PROGRAMMING FOR PROBLEM SOLVING

THEORY, PROBLEMS AND APPLICATIONS USING C

Aimed @ Empowering
Budding IT Professionals of 21st Century

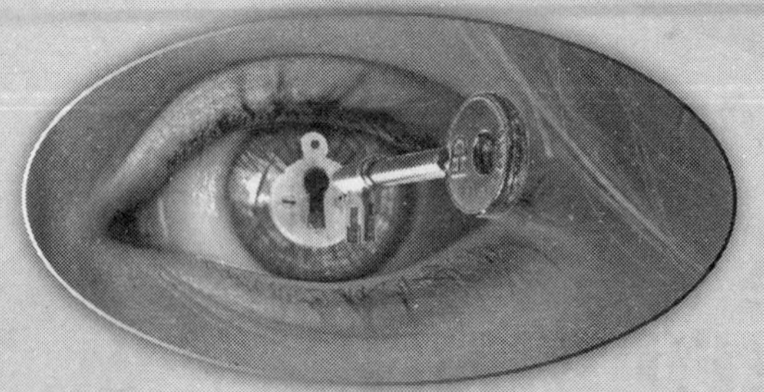

An Eye Opener

An eye opener is something that makes you see a situation or the world differently, that makes you come to a realization or that forces you to change your perspective.

What is Programming?

Programming is a way to "instruct the computer to perform various tasks".

"**Instruct the computer**": this basically means that you provide the computer a set of instructions that are written in a language that the computer can understand.

"**Perform various tasks**": the tasks could be simple involving few instructions or complex involving hundreds of instructions.

Hence, in brief, "*Programming is a way to tell computers to do a specific task*".

Now, a million dollars question here is that you need to address to yourself – "Do you know how to perform the task in hand?

If the answer is NO, then how can you instruct the computer?

Remember that

"*Programming is a Science of Sciences & Art of Arts*"

When we say Science, it means it is methodology that needs to be understood, and when we say Art, it means it is a skill that needs to be practiced.

As rightly said,

"*Practice makes a person perfect*"

Therefore, practice, practice, practice, . . ., and practice, but with understanding.

The book in your hand will provide the answer. It will provide you all the requisite knowledge you need and will empower you to become not only skilled rather sharp task master (problem solver).

. . . A Word of Promise

PROGRAMMING FOR PROBLEM SOLVING

THEORY, PROBLEMS AND APPLICATIONS USING C

Special Edition for the States of Telangana & Andhra Pradesh
Covers Syllabus of all Public & Private Universities

Prof. R. S. Salaria
Professor
Computer Science & Engineering
Head
Centre of Excellence
Coding & Code Optimization
Guru Nanak Institutions, Hyderabad, Telangana.

Prof. H. S. Saini
Professor
Computer Science & Engineering
Managing Director
Guru Nanak Institutions,
Hyderabad

Prof. Sunil Tekale
Professor
Computer Science & Engineering;
Dean Academics
Malla Reddy College of Engineering,
Hyderabad

KHANNA BOOK PUBLISHING CO. (P) LTD.
Publisher of Engineering and Computer Books
4C/4344, Ansari Road, Darya Ganj, New Delhi-110002
Phone : 011-23244447-48 **Mobile:** +91-9910909320
E-mail : contact@khannabooks.com
Website : www.khannabooks.com

Dear Readers,

To prevent the piracy, this book is secured with HIGH SECURITY HOLOGRAM on the front title cover. In case you don't find the hologram on the front cover title, please write us to at *contact@khannabooks.com* or whatsapp us at +91-99109 09320 and avail special gift voucher for yourself.

Specimen of Hologram on front Cover title:

Moreover, there is a SPECIAL DISCOUNT COUPON for you with EVERY HOLOGRAM.

How to avail this SPECIAL DISCOUNT:

Step 1: Scratch the hologram
Step 2: Under the scratch area, your "coupon code" is available
Step 3: Logon to *www.khannabooks.com*
Step 4: Use your "coupon code" in the shopping cart and get your copy at a special discount
Step 5: Enjoy your reading!

Thanks,
Khanna Publishing House

Price: ₹ 450.00

PROGRAMMING FOR PROBLEM SOLVING

Prof. R. S. Salaria, Prof. H. S. Saini, Prof. Sunil Tekale

ISBN: 978-93-89139-12-9

Reprint: 2022

Published by:

Khanna Book Publishing Co. (P) Ltd.
4C/4344, Ansari Road, Darya Ganj, New Delhi-110 002
Phone: 011-23244447-48 **Mobile:** +91-9910909320
E-mail: contact@khannabooks.com

Printed by: SPS Printers

About the Authors

Prof. R. S. Salaria, MTech (IIT/Delhi), PhD*

- Professor, Computer Science & Engineering
- Hard Core Educator, Prolific Author, Excellent Motivator & Social Reformer
- Alumnus of IIT, Delhi
- President, Innovative Foundation for Holistic Education (IFHE)
- Certified Software Quality Professional, MoCIT, Govt. of India
- Certified Programmer & Trainer, SUN Microsystems, USA
- Former Head, Department of Computer Science & Engineering, Thapar University, Patiala, Punjab
- Authored 25 books on Computer Science, Engineering Mathematics & Communication skills
- Delivered training programmes to various IT industries
- Conducted 60+ workshops for faculty and 85+ workshops for the students
- Strong corporate connect

Prof. H. S. Saini, MTech, PhD

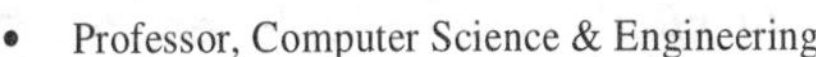

- Professor, Computer Science & Engineering
- Managing Director, Guru Nanak Group of Institutes, Hyderabad
- Taught for University of Westminster and Oxford Brookes University, UK, at their off-shore centres
- Founder and Chief Editor of Journal of Innovation in CSE/ECE/IT
- Recipient of Eminent Academia Award from Indus Foundation
- Mentored half a dozen startups in the state of Telangana
- Key member on the panel of NBA & NAAC expert committees
- Expert in the field of Fuzzy Systems, AI, Knowledge Based Systems and Web Technologies
- Established Industry-Academia Collaborations, Foreign University Collaborations and Centers of Excellence in Cutting Edge Technologies
- Authored 07 books and published 54+ Innovative Research papers
- Fellow of Computer Society of India and Institution of Engineers
- Strong corporate connect

Prof. Sunil Tekale, MTech, PhD

- Professor, Computer Science & Engineering
- Started career by imparting Training on Y2K problem
- Certified as Quality Management Educator by PHILIP CROSBY, Guru of Quality Management
- Certified Trainer, SUN Microsystems and Microsoft
- Former Managing Director NIIT Pvt. Ltd., Franchisee of NIIT Ltd.
- 15 years of experience in training corporate clients and 10 years in Academics
- Registered 3 patents on innovative products
- Delivered training programmes to various professionals in Banking Sector and Government Employees including Top Level Administrative officers
- Conducted multiple workshops, Faculty Development programs (FDP) for faculty and workshops/seminars for the students
- Trained more than 10,000 students in various Technologies
- Strong corporate connect

About the Book

This book is specially conceived with an aim to serve not only as a textbook to cover the syllabus of public & public universities, rather a intelligent book (*i*-book) that empowers readers to acquire 21st century problem solving skills and sharp coding skills.

Therefore, keeping in mind the requirements of 21st Century, a special emphasis is laid on the following aspects:

- Designed keeping in mind the objectives of **OUTCOME BASED EDUCTION**.
- Simple and lucid language that enables even an average reader to grasp the fundamental concepts of the subject.
- Illustrative examples (*165+ in number*) to demonstrate the application of the concepts.
- Multiple choice questions (*225+ in number*) to provide an opportunity for self-assessment of the fundamental concepts learned.
- Short-answer type questions (*192+ in number*) to provide an opportunity to synthesize the fundamental concepts learned to answer precisely.
- Programming exercises (*112+ in number*) to provide an opportunity to harness their coding skills.
- Coding problems (*42 in number*) from the IT/IT-enabled industry perspective to conquer the screening phase of the placement process.
- Frequently asked questions (*43 in number*) to conquer the technical round of the **CAMPUS PLACEMENT PROCESS** with flying colors.
- The whole content is designed and organized as per the **BLOOM'S TAXONOMY.**

We are sure that our efforts in preparing a wonder piece of work will bring outstanding results not only in university examinations but also in bagging handsome package in campus placement in top notch IT giants that include **Microsoft, Google, TCS, IBM, Wipro, Amazon, Facebook**.

For faculty handling the subject of Programming, the book will be a great resource. It will help in delivering the content effectively, designing the challenging assignments, head scratching MCQs, absorbing programming assignment, that will make learning problem solving & programming a great fun for the students.

Further, as a special gesture, we have decided to provide free assistance to the readers during the camps placements, free of cost

Looking forward to your honest feedback that will help us to fine tune the content to meet your expectations.

. . . Prof. R. S. Salaria
Email: *rssalaria@yahoo.com*

. . . Prof. H. S. Saini
Email: *hssaini@yahoo.com*

. . . Prof. Sunil Tekale
Email:*sunil.tekale2010@gmail.com*

Contents at a Glance

Web Resources

To empower and boost the confidence of the faculty in delivering the content on the topic of Programing for Problem Solving effectively, web resources in the form of PPTs and solutions of the selected exercises are provided, which are not to be shared with the students without the permission of the author.

Link: *www.khannabooks.com/RSSalaria/PPS/Faculty*

Every Sunday @ 11:00AM, we will have webinar for faculty to address their issues & concerns, which they can post on the above link. The link for the same will be provided on the same day by 10:45AM.

~~~~~~~~~~~~~~~~~~~~~~~~~~~~~~~~~

To empower students to learn & master the subject of Programing for Problem Solving, a web link is created to address student's queries, doubts, and problems. Students can upload their queries, doubts, and problems on the following link:

Link: *www.khannabooks.com/RSSalaria/PPS/Students*

Every Sunday @ 04:00PM, we will also have webinar for students to clear their queries, doubts, and problems. The link for the same will be provided on the same day by 03:45PM.
~~~~~~~~~~~~~~~~~~~~~~~~~~~~~~~~~

Table of Contents

Chapter 4: Data Types, Variables and Constants.......... 117-136

Chapter 5: Operators and Expressions.......... 137-170

Chapter 9: User-Defined Functions 245-282

Chapter 10: Arrays and Strings 283-324

Chapter 0

Coding/Programming Problems for Campus Placements

The book in your hand is the only text book that gives you the exposure regarding the kind of problem asked and the format in which they are asked during the placement process by majority of the IT companies. You are required to strictly adhere to the instructions otherwise your submission of the solution can be rejected.

The problems listed here, are just indicative, and will be expanded further in future editions. The problems listed and similar had been asked in the previous placements in the coding/programming section of first phase (screening) by majority of the IT companies including TCS, Wipro, Cognizant, Accenture, Amazon, Facebook, etc. While going through the content of the book, try to solve these problems, and this practice will help you to harness your problem solving and coding skills to crack the placement with flying colors.

As a special gesture, the readers of this book will be provided with placement assistance free of cost, for which you may be required to register online.

Best wishes for your great future.

1. Area of Circle

Write a program to find area of circle. (Assume pi=3.14)

Input Format:
Input consists of float value which corresponds to radius of circle.

Output Format:
Output consist of area of circle in mtsq correct to 2 decimal places.

Sample Input:
5.5

Sample Output:
94.99

2. Leap Year

A year will be a leap year if it is divisible by 4 but not by 100. If a year is divisible by 4 and by 100, it is not a leap year unless it is also divisible by 400.

Write a program to check whether a given year is a leap year or not.

Input Format:
Input consists of a single integer.

Output Format:
Output consists of a single line. Refer sample output for details.

Sample Input 1:
2020

Sample Output 1:
2020 is a leap year

Sample Input 2:
2019
Sample Output 2:
2019 is not a leap year

3. Validity of Date

Write a program to check whether the given date in format *dd/mm/yyyy* is valid or not.

Input Format:
Input consists of date in given format.

Output Format:
Output consists of a single line. Refer sample output for details.

Sample Input 1:
Enter date in format dd/mm/yyyy: 19/11/2020
Sample Output 1:
19/11/2020 is valid date.

Sample Input 2:
Enter date in format dd/mm/yyyy: 29/02/2019
Sample Output 2:
29/02/2019 invalid date.

4. Find Your Day of Birth

Write a program to find the day on which you were born. Your date-of-birth is given in the format *dd/mm/yyyy*.

Input Format:
Input consists of date in given format.
Output Format:
Output consists of a single line. Refer sample output for details.

Sample Input 1:
Enter your date of birth dd/mm/yyyy: 28/10/1955

Sample Output 1:
You was born on Friday.

**Bill Gate, Founder of Microsoft, was born on this day.*

5. Armstrong Number

An Armstrong number is a 3-digit integer such that the sum of the cubes of its digits is equal to the number itself.

For example, 371 is an Armstrong number, since

3^3 + 7^3 + 1^3 = 371

Write a program to find whether a given 3-digit number is an Armstrong number or not. Restrict your input to 3-didit number otherwise display message "Wrong input".

Input Format:
Input consists of a single integer.

Output Format:
Refer sample output for details.

Sample Input 1:
153
Sample Output 1:
Armstrong Number

Sample Input 2:
135
Sample Output 2:
Not an Armstrong Number

6. Prime Number

A prime number is a natural number that is divisible by only 1 and itself only.

Write a program to find whether the given natural number 'n' is a prime number or not.

Input Format:
Input consists of a single integer.

Output Format:
Refer sample output for details.

Sample Input and Output 1:
Enter the given natural number: 17
15 is a prime number

Sample Input and Output 2:
Enter the given natural number: 145
145 is not a prime number

7. Sum of digits

Write a program to find the sum of digits of a given natural number 'n'.

Input format:
Input consists of an integer value.

Output Format:
Refer sample output for details.

Sample Input and Output 1:
Enter given natural number: **1345**
Sum of digits in 1345 is 13.

Sample Input and Output 2:
Enter given natural number: **1111111111**
Sum of digits in **1111111111** is 10.

8. Factorial

Write a program to find the factorial of a given natural number 'n'.

Input format:
Input consists of an integer value.

Output Format:
Refer sample output for details.

Sample Input and Output 1:
Enter given natural number: 4
4! = 24

Sample Input and Output 2:
Enter given natural number: 6
6! = 720

9. Palindrome Number

A Palindrome number is one which reads same from both the ends. For example, 1881, 232454.

Write a program to find whether the given number 'n' is a palindrome or not.

Input Format:
Input consists of a single integer.

Output Format:
Refer sample output for details.

Sample Input and Output 1:
Enter the given natural number: 1001
1001 is a palindrome.

Sample Input and Output 2:
Enter the given natural number: 1100
1100 is not a palindrome.

10. Fibonacci Sequence

Write a program to print first 'n' terns of the Fibonacci sequence.

For example, first 7 terms of Fibonacci sequence

0 1 1 2 3 5 8

Input format:
Input consists of an integer value.

Output Format:
Refer sample output for details.

Sample Input and Output:
Enter value for n: **10**
First 10 terms of Fibonacci sequence are
0 1 1 2 3 5 8 13 21 34

11. Perfect Number

A perfect number is a number in which the sum of its proper divisors is equal to the number itself. Proper divisors of a number are all divisors of a number excluding itself.

Write a program to find whether a given number 'n' is a perfect number or not.

Input format:
Input consists of an integer value.

Output Format:
Refer sample output for details.

Sample Input and Output 1:
Enter given natural number: 6
6 is a perfect number.
Sample Input and Output 2:
Enter given natural number: 12
12 is not a perfect number.

12. Strong Number

A strong number is a number whose sum of the factorial of its each digit equals to the number itself.

For example, 145 is a strong number since

1! + 4! + 5! = 1 + 24 + 120 = 145

Write a program to find whether a given number 'n' is a strong number or not.

Input Format:
Input consists of a single integer.

Output Format:
Refer sample output for details.

Sample Input and Output 1:
Enter the given number: 145
145 is a strong number.

Sample Input and Output 2:
Enter the given number: 110
110 is not a strong number.

13. Special Number

A 2-digit number is said to be a special number if the sum of the sum of its digits and the products of its digits is equal to the number itself.

For example, 19 is a special number. The digits in 19 are 1 and 9. The sum of the digits is 10 and the product of the digits is 9, sum of sum of digits and product of digits is 19.

Write a program to find all special numbers.

Input Format:
Nil

Output Format:
List of all special numbers separated by one or more spaces. Refer to sample output for more details.

Output:
Here is the list of special numbers.
19 29 39 49 59 69 79 89 99

14. GCD of numbers

Write a program to find the greater common divisor (GCD)/highest common factor (HCF) of two natural numbers 'm' and 'n'.

Input Format:
Input consists of 2 natural numbers.

Output Format:
Refer sample input and output for formatting specifications.

Sample Input and Output:
Enter first natural number 'm' : 36
Enter second natural number 'n' : 27

GCD(36,27) = 9

15. LCM of numbers

Write a program to find the least common multiplier (LCM) of two natural numbers 'm' and 'n'.

Input Format:
Input consists of 2 natural numbers.

Output Format:
Refer sample input and output for formatting specifications.

Sample Input and Output:
Enter first natural number 'm' : 40
Enter second natural number 'n' : 130

LCM(35,120) = 260

16. Series Problem #1

Consider the following series:

1, 1, 2, 3, 4, 9, 8, 27, 16, 81, . . .

If you note carefully, you will find that even term represent one GP and odd term represent second GP.

Write a program to generate first 'n' terms of this series when common ratios of these two GPs are given as 'r1' and 'r2', respectively.

Input Format:
Input consists of three integers, 'n', 'r1' and 'r2', in order, on separate lines.

Output Format:
Display the first 'n' terms of the series.

Sample Input:
12
2
3
Sample Output:
1 1 2 3 4 9 8 27 16 81 32 243

17. Series Problem #2

Consider the following series:

1, 1, 2, 5, 4, 9, 8, 13, 16, 17, . . .

If you note carefully, you will find that odd term represent GP with common ratio of 2, and even term represent AP with a common difference of 4.

Write a program to generate first 'n' terms of this series when 'r' as common ratio of GP, and 'd' is the common difference of AP.

Input Format:
Input consists of three integers, n, r and d, in order, on separate lines.

Output Format:
Display the first 'n' terms of the series.

Sample Input:
12
2
4
Sample Output:
1 1 2 5 4 9 8 13 16 17 32 21

18. Series Problem #3

Consider the following series:

1, 1, 1, 5, 2, 9, 3, 13, 5, 17, 8, 21, 13,. . .

If you note carefully, you will find that odd term represent terms of Fibonacci sequence, and even term represent AP with a common difference of 4.

Write a program to generate first 'n' terms of this series with 'd' as the common difference of AP.

Input Format:
Input consists of two integers, 'n' and 'd' on separate lines.

Output Format:
Display the first 'n' terms of the series.

Sample Input:
10
4
Sample Output:
1 1 1 5 2 9 3 13 5 17

19. Series Problem #4

Consider the following series:

20, 1, 16, 1, 12, 2, 8, 3, 4, 5, 0, 8, -4, 13, -8,. . .

If you note carefully, you will find that the odd terms represent AP with a common difference of -4, and even terms represent term of Fibonacci sequence.

Write a program to generate first 'n' terms of this series with 'd' as the common difference of AP.

Input Format:
Input consists of two integers, 'n' and 'd' on separate lines.

Output Format:
Display the first 'n' terms of the series.

Sample Input:
10
-4
Sample Output:
20, 1, 16, 1, 12, 2, 8, 3, 4, 5.

20. Decimal to Binary Conversion

Write a program to perform decimal to binary conversion.

Input Format:
Input consists of a single decimal number.

Output Format:
Print the equivalent binary number.

Sample Input and Output:
Enter the decimal number: 10
Binary equivalent of decimal number 10 is 1010.

21. Binary to decimal Conversion

Write a program to perform binary to decimal conversion.

Input Format:
Input consists of a single binary number.

Output Format:
Print the equivalent decimal number.

Sample Input and Output:
Enter the binary number: 1010
Decimal equivalent of binary number 1010 is 10.

22. Sorting

Write a program to sort elements of an array named 'a' of size 'n'. Elements of array are of type 'int'.

Input Format:
Input consists of 'n+1' integers. The first integer corresponds to 'n'. The next 'n' integers correspond to the elements in the array.

Output Format:
Display the elements of the sorted array.

Sample Input:
5
3
2
6
8
1
Sample Output:
1 2 3 6 8

23. First 3-Largest Numbers

Write a program to find the first 3-largest numbers from an array named 'a' of size 'n', without sorting the array. The elements of the array are of type 'int'.

Input Format:
Input consists of 'n+1' integers. The first integer corresponds to 'n'. The next 'n' integers correspond to the elements in the array.

Output Format:
Display the first 3-largest numbers in order.

Sample Input:
10
35 20 61 80 19 40 100 75 57 60

Sample Output:
100
80
75

24. Leaders

An element is a leader if it is greater than all the elements to its right side, i.e., following it. And the rightmost element is always a leader.

For example in the array {16, 17, 4, 3, 5, 2}, leaders are 17, 5 and 2.

Write a program to print all the leaders in the array.

Input Format:
The first line of the input consists of an integer 'n' that corresponds to the number of elements in the input array.
The second line consists of elements of array separated by space.

Output Format:
List of numbers that are leaders, one single line, separated by space.

Sample Input 1:
6
16 17 4 3 5 2
Sample Output 1:
17 5 2

Sample Input 2:
6
16 10 4 3 18 19
Sample Output 2:
19

25. Majority Element

A majority element in an array of size 'n' is an element that appears more than 'n/2' times.

1. If there exists a majority element, then display that element.
2. If such element doesn't exist, then display "No Majority Element exists".

Write a program that prints the majority element in an array.

Input Format:
The first line of the input consists of an integer 'n' that corresponds to the number of elements in the input array.
The second line consists of elements of array separated by space.

Output Format:
Refer sample output formatting specifications.

Sample Input 1:
9
3 3 4 2 4 4 2 4 4
Sample Output 1:
4

Sample Input 2:
8
3 3 4 2 5 4 2 4 6
Sample Output 2:
No Majority Element exists.

26. Sorting a String

Write a program to sort a string in alphabetical (dictionary) order.

Input Format:
Input consists of a string. Assume that the input string consists of only letters and the maximum length of the string is 50.

Output Format:
A string whose letters are sorted in alphabetical order.

Sample Input:
Anitha

Sample Output:
Aahint

27. Duplicate Elements

Given an array of 'n<=50' elements which contains elements from 1 to n, generated randomly, with any of these numbers appearing any number of times.

Write a program to find these repeating numbers.

Example:
Let A = {1, 2, 3, 1, 3, 6, 6, 3}
Duplicate elements = 1 3 6

Input Format:
A single number 'n' that corresponds to the number of elements in the array.

Output Format:
Refer sample output formatting specifications.

Sample Input 1:
8

Let us assume that following are the elements generated randomly:

3 1 5 3 2 1 2 6

Sample Output 1:
3 1 2

Sample Input 2:
5

Let us assume that following are the elements generated randomly:

3 1 5 4 2

Sample Output 2:
No duplicates

28. Distinct Elements

Given an array of 'n<=50' elements which contains elements from 1 to n, generated randomly, with any of these numbers appearing any number of times.

Write a program to find these repeating numbers.

Example:
Let A = {1, 2, 3, 1, 3, 6, 6, 5}
Distinct elements = 2 5

Input Format:
A single number 'n' that corresponds to the number of elements in the array.

Output Format:
Refer sample output formatting specifications.

Sample Input 1:
8

Let us assume that following are the elements generated randomly:

1, 2, 3, 1, 3, 6, 6, 5

Sample Output 1:
2 5

Sample Input 2:
6

Let us assume that following are the elements generated randomly:

3 2 5 3 2 5

Sample Output 2:
No distinct element.

29. To Remove Duplicate Elements

Given an array of 'n<=50' elements. Write a program to remove duplicate elements from the array.

Input Format:
The first line of the input consists of an integer 'n' that corresponds to the number of elements in the input array.
The next 'n' lines in the input correspond to the elements in the array.

Output Format:
Refer sample output formatting specifications.

Sample Input:
10
10
22
31
10
31
16
26
45
22
31

Sample Output 1:
10
22
31
16
26
45

30. Missing Number

Write a program to find the missing integer among the array of 'n<=9' integers that are in the range 1 to (n+1).

Input Format:
The first line of the input consists of an integer 'n' that corresponds to the number of elements in the input array.

The second line consists of 'n' integers in the range 1 to (n+1), separated by space.

Output Format:
List of missing number(s).

Sample Input 1:
5
1 5 3 2 6

Sample Output 1:
4

31. Sum

Given an unsorted array 'a' of size 'n' of non-negative integers, find a continuous sub-array which adds to a given number sum.

Input Format:
The first line contains an integer 'n' denoting the size of the array.
The second line contains integers denoting the elements
of the array separated by space.
The last line contains an integer, denoting the sum.

Output Format:
The output line contains integers denoting the indexes of the sub-array.

Sample Input 1:
6
1 4 3 0 5 10
7

Sample Output 1:
Sum found between indexes 1 and 4

Sample Input 2:
7
1 4 3 0 5 10
9

Sample Output 2:
No sub-array found.

32. Equilibrium

Equilibrium position in an array is a position such that the sum of elements before of it is equal to the sum of elements after it.

Given an array 'a' of size 'n' of non-negative integers, find the index in the array where equilibrium occurs first in the array.

Input Format:
The first line contains an integer 'n' denoting the size of the array.
The second line contains integers denoting the elements of the array separated by space.

Output Format:
The output line contains integers denoting the equilibrium position or -1 if no equilibrium exists.

Sample Input 1:
7
-7 1 5 2 -4 2 1

Sample Output 1:
3

Sample Input 2:
7
1 4 3 0 5 10

Sample Output 2:
-1

33. Reverse Sub-array

Given an array 'a' of size 'n' of non-negative integers. Write a program to reverse every sub-array of size 'k', where 'n' is multiple of 'k'.

Input Format:
The first line contains an integer 'n' denoting the size of the array.
The second line contains integers denoting the elements of the array separated by space.
The last line contains an integer 'k' denoting the size of the sub-array.

Output Format:
The output line contains integers denoting the elements of the array, with elements of each sub-array reversed..

Sample Input:
6
-7 1 5 6 -4 2

Sample Output:
5 1 -7 2 -4 6

34. ASCII to Integer Conversion

Your task is to implement the function atoi. The function takes a string 'str' denoting a decimal number string, converts it to integer and returns it.

Input Format:
A numeric string.

Output Format:
Integer value

Sample Input:
"8671"

Sample Output:
8671

35. Triplets

Given an array 'a' of size 'n'. Write a program to find all the triplets such that sum of any two elements is equal to the third element.

Input Format:
The first line contains single integers 'n' denoting the size of the array.
The next line contains the elements of the array, separated by space.

Output Format:
The output line contains all those triplets whose sum of any two elements is equal to the third element.

Sample Input:
8
7 2 5 6 -4 6 3 9 8

Sample Output:
(7,2,5) (6,3,9)

36. Counting Bits

Write a program that counts the bits that are 1 in the binary representation of natural number 'n'.

Input Format:
Input consists of a single natural number 'n'.

Constraint:
20 <= N <=1000

Output Format:
Output consists of the counts of 1s in the binary representation of 'n'.

Sample Input 1:
127 (01111111)

Sample Output 1:
Number of 1-bits in 127 = 7

Sample Input 2:
32 (100000)

Sample Output 2:
Number of 1-bits in 32 = 1

37. Left Rotation

A *left rotation* operation on an array of size *n* shifts each of the array's elements 1 unit to the left. Given an integer, *d*, rotate the array that many steps left.
Example,
$d = 2$
$arr = [1,2,3,4,5]$
After 2 rotations, $arr' = [3,4,5,1,2]$.

Input Format:
The first line contains two space-separated integers that denote the number of integers, and the number of left rotations to perform.
The second line contains space-separated integers that represent the elements of the array.

Output Format:
List of elements of left rotated array.

Constraints
- $1 \leq n \leq 10^5$
- $1 \leq d \leq n$
- $1 \leq a[i] \leq 10^6$

Sample Input:
5 4
1 2 3 4 5

Sample Output:
5 1 2 3 4

38. Magic Triplets

Given an array 'a' of 'n' integers, a triplet (a[i], a[j], a[k]) is called Magic Triplet if a[i] < a[j] < a[k] and i < j < k. The task is to count number of magic triplets in a given array.

Input Format:
The first line contains an integer 'n' denoting the size of array.

The second line contains 'n' space-separated integers denoting the elements of the array.

Output Format:
Single number representing the number of triplets.

Constraints:
$1 <= n <= 10^3$
$1 <= a[i] <= 5000$

Sample Input 1:
3
3 2 1
Sample Output 1:
0

Sample Input 2:
4
1 2 3 4
Sample Output 2:
2

39. Numbers with alternate bits

Input Format:
Input consists of a single natural number 'n'.

Output Format:
Refer sample output formatting specifications.

Sample Input 1:
5 (101)

Sample Output 1:
Number has alternate pattern of 0s and 1s.

Sample Input 2:
15 (1111)

Sample Output 2:
Number doesn't have alternate pattern of 0s and 1s.

40. EVEN or ODD

Write a program to check whether the given natural number is EVEN or ODD.

Input Format:
Input consists of a single natural number 'n'.

Constraints:
1. Modulus operation by 2 as well 10 is not
2. Use of *if* statement is not permitted.

Output Format:
Refer sample output formatting specifications.

Sample Input 1:
Enter natural number : 35

Sample Output 1:
35 is odd number.

Sample Input 2:
Enter natural number : 120

Sample Output 2:
120 is even number.

41. Swapping

Write a program to swap two natural numbers.

Input Format:
Input consists of a two natural number *a* & *b* separated by space.

Constraints:
1. Use of third variable is not allowed
2. Only single statement is permitted to facilitate the swapping.
3. $0 < a, b \leq 10\text{^}7$

Output Format:
Refer sample output formatting specifications.

Sample Input:
Enter two natural number a & b : 20 25

Sample Output:
After swapping
a = 25, b = 20

42. Forming Largest Number

Write a program to find the largest number that can be formed from the digits of a given number *n*.

Input Format:
Input consists of a single positive number *n*.

Constraints:
$0 < n \leq 10\text{^}7$

Output Format:
Refer sample output formatting specifications.

Sample Input:
Enter given n : 182735

Sample Output:
Largest number that can be formed from the digits of number 182735 is 875321.

❑ ❑ ❑

Chapter 1

Introduction to Computers

Learning Outcomes

After reading this chapter, students will be able to

- describe a computer and its operations
- differentiate hardware and software
- differentiate between data and information
- explain characteristics of a computer
- explain the functional diagram of a computer
- describe various I/O devices, storage devices and I/O ports
- explain the concept of booting
- describe various essential aspects of software
- describe various number systems
- covert numbers from one number system to another number system
- explain the factors that affect the performance of a computer
- explain the meaning of specification of a computer system

1.1 INTRODUCTION

The human civilization has seen many revolutions, which changed its life style many times. With the discovery of fire, man started eating cooked food. The invention of steam engine led to the industrial revolution and invention of telephone led to the communication revolution. Now, we are already going through information revolution brought about by the wonder machine known as a *Computer*. At present everyone is talking about computer as every one is getting benefits of it directly or indirectly.

In this chapter, we will discuss about basic aspects of computers - *the wonderful machine.*

1.2 WHAT IS A COMPUTER?

Computer is an electronic machine that performs tasks or computations according to a set of instructions called *programs*.

The first fully electronic computers, introduced in the 1940s, were huge machines that required teams of people to operate. Compared to those early machines, today's computers are amazing - they are thousands of times faster and can fit on your desk, in your lap, or even in your pocket.

A computer, in general, is capable of

- Receiving the data and instructions in various forms,
- Storing the data and instructions in its memory,
- Retrieving the data already stored on secondary storage,
- Performing arithmetic and logical operations, according to the specified instructions, with very high speed and greater accuracy,
- Producing the results in many forms, and
- Communicating with other computers.

1.2.1 Hardware and Software

Computers work through an interaction of *hardware* and *software*.

Hardware refers to the parts of a computer that you can see and touch, including the case and everything inside it. The most important piece of hardware is a tiny rectangular chip inside your computer called the *microprocessor*. It's the "brain" of your computer—the part that translates instructions and performs calculations. Hardware items such as your monitor, keyboard, mouse, printer, and other components are often called *hardware devices*, or *devices*.

Software refers to the instructions, or programs, that tell the hardware what to do. A word processing program using which you can write letters on your computer is a type of software. The most important example of software is an *operating system* (OS) that manages all the resources (such as memories, input/output devices, etc.) of your computer system, and provides an interface using which the user can interact with the computer to perform various tasks. Two well-known operating systems are Windows and Linux.

1.2.2 Data and Information

The terms *data* and *information* have been used to mean the same thing. But actually information is more than just data.

Data – The term *data* simply refers to a value or set of values. These values may represent some observations from an experiment, some figures collected during some survey (such as *census*, *exit polls*, etc.), marks obtained by a student in an examination, etc.

Information – In simple terms, information is a processed data. Data is just a collection of values (raw data), from which no conclusions can be drawn. Thus, data as such is not useful for decision making. When the data is processed, by applying certain rules, new generated data becomes information. This new generated data (information) conveys some meaning and hence can be used for decision-making. For more clarity, consider the next example.

An agricultural scientist wants to study the effect of a particular pesticide on a new variety of wheat. He uses a different amount of pesticide for different fields of wheat and notes down the yield of wheat in each field. Here, the amount of pesticide used and yield of wheat in each field represents data.

As such these values do not convey anything regarding the effect of pesticide on the yield. Next he applies the technique of correlation analysis to determine the correlation coefficient whose value lies in the range −1 to +1. This coefficient, which is also a value, conveys valuable information about the effect of pesticide on yield of wheat. If correlation coefficient is greater than 0, it means increased amount of pesticide has positive effect on the yield. If correlation coefficient is less than 0, it means increased amount of pesticide has negative effect on the yield. If correlation coefficient is equal to 0, it means increased amount of pesticide has no effect on the yield.

1.3 CHARACTERISTICS OF A COMPUTER

A computer is a very powerful and versatile machine. The key characteristics of a computer include the following:

Speed - A computer can perform arithmetic as well logical operations at very high speed. The speed of a computer is usually measured in Million Instructions Per Second (MIPS) or Millions of Floating-point Operations Per Second (MFLOPS). For example, a simple microcomputer (also known as PC) can have a speed of about 1 MFLOPS, where as a super computer can have a speed more than 80 MFLOPS.

An average computer executes instructions at 20 MIPS. Even such a high unit to measure the speed of computers, becomes so small for very high speed parallel computers. For example, an ordinary computer can add two 18 digits numbers in just 3×10^{-4} seconds.

Accuracy - If the data and instructions given are correct, computers will always give the correct results. The errors in the calculations are due to humans, who either don't give the correct data or instructions to the computer or ignore the data handling limitations of the computer.

Capacity - Information is useful only if it is available on time and computer is the only solution for it. Computer can handle very large amount of data. Whole encyclopedia, which is on many volumes of books, can be stored on a single disk of computer and any information can be accessed in seconds by just a push of button.

Reliability - Being a machine, computers never get tired or bored. Computer can perform millions of similar type of operations without getting bored or tired, for number of days together without showing any sign of fatigue. It does not require even tea or lunch break. Computer is made of mostly electronic components, which don't have any wear and tear; therefore computer can work for years with requiring any maintenance.

Versatility - The basic operation of computer is to receive the data, process it and produce the accurate result at a very high speed. This quality makes the computer a very versatile device which can be used for number of applications like word-processing, accounting, scientific calculations, designing, training, controlling and entertainment.

Automatic - Computer works automatically without any aid from the humans. It requires the data and instructions only in the beginning, later it works automatically according to the instructions.

Communication - Computers can communicate with each other using telephones and modems. This enables the computer to share the information and resources at much lower price. The large problems can be distributed among the various computers to have fast processing of data.

Diligence - A computer is free from tiredness, lack of concentration, fatigue, etc. It can work for hours without creating any error. It does not require any tea breaks. If millions of calculations are to be performed, a computer will perform every calculation with the same accuracy. Due to this capability, it overpowers human being in routine type of work.

No IQ - Computer is a dumb machine and it cannot do any work without instruction from the user. It performs the instructions at tremendous speed and with accuracy. It is you to decide what you want to do and in what sequence. So a computer cannot take its own decision as you can.

No Feeling - It does not have feelings or emotion, taste, knowledge and experience. Thus, it does not get tired even after long hours of work. It does not distinguish between users.

1.4 LIMITATIONS OF A COMPUTER

The only limitation of the computer is that it does not have any intelligence of its own. It can't work on its own. It can't learn from its mistakes or previous experiences. It can't think or innovate. The research is going on to render artificial intelligence to computers.

1.5 FUNCTIONAL COMPONENTS OF A COMPUTER

A computer is a combination of hardware and software. Hardware represent the physical components of a computer like motherboard, memory devices, monitor, keyboard, etc., while software represents the set of programs or instructions. Both hardware and software together make the computer system function.

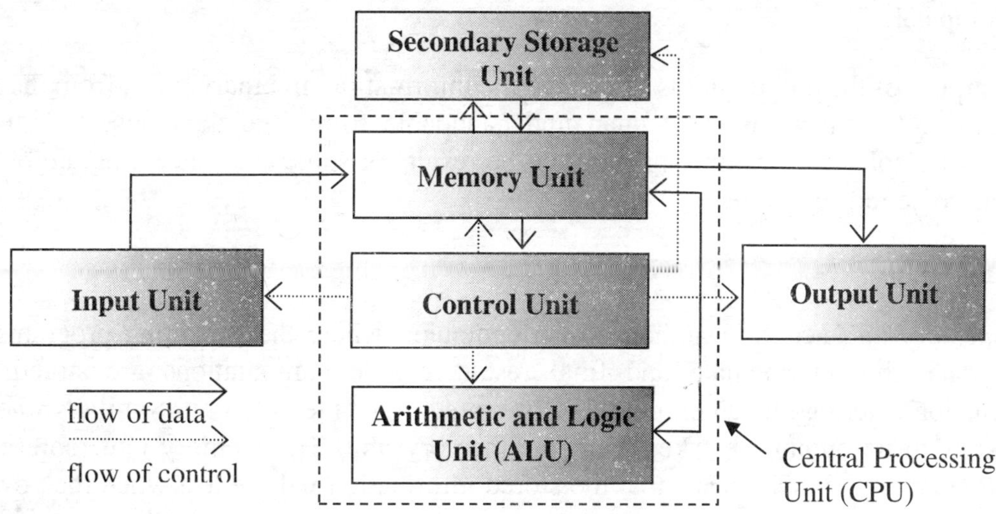

Figure 1.1: Functional components of a Computer

Every task given to a computer follows an Input- Process-Output Cycle (IPO cycle). It needs certain input, processes that input and produces the desired output. The input unit takes the input, the central processing unit does the processing of data and the output unit produces the output. The memory unit holds the data and instructions during the processing.

Let us have a look at the functional components of a computer.

Input Unit

The *input unit* consists of various input devices that can be connected to a computer, such as *keyboard*, *mouse*, *light pen*, *microphone*, etc. The standard input device is *keyboard*, which will come with every new computer. At any time, more than one input device can be connected to a computer.

The purpose of the input unit is to receive the information from the outside world, convert it into binary form, and then transfer it to the main memory. It may consist of entering data to be processed by some program, recording of speech, capturing of a scene or photograph, or selecting some objects on the display.

Output Unit

The *output unit* consists of various output devices that can be connected to a computer, such as *visual display unit* (VDU) or simply known as *monitor*, *printers*, *plotters*, *audio systems* etc. The standard output device is a *monitor*, which will come with every new computer. At any time, like input devices, more than one output device can be connected to a computer.

The purpose of the output unit is to receive the information, in binary form, from the main memory unit, convert it into human understandable form, and then output it. It may consist of displaying, printing, or plotting the results of some computer program or even playing some music or movie.

Memory Unit

The *memory unit* acts as warehouse of a computer, where the executing programs, the input data, the intermediate and final results of the computations are stored. The information entering into a computer is first stored in the memory unit. Similarly, the data or results to be output are taken from the memory unit. The memory unit represents a *volatile* memory, which means that the stored information will be lost when the power is switched off.

Arithmetic and Logic Unit

The arithmetic and logic unit (ALU), as its name implies, performs arithmetic and logical operations. Arithmetic operations include addition, subtraction, multiplication, and division. Logical operations are used to compare two data items, and include operations such as less than ($<$), less than or equal to ($\leq$), greater than ($>$), greater than or equal to ($\geq$), equal to ($=$), and not equal to ($\neq$).

Control Unit

The control unit co-ordinates and controls all the other parts of the computer system. Under the direction of a program, the control unit performs four basic operations:

- *Fetch* - getting the next program instruction from the computer's memory.
- *Decode* - figuring out what the program is telling the computer to do?
- *Execute* - performing the requested action, such as adding two numbers or deciding whether one of them is larger.
- *Write-Back* - writing the results back to memory.

This four-step process is called *a machine cycle*, or *a processing cycle*, and consists of two phases: *the instruction cycle* (*fetch* and *decode*) and *the execution cycle* (*execute* and *write-back*).

Secondary Storage Unit

The *secondary storage unit* provides the long-term storage for important data for future use. The data for the secondary storage unit is transferred from the memory unit to the secondary storage unit. In order to process the data residing in secondary storage, it is transferred from the secondary storage to the memory. Please remember that data stored in secondary storage can not be processed directly in secondary storage itself. The data can only be processed if it is in the memory.

The various devices used in secondary storage unit are, also called *storage devices*, are *floppy disks*, *hard disks*, *tapes*, CDs. DVDs, pen drives, etc. The floppy disks, tapes, CDs, DVDs, pen drives are removable storage devices, where as hard disks are (mostly) fixed.

The device to read and write data onto a floppy disk is known as *floppy drive*, for tape it is *tape drive*, for CD it is CD drive, for DVD it is DVD drive, etc. The hard disk and its drive are embedded in a single unit, which is fixed inside the system box.

- The *control unit*, the *arithmetic and logic unit*, and the *main memory unit*, collectively, are known as *Central Processing Unit* (CPU).
- All the external devices, such as *input device* and *output devices*, connected to the computer are known by the common name as *peripherals*.

Figure 1.2 shows the block diagram of a personal computer (microcomputer). With the advancement in the field of materials and microelectronics, it became possible to embed the *control unit*, *arithmetic and logical unit*, and limited high speed memory in the form of few *registers*, inside the single *Very Large Scale Integration* (VLSI) chip, known as a *microprocessor*.

Formally, a *microprocessor*, commonly referred as *processor*, is a general-purpose programmable device, which can receive the information in binary form, process it and transmit the results. It can be programmed to solve a wide variety of problems using the instructions that it understands. The set of all those instructions that it can understand is known as the *instruction set* of the processor.

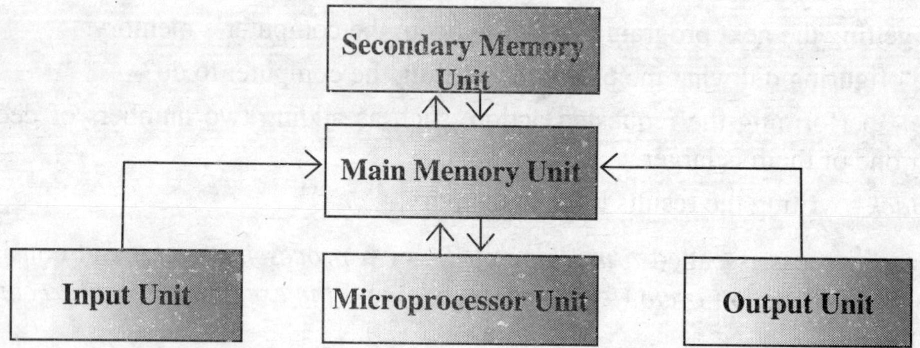

Figure 1.2: Functional diagram of a Personal Computer

The role of the other units remains the same as in the case of a general digital computer. Majority of the present PCs including laptops are built around the *Intel* or *AMD* based microprocessors.

1.6 CONCEPT OF BOOTING

In simple terms, *booting* is a process in which your computer gets initialized. This process includes initializing all your hardware components in your computer and getting them to work together and to load your default operating system which will make your computer operational. The technical details of booting process can be summarized as follows:

1. When the computer is switched on, a copy of boot program is brought from ROM into the main memory.
2. Then CPU runs a jump instruction that transfers to BIOS (Basic Input output System) and it starts executing.
3. BIOS conduct a series of self diagnostic tests called POST (Power On Self Test). These tests include memory test, configuring and starting video circuitry, configuring system's hardware and checking other devices that help computer to function properly.
4. Thereafter, the BIOS locate a bootable drive to load the boot sector. The Boot Strap Loader program loads and executes the operating system.

The booting process is of two types:

- **Cold booting:** When the system starts from initial state, *i.e.*, it is switched on. It is also called *hard booting*. The system goes through a number of stages as explained above

- **Warm booting:** When the Reset button or *Ctrl+Alt+Del* key combination is used to restart the system. It is also called *soft booting*. Here the system does not start from the initial state and therefore the diagnostic tests are not carried out in this case.

1.7 COMPUTER HARDWRE

Hardware refers to the parts of a computer that you can see and touch, including the case and everything inside it. In this chapter, we will learn about various components of the computer system that constitutes its hardware.

1.7.1 System Unit

The system unit of the PC is the case that houses the computer's main components that includes the CPU and the memory.

Figure 1.3: Typical home PC

1.7.1.1 Outside View

Figure 1.4: (*a*) Back panel of system (*b*) Front panel of system unit

- *Power Switch* – Sometimes this is located in the rear of the computer so you can't hit it by accident, which might cause you to lose your work.
- *Ports* – A port is a receptacle that enables you to plug in peripherals such as a printer, a keyboard, and mouse to the connectors, which are usually on the back of the computer.

- *Front Panel* – On most of the computer's you'll find *reset switch* that enables you to restart your computer in the event of a failure, a *drive activity light* (a light that tells when your hard disk is accessing data), and a *power-on light* (a light that tells you whether the power is on). You may also find a *key lock* that enables you to prevent others from operating the machine.

1.7.1.2 Inside View

- *Motherboard* – also called main board, is a large printed circuit board on which complex patterns of copper pathways have been created by means of an etching process. The pathways link the semiconductor chips, including the microprocessor and the chips that provide the computer's memory. They also link the motherboard's expansion slots, which are receptacles designed to fit expansion boards, also called *adapter cards*.

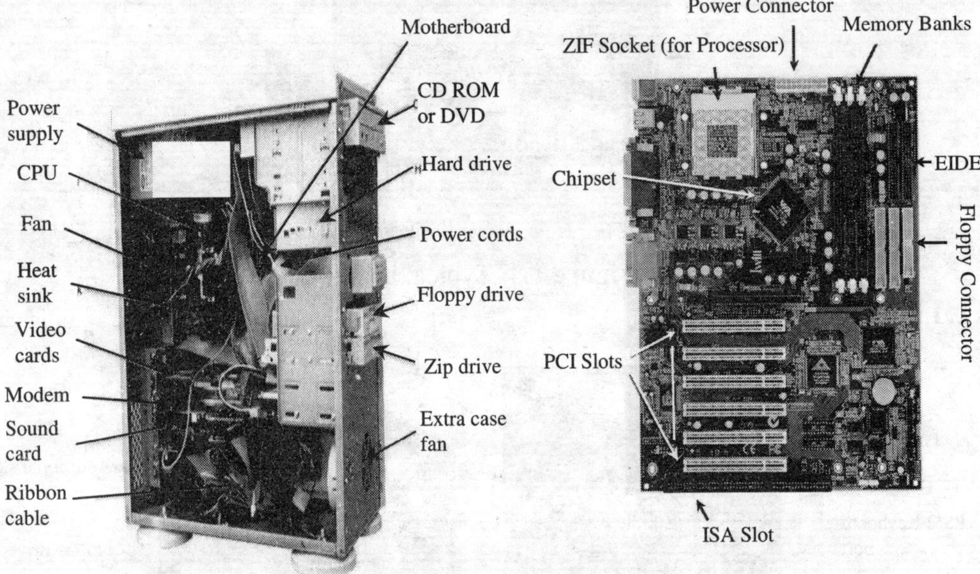

Figure 1.5: (*a*) Inside view of the system (*b*) Typical Motherboard

- *Power Supply* – it transforms alternating current into the direct current needed for the computer's operation and also steps the voltage down to the low voltage required by the motherboard.
- *Drive bays* – to accommodate the computer's *floppy disk drive*, *CD or DVD drive*, and a removable media drive such as a *Zip drive*. Some of the bays may be available for system upgrades.
- *Hard disk* – it may be mounted in a covered drive bay or deeper in the system unit.
- *Floppy disk drive* – it uses floppy disks, usually 3½-inch magnetic disks.
- *CD-ROM or DVD-ROM drive* – it uses optical CD or DVD discs.
- *Removable media drive* – it uses high-capacity removable disks.

Of all these components, by far the most important is the CPU, or microprocessor. As you've learned, in a strict sense, the CPU is the computer. More than any other factor, the CPU's characteristics determine the performance of the overall system.

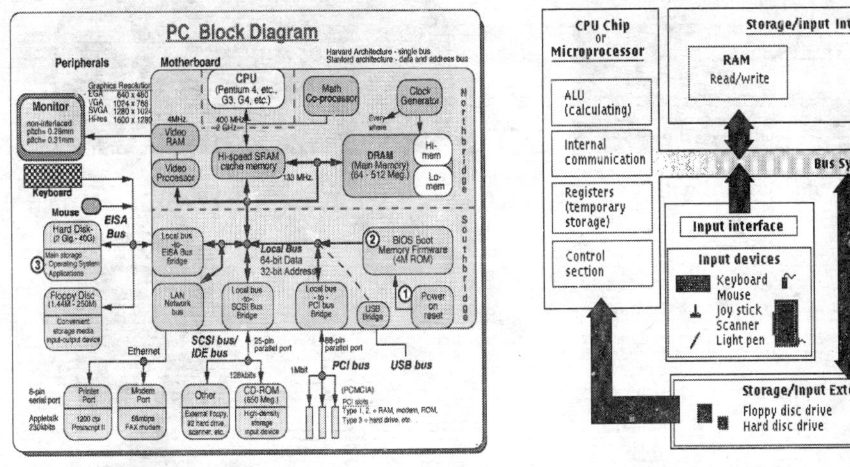

(*a*) PC Block Diagram

(*b*) Interfacing of various components

Figure 1.6: Block diagram and interfacing of various components of a PC

1.7.2 Input/Output Devices

Computers will of no use unless it can interact and communicate with the outside world. Input/output devices are attached (connected) to the computer so that it can interaction and communication with the outside world.

The input/output devices are also known as *peripherals* since they surround the CPU (processor) and memory of a computer system.

1.7.2.1 Input Devices

Input devices take the information from the outside world in human readable form, convert them into a form that a computer can understand (in 1s and 0s, i.e., binary form) and passes it to the computer for further processing. This may represent data to be processed by the computer or commands that are to be executed by the computer.

The typical input devices used in almost all computers are keyboard, mouse, microphone, and webcam, etc.

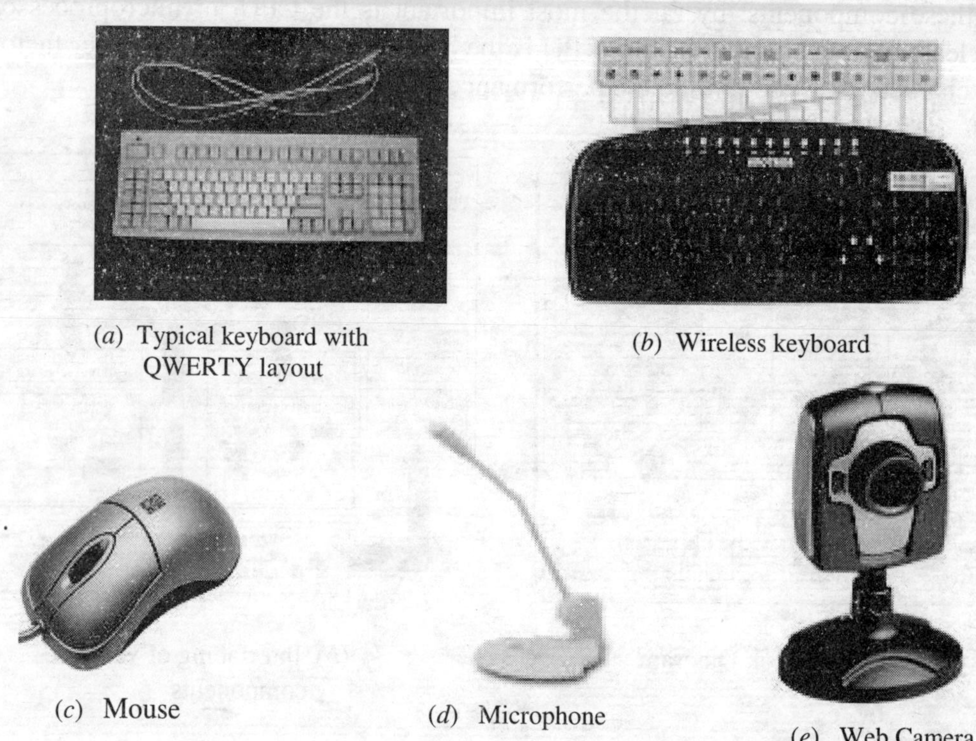

(*a*) Typical keyboard with QWERTY layout

(*b*) Wireless keyboard

(*c*) Mouse

(*d*) Microphone

(*e*) Web Camera

Figure 1.7: Common input devices

1.7.2.2 Output Devices

Output devices take the information from the computer in computer readable form, convert them into human readable form and pass onto the user in appropriate form (textual, graphical or multimedia) on a appropriate medium (display on screen, print on a paper, etc.).

The typical output devices used are visual display (monitor/screen), printer, plotter, and speaker, etc.

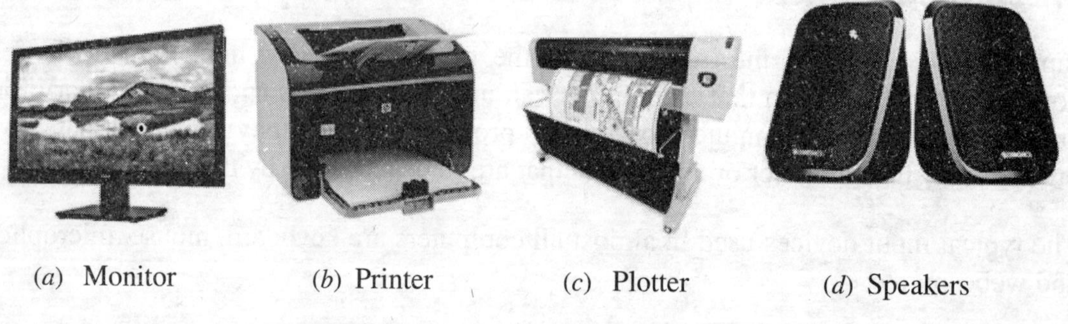

(*a*) Monitor

(*b*) Printer

(*c*) Plotter

(*d*) Speakers

Figure 1.8: Common output devices

1.7.3 Memories and Storage Devices

As you know, by this time, that hardware and software are integral parts of a computer system. Software constitutes that part of a computer system that drives the hardware to get the task done. Also software is generic name for any kind of program(s). All the programs process the relevant data to generate the desired results.

The program(s) and the data need to be stored in the computer system before they can be used. For this purpose, the memory (called *main memory* or *primary memory*) contained in the memory unit of the CPU is utilized.

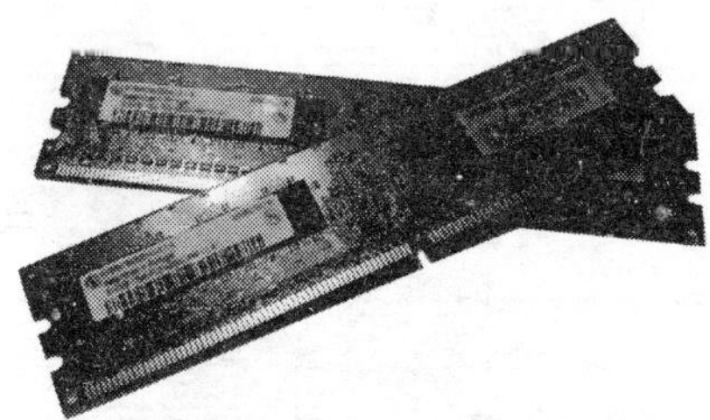

Figure 1.9: Memory chips

However, this memory loses its contents when the power is switched off.

Since the program(s) are to be used in future too, they need to be stored a kind of memory that does not lose its contents when the power is switched off. Likewise, the data entered into the computer system and the results produced, if needed in future, also need to be stored in such memory. For this purpose, additional devices, called *storage devices* (such as hard disk, floppy disk, CD or DVD, pen drive), are attached to the computer system which supplements the primary memory. The total memory that results from the various storage devices is called *secondary memory.*

(*a*) Hard disk (*b*) CD/DVD (*c*) Pen drive

Figure 1.10: Common storage devices

1.8 INPUT/OUTPUT PORTS

A *port* is a connection point that acts as interface between the computer and the internal device or external device.

A port that connects internal device (such as hard disk drive, CD ROM or DVD drive, etc.) is known as *internal port*.

On the other hand, a port that connects an external device (such as monitor, mouse, printer, modem, etc.) is known as *external port*.

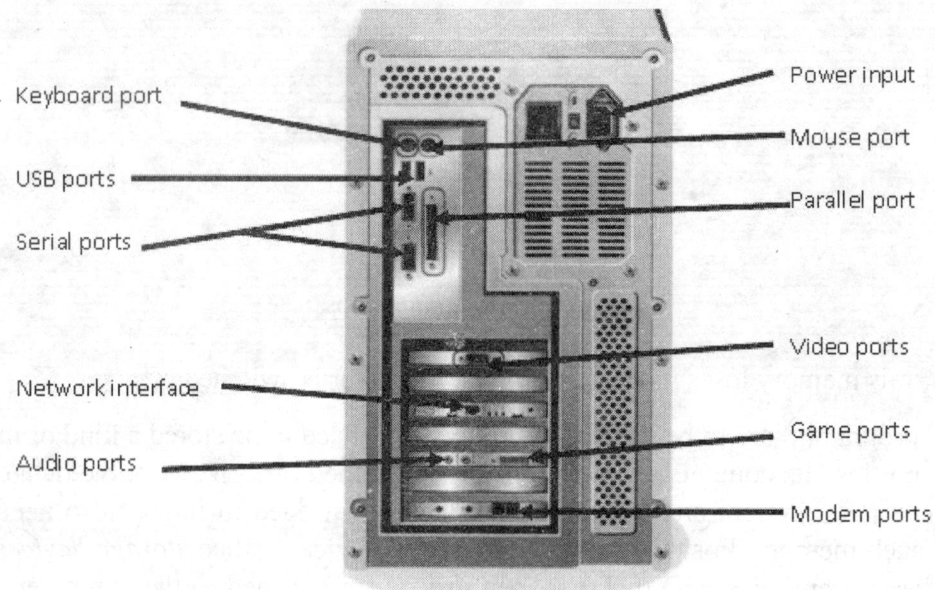

Figure 1.11: Back panel of system box showing various ports

Ports are classified as *serial ports* and *parallel ports*.

- **Serial Port** – A *serial port* is a port that transmit data sequentially one bit at a time. So they need only one wire to transmit 8 bits. The disadvantage is that it takes 8 times longer to transmit a byte. Also it is necessary to send a start bit (to mark the beginning of the byte) before each byte of data, a stop bit (to mark the end of the byte) after each byte of data, and a parity bit to help check the integrity of data. Serial ports are usually 9-pin or 25-pin male connectors.
- **Parallel Port** – A *parallel port*, in contrast to serial port, is a port that can transmit 8 bits or 1 byte at a time. The main advantage of parallel port is that it transfers data much faster than a serial port. Parallel ports come in form of 25-pin female connectors.

1.9 COMPUTER SOFTWARE

Software, in its most general sense, is a set of instructions or programs that tell the hardware what to do. Software can be difficult to describe because it is *virtual* not physical like computer hardware. Instead, software consists of lines of code (instructions) written by computer programmers that have been compiled into a computer program. Software programs are stored as binary data that is copied to a computer's hard drive, when it is installed. Since software is virtual and does not take up any physical space, it is much easier (and often cheaper) to upgrade than computer hardware.

The most important example of software is an *operating system* (OS) that manages the resources of the computer system, and provides an interface using which the user can interact with the computer to perform various tasks.

It is important to remember that hardware and software are integral parts of a computer system — *hardware is of no use without software and software cannot be used without hardware.* It is the software that drives hardware, i.e., it is the software that gives hardware its capability.

Software can be broadly are categorized as

- **System software** – It is a type of software that is designed to run a computer's hardware and application programs. If we think of the computer system as a layered model, the system software is the interface between the hardware and user applications. The *operating system* and *language translators* (*assembler*, *compiler* and *interpreter*), *linker* ad *loader* are examples of system software.
- **Application software** – It is a type of software that is designed for a specific purpose and is used by end users. Word processor, database software, spreadsheet software, entertainment software, reservation software, inventory management software, etc., are examples of application software.
- **Utility software** – It is a type of software that is designed to help analyze, configure, optimize or maintain a computer. It is used to support the computer infrastructure. Antivirus software, compression tools, disk management tools, etc., are examples of utility software.

1.9.1 System Software

System Software is the software that is directly related to coordinating computer operations and performs tasks associated with controlling and utilizing computer hardware. These programs assist in running application programs and are designed to control the operation of a computer system.

Examples of system software include operating system, assembler, compiler, interpreter Linker, loader, etc.

Compiler, interpreter and assembler are called language translators/processors. They translate source code (program written in assembly language or high-level language) to object code (machine language) which computer can understand and execute on hardware.

1.9.1.1 Operating System

The most important program on any computer is the *Operating System* or simply OS. The OS is a large program made up of many smaller programs. It acts as resource manager of the computer system and provides an interface between the users and the hardware.

The resources of a computer system include processor, memories and I/O devices, users, etc. As a manager it coordinates the activities of various resources so as to provide the users with the required information.

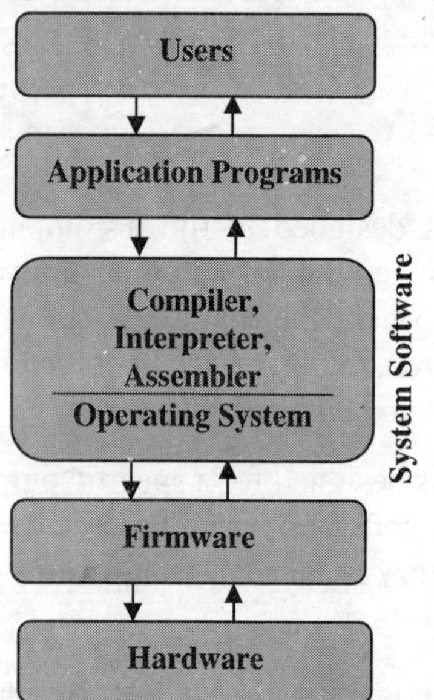

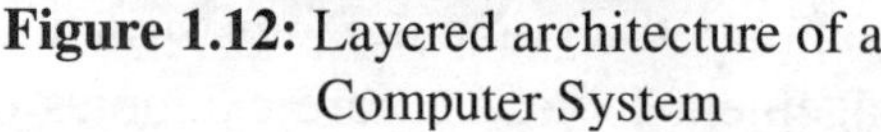

Figure 1.12: Layered architecture of a Computer System

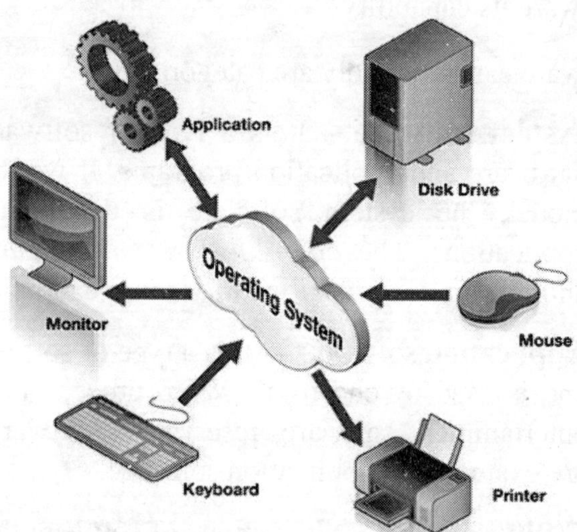

Figure 1.13: Operating System managing various resources of a Computer System

Commonly Used Operating Systems for PCs/Laptops

Some of the commonly used operating systems are discussed below:

Unix — It is a multi-user operating system. It was developed in the early 1970s at Bell Laboratories by Ken Thomson and Dennis Ritchie. It was the first operating system developed using C language.

Unix operating system and its variants are very rich in features handling system security and networking. Because of these features, Unix is a natural choice for large computer systems and for managing servers providing critical services.

Linux — Linux is a very popular open source operating system enhanced and backed by thousands of programmers worldwide. The name Linux was derived from its inventor Linus Torvalds. Linus, as a student, wrote the first version of Unix like operating system, as a toy project. Later, he posted the source code on Internet and asked the programmers across the world to help him build it into a working system, and as result the Linux was born. Linux has the ability to manage thousands of tasks at the same time. Like Unix, the Linux is also very rich in system security and networking features.

Windows — Windows is an operating systems developed by Microsoft to overcome the limitations of its own operating system – *the MS DOS.*

Windows operating system provides an easy to use graphical user interface (GUI). As part of the GUI, it shows a desktop where the various tasks are represented by small pictures, called *icons*. The user only needs to point at an icon and click into order to select and double click to launch the program/application.

Microsoft launched Windows 1.0 operating system in 1985 and since then Windows has ruled the world's software market, and launched various versions of Windows line Windows 95, Windows 98, Window NT, Windows XP, Windows 7, Windows 8, and the latest being Windows 10 (at the time of publishing this book).

- **Solaris** solaris — Solaris is a Unix based operating system originally developed by SUN Microsystems. In 2010, after SUN's acquisition by Oracle, it was renamed Oracle Solaris.

  It is known for its scalability. It supports SPARC-based and x86-based workstations and servers from Oracle and other vendors. Efforts are on to port to additional platforms.

- **BOSS (Bharat Operating System Solutions)** BOSS Bharat Operating System Solutions — It is a Linux distribution developed by C-DAC, Chennai, in order to benefit the usage of Free/Open Source Software in India. BOSS Linux, derived from Debian, is a key deliverable of National Resource Centre for Free/Open Source Software (NRCFOSS). It has enhanced Desktop Environment integrated with Indian language support and other software.

  It has been endorsed by the Government of India for adoption and implementation on a national scale.

Commonly Used Operating Systems for Mobile/Hand Held Devices

Android — Android is an open source and Linux-based mobile operating system for mobile devices such as smart phones and tablets.

Android is one of the most user friendly, i.e., easy to use. Its user interface is mainly based on direct manipulation, using touch gestures that loosely correspond to real-world actions, such as swiping and tapping, to manipulate on-screen objects, along with a virtual keyboard for text input. Initially it was originally developed by Android Inc., and was later bought by Google in 2005.

iOS — Originally iPhone OS, is a mobile operating system created and developed by Apple Inc., and distributed exclusively for Apple hardware. It is the operating system that presently powers many of the company's mobile devices, including the iPhone, iPad, and iPod touch. It is the second most popular mobile operating system in the world by sales, after Android.

Symbian — Symbian is a mobile operating system targeted at mobile phones that offers a high-level of integration with communication and personal information management (PIM) functionality.

It combines middleware with wireless communications through an integrated mailbox and the integration of Java and PIM functionality.

Nokia has made the Symbian platform available under an alternative, open and direct model, to work with some Original Equipment Manufacturers (OEMs) and the small community of platform development collaborators.

1.9.1.2 Language Translators

Language translators are also known as *language processors*. It is pertinent to mention here that while machine language requires no translation, the assembly language and high-level languages requires translators to translate instructions written in assembly language and machine language before they can be executed.

A program written using assembly language or high-level language is known as *source code*. The translated version of the source code is a program in machine language is known as *object code*.

The language translator for assembly language is known as an *assembler*, and the language translators for high-level languages are known as *compilers* and *interpreters*.

Assembler

An assembler is a program that translates a program in assembly language (*source code*) to machine language (*object code*).

As there is one-to-one correspondence between instructions in machine language and assembly language, the assembler translates each instruction in assembly language to corresponding instruction in machine language.

Compiler

A compiler is a program that translates (*compiles*) the program written in a high-level language (*source code*) to machine language (*object code*).

This process of translation is known as *compilation*. During compilation, compiler reads the source code, line by line, and checks for syntax errors. Note that any violation of the rules of the programming language (also known as the grammar of the language), is known as *syntax error*.

If there are syntax errors, then it displays the error messages identifying the error locations and prints a program listing that highlights the locations and the likely causes of the errors, and no translation takes place. However, if there are no syntax errors in the source code, the compiler re-reads the source code and does the translation and creates an object code which then it writes onto another file, usually with extension *obj* (for object).

Once translation is done, the compiler has no further role in the subsequent steps. However, if the source code is modified, we need to recompile it to incorporate the modification in the object code.

Interpreter

An interpreter is another type of translator used for translating high-level languages. However, interpreter doesn't produce object code. Instead, it translates one line of the source code at a time and executes the translated instruction. It continues till all the instructions are translated an executed. Note that if any instruction is to be executed repeatedly, it will be translated each time.

Note that once the source code is translated into the object code by the compiler, which is one time task, the object code, after further processing (linking), is executed directly which result in faster execution of the program. However, the interpreter first understands the code and then executes the instruction, which lead to slow execution of the program. In addition, every time the program is executed using interpreter, the translation will be repeated.

Although interpreters run programs more slowly than programs compiled into object code, they are helpful tools for learning language for the beginners and also for debugging the programs. As the program executes line by line the programmer can see exactly what each line does.

Here it is important to mention that in order to execute a program using interpreter, first we need to run the interpreter, then only then we can instruct the interpreter the execute the user program.

Therefore, every time to run the program, you need to run the interpreter as well.

For most programming languages, a programmer can use both an interpreter and a compiler. The interpreter comes in handy during program development. When the program development is finished, it is compiled for efficient use.

Table 1.1: Comparison of Compiler and Interpreter

Compiler	Interpreter
It takes entire program as unit of translation and does not execute the program.	It translates one instruction at a time and executes it.
After translation, it writes the object code onto a disk file.	It translates instruction by instruction and does not create a permanent object code.
It does not execute the program after translation.	It executes every instruction after translation and continues doing so till all instructions are finished.
After translation, neither the source code nor the compiler is required for executing the program.	Both, interpreter as well the source code are required for execution of the program. In fact, first interpreter program is executed, and then it is instructed to execute the program under consideration.
It takes entire program as unit of translation and does not execute the program.	It translates one instruction at a time and executes it.
After translation, it writes the object code onto a disk file.	It translates instruction by instruction and does not create a permanent object code.
It does not execute the program after translation.	It executes every instruction after translation and continues doing so till all instructions are finished.
After translation, neither the source code nor the compiler is required for executing the program.	Both, interpreter as well the source code are required for execution of the program. In fact, first interpreter program is executed, and then it is instructed to execute the program under consideration.
Program execution is very fast since the object code is directly executed.	Program execution is slow since each instruction is translated first and then executed.
Compilers play almost no role in debugging, since after compilation, the compiler is not in picture and have no role during program execution.	Debugging is very easy since each instruction is executed one-by-one, programmer and/or user can see what each instruction does.
To execute the program on a different computer, you need to compile the program on that computer. It will be essential sometimes since it may have different machine language	However, if you have a interpreter on another machine, the program can be executed as such since translation takes place during execution, and will take place automatically.
Well suited for application where high performance is the need.	Well suited for the beginners who are just learning the particular high-level language.

1.9.1.3 Programming Environment

Programming environment is an environment in which programs are created and tested, and it has less influence on the design of the language that the operating environment on which the programs are expected to be run.

A programming environment consists of support tools and a command language (set of commands) to invoke them. Each support tool is another program that may be used by the programmer as an aid during one or more stages of creation of a program.

Typical tools in a programming environment include *editors, language translators, linkers, loaders, test data generators*, etc. Language translators are explained in the earlier section, the remaining tools are described here.

Editor – *Editor* is a program that allows creating and editing of files. Further, there are editors that allow working with text files only, and are called *text editors*. Examples of popular text editors are *notepad* and *edit* of *Windows* operating system, *vi* of *Unix* and *Linux* operating systems. Even popular word processing programs such as MS Word, allows to create text files in addition to word documents.

Linker – Note that the object code produced by the compiler is not in a form that can be executed by the computer. Some more information needed to be pumped into the object code to make it in a form that can be executed by the computer.

When you write a program, you may be referring to various library functions for doing input/output and for mathematical computation (*sqrt, sin, cos, abs, exp* to name a few frequently used mathematical library functions). The instructions defining the task to be performed by these library function are in the machine language (binary) form and are contained in system libraries (files with extension LIB) supplied with the compiler. Plus, you may be referring to some user-defined functions which are written in a separate program file, and whose object code is lying separately than the object code under consideration. Therefore, there is a need to combine the object code of the library as well as user-defined functions with the object code under consideration, and this task is accomplished by the program called *linker.*

Linker is a program that combines the object of the program with the object codes of the library and user-defined functions, in the desired manner, and writes the same in a separate program file with extension, usually EXE under Windows. This file produced by the linker is known as *executable file* or simply as *run file*, and is the file that is used to execute the program.

Loader – *Loader* is a program that is responsible for transferring the executable file from the secondary storage to memory. On execution, the loader, with the assistance of the operating system, first finds the free memory of the requisite amount and then transfers the executable file into that part of memory, takes appropriates steps such as address translation, and then initiates its execution. From that point onwards, the role of the loader is over, and the control is taken over by the program. Instruction of the program are executed to perform the desired task, but under the overall supervision of operating system. As soon as the program finishes its execution, it is removed from the memory.

Debugger – All sort of errors that are encountered during program development (syntax errors, logical error and runtime/execution errors) are known as bugs, and the process of identifying those errors and fixing (correcting) them is known as *debugging*. The logical errors are most difficult to track. To catch logical errors, we need to observe the pattern in which instructions of program are executed, intermediate and final results are produced.

Debugger is a program that, in general, allows user to monitor the trace the execution of executable version of program.

Test Data Generator – *Test data generator* is a powerful and simple automatic tool that allows you to generate test data for testing a programs. The testing data is generating according to the internal structure of the program. Note that here by test data we mean that data for which the desired output is known. Testing is done to ensure that the program is performing its intended task.

Popular Integrated Developments Environments (IDEs), such as Turbo C/C++ and Borland C++, Visual Studio, Dev C++, CodeBlocks, Netbeans, Eclipse, *etc.*, integrate various tools (editor, compiler, linker, debugger, etc.) in a unified manner, so that a programmer can work with these tools with much ease and more productively.

1.9.2 Application Software

Application software is a term which is used for software created for a specific purpose. It is generally a program or collection of programs used by the end users. It can be called an *application* or simply an *app*.

Application software can be of two types – *general purpose* and *customized.*

Office tools come under the category of general purpose application software where as *domain specific tools* come under the category of customized application software.

1.9.2.1 Office Tools

Office tools represent that category of application software that is developed for the common users for their day-to-day applications and uses. The users may use them in the manner they want.

Some of the commonly used office tools are *word processor software* (for example, *Microsoft Word), presentation tool/software* (for example, *Microsoft PowerPoint), spreadsheet software* (for example, *Microsoft Excel),* and *database management system* (for example, *MySQL, Oracle*), *etc*.

1.9.2.1 Domain Specific Software/Tools

Domain specific software/tools represent that category of application software that is developed as per the user's requirement. Such type of software is customer specific. It is made keeping in mind the individual needs of the user. Such software cannot be installed and used by any other user/customer since the requirements may differ.

Some of the commonly used domain specific tools are *school management system, inventory management system, payroll system, financial accounting, hotel management, reservation system*, and *weather forecasting system, etc.*

1.9.3 Utility Software

Utility software is software designed to perform specific tasks related to the *file management, file compression, disk management, virus protection*, etc.

It is used to support the computer infrastructure in contrast to application software, which is aimed at directly performing tasks that benefit ordinary users.

Nowadays, utility software is part of the operating system itself.

Commonly used utility software include *anti virus, file management tools, compression tools* and *disk management tools* such as *disk cleanup, disk defragmenter, backup.*

1.9.4 Open Source Concepts

Software is mainly categorized into the following categories based on their licenses:

- Open Source Software
- Freeware
- Shareware
- Proprietary Software

❑ Open Source Software

The term *open source* refers to something that can be modified and shared because its design is publicly accessible.

Open source software (OSS) is software with source code that anyone can use, inspect, modify, enhance and redistribute.

Source code is the part of software that programmers can improve that software by adding features to it or fixing parts that don't always work correctly.

Linux, PHP, MySQL, Java, and Python, etc., are few examples of Open Source Software.

- **Freeware**

Freeware is any copyrighted software that may be freely downloaded, installed, used and shared. Such programs are available for use at no cost to general end users. Freeware differs from free software, as the latter allows a user to modify source code for republishing or integration with other software.

Generally, freeware is a concise and limited version of a larger and paid software program. Vendors publish freeware to provide a limited but free version of software to prospective buyers prior to purchase. Additionally, large independent software vendors publish freeware to enhance brand buzz and market reputation.

- **Shareware**

Shareware is also copyrighted software that is distributed free, for specified period of time, on a trial basis with the intention that the user may buy it later. When the trial period ends, the software must be purchased or uninstalled.

Some shareware come with a built-in expiration date such as after 30 days, the user can no longer get access to the software.

- **Proprietary Software**

Proprietary means something owned by a specific company or an individual.

Therefore, proprietary software is either owned by a company or an individual. It is commercial software that can be bought, leased or licensed from its vendor/developer.

The restrictions or conditions imposed by the vendor/developer on the use of proprietary software are elaborated within the software's end-user license agreement (EULA), terms of service agreement (TOS) or other related use agreements. The user/organization must accept the agreement prior to installing or using the software. The software developer/vendor can take legal action against the end-user/organization for violating the EULA or TOS.

1.10 PROGRAMMING LANGUAGES

You are aware with the term language. A language is nothing but a means of communication. As you know, we communicate with each other either by speaking or by using gestures.

When we communicate by speaking, we are doing so by using languages that we know such as English, Punjabi, Hindi, etc. The languages spoken by the human beings are called *natural languages*.

When we communicate by gestures, we are doing so by the involvement of our physical organs. This language of communication is called *the body language*.

Similarly, the languages used for the purpose of instructing the computer to do a given task are called *computer programming languages* or simply as *programming languages.* Since the evolution of the computer, *a magical machine*, many programming languages have been developed.

1.10.1 Classification of Programming Languages

There are various classifications of programming languages. Although it is not possible to sort all languages into convenient groups, some attempt to categorize them is still possible and indeed helpful because some of them require very different programming techniques.

The oldest approach to classification of programming languages is according to the level of interaction and their dependence on the organization and architecture of the underlying processor. This approach divides programming languages into two major categories – *low-level languages* and *high-level languages*.

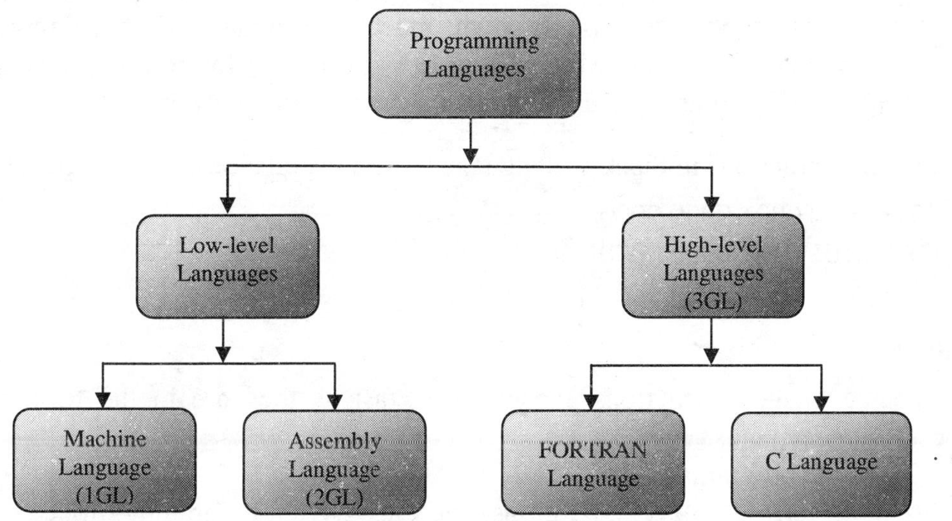

Figure 1.12: Classification of programming languages

1.10.1.1 Low-level Languages

Using low-level languages, you can have total control of the processor and can instruct the processor to perform all types of tasks that a processor can perform. However, to use low-level languages, you need to have through knowledge of the architecture of the processor, *i.e.*, internal functional blocks and set of instruction that they follow.

The low-level languages are further classified as follows:

- Machine Language
- Assembly Language

1.10.1.1.1 Machine Language

Machine language, as its name suggests, is the language that a machine (computer, hence *the processor*) can understand. Since, it was the first and the only language used when computer were invented, it was also known as first generation programming language (1GL).

The machine language involves expressing the instructions using 1s and 0s only. Each instruction can be divided into parts, where first part is used to instruct the computer *what to do*, called *the operation code*, and the remaining parts are used to represent the data on which the computer will perform the specified the operation.

Since, programming using machine language require an in-depth knowledge of the hardware, only people with sound knowledge in the disciplines of Physics, Electrical & Electronics Engineering were able to program the early computers.

Further, since the hardware design of the processor varies from manufacturer to manufacturer, each processor has its own machine language. For example, Intel processors from Intel Company, Motorola processors from Motorola Company, AMD processors from AMD Company, etc., have their own machine language.

Example 1.1: Instructions in machine language to find average of numbers 10 and 50.

```
1011 1000 0001 0000 0000
0000 0101 0101 0000 0000
1101 0001 1110 1000
```

Advantages

1. Machine language instructions execute very fast as they are in the form that the processor understands.
2. The size of the programs is very small.
3. Ability to perform low-level operations, and thus gives a programmer total control of the hardware.

Disadvantages

1. Writing instructions in a machine language is very time-consuming and error-prone task. In addition, the programmer needs to have detailed knowledge of the hardware.
2. Program written in machine language is very difficult to read, understand and modify.
3. Machine language is dependent on the processor. Therefore, you can't use the same program on a computer with different processor, *i.e.*, machine language is not portable. You have to write it again from scratch, which a wastage of time and effort.
4. The programmer has to remember a lot of operation codes to write a program, which results in program errors.
5. High programming skill is required to do programming in machine language.
6. It is difficult to debug (locate and correct the errors) the program.

1.10.1.1.2 Assembly Language

The name given to this language is *the assembly language*. The use of mnemonics in preparing the instructions made the task of programming a much easy as compared to machine language. However, since computer can understand the machine language only, computer instructions written using assembly language need to be translated to the machine language. To do this translation, a translator program, called *assembler* was developed. The assembler will take the instructions written using assembly language and translates them to the machine language.

Table 1.2: Common mnemonics used in assembly language

Mnemonic	Description
MOV	To assign (copy) source operand to destination operand.
ADD	To perform addition operation.
SUB	To perform subtraction operation.
DIV	To perform division operation.
MUL	To perform multiplication operation.
JMP	To transfer program control from one section of program to another section.
CMP	To perform the comparison operation for decision making.
SHR	To shift the bits of the operand from left to right.
SHL	To shift the bits of the operand from right to left.

Note that, like the machine language, there is one and only one assembly language for a given computer.

Example 1.2: Instructions in assembly language to find average of numbers 10 and 50.

```
MOV AX, 10
ADD AX, 50
SHR AX, 1
```

Advantages

1. The symbolic programming of assembly language is easier to understand and saves a lot of time and effort of the programmer.
2. It is easier to correct errors and modify program instructions.
3. Assembly language has the same efficiency of execution as the machine level language, because this is one-to-one correspondence between instruction in assembly language program and machine language.

Disadvantages

1. Assembly language is also machine dependent.
2. Even to do a simple task, you have to write a lot of instructions.
3. High programming skill is also required to do programming in assembly language.

1.10.1.2 High-level Languages

Though, the assembly language made the task of programming bit easy as compared to machine language, but still to work with assembly language you require deep knowledge of the computer hardware. Also, the assembly language is machine dependent, and hence the program written in assembly language can't be used on other computers.

Therefore, further need was felt to have programming languages, where you don't require the knowledge of computer hardware, and also the program once developed can be executed (run) almost on any computer. The quest to fulfill such a need lead to the development many languages which uses an English language like styles to express instructions. These languages were given the name of *high-level languages (HLLs).*

Like, assembly language, the instructions expressed using high-level languages need to be translated into machine language. This translation can be done before the computer executes the instructions or during execution of the instructions. The translator that translates the instructions before they are executed is called *compiler*, and the translator that translates the instructions during execution is called *interpreter*.

You need separate compiler and/or interpreter for each high-level language, and for each computer. However, once you write a program in majority of high-level languages, it can be used on any computer without any modification or with minimum modifications, and thus it saves lot of valuable time and effort.

Example 1.3: Instruction in C, C++, and Java languages, to find average of numbers 10 and 50.

```
average = ( 10 + 50 ) / 2;
```

Advantages

1. High-level languages are easy to learn and use. It is because that they are similar to the languages used by us in our day to day life.
2. The effort required to write instructions in high-level languages is very less as compared to machine language and assembly language.
3. High-level languages are machine independent, and thus programs written using high-level languages can be used on any computer.
4. Program written in high-level languages can be easily understood, and hence modification can be done with quite ease.

Disadvantages

1. Programs developed in high-level languages are large, and thus occupy more main memory and take more time to execute, *i.e.*, high-level languages are less efficient.
2. Majority of the high-level language lack in features for performing low-level operations, with an exception of C/C++ languages.

Table 1.3: Description of some modern HLLs

HLL	Targeted Application
C	Well suited for creating system programs such as operating systems, language translators, etc.
C++	Well suited for large and complex applications. Uses object-oriented approach to manage size and complexity.
Java	Well suited for mobile & distributed applications. Uses object-oriented approach to manage size and complexity. At present around 3 billion devices run on Java.
Python	Well suited for artificial intelligence based applications such as Machine Learning.
R	Well suited for statistical computing and graphical applications.
Kotlin	A new language has all features of Java but is very concise as compared to Java. Kotlin is now Google's preferred language for Android app development.

1.10.1.3 Fourth Generation Languages (4GLs)

Though, high-level languages makes task of programming quite easy, still we have to develop a procedure (a set of instructions) to complete an entire task. Because of this, majority of the high-level languages fall into the category of what we call as *procedural languages*.

In order to handle the routine type of tasks, a need was felt to have a kind of languages where the user of the data don't have to write a detailed procedure, which lead to the development of non-procedural languages. One of the most popular languages with this capability is Structured Query Language (SQL). SQL allow the user to insert new data into the database, delete data from the database, and retrieve data from the database without writing the detailed procedure.

For example, the following simple statement in SQL

```
SELECT * FROM EMPLOYEE
```

will retrieve the information of all the employees from the table named *EMPLOYEE* in the given database.

Table in a database, in lay man's language, is something like a file containing the information of the employees in an organization. Database is something like collection of all the files containing information about the entire organization.

1.11 DATA REPRESENTATION AND NUMBER SYSTEMS

In computers, data as well as instructions to manipulate the data to achieve the desired results, is represented using 1s and 0s, i.e., it uses binary number system. Data representation refers to the form in which the data is stored, processed, and transmitted. Numeric data consists of numbers that can be used in arithmetic operations.

In simple terms, a *number system* is a way to represent numbers.

Formally, a number system is defined as a set of

1. Symbols or digits,
2. Rules describing the ways numbers are formed from these digits, and
3. Rules specifying the operations performed on these numbers.

The total number of digits in a number system is called *base* or *radix* of the number system.

Computer system supports following number systems.

- Decimal number system
- Binary number system
- Octal number system
- Hexadecimal (hex) number system

Table 1.4: Summary of number systems

Number System	Radix	Symbols Used	Example
Binary	2	0, 1	$(1010011)_2$
Octal	8	0, 1, 2, 3, 4, 5, 6, 7	$(235)_8$
Decimal	10	0, 1, 2, 3, 4, 5, 6, 7, 8, 9	$(1200)_{10}$ or simply as 1200
Hexadecimal	16	0, 1, 2, 3, 4, 5, 6, 7, 8, 9, A, B, C, D, E, F where A = 10, B = 11, C = 12, D = 13, E = 14, F = 15	$(10A3)_{16}$

Since binary digits are contained in octal digits, which in turn are contained in decimal digits, which further are contained in hexadecimal digits, therefore, if we are working with many number systems then we need some notation to distinguish the number expressed in these number systems. Therefore, as a general notation, if $d_1, d_2, d_3, \ldots, d_n$ are distinct symbols used in a particular number system with base *r*. Then a number with digits d_1, d_3, d_7 will be expressed as $(d_1d_3d_7)_r$.

For example, number

$(1245)_{10}$ is a decimal number, $(10101)_2$ is a binary number,
$(124)_8$ is an octal number, and $(101)_{16}$ is a hexadecimal number.

However, if the base is omitted in expressing the number, the number will be treated as decimal number, by default.

1.11.1 Decimal Number System

The word decimal is derived from the Latin word *deci*, meaning ten. Decimal number system uses 10 symbols to represent quantitative values: 0, 1, 2, 3, 4, 5, 6, 7, 8, and 9. In decimal system, each positional value (each weight) is 10 raised to the power of its position in the number.

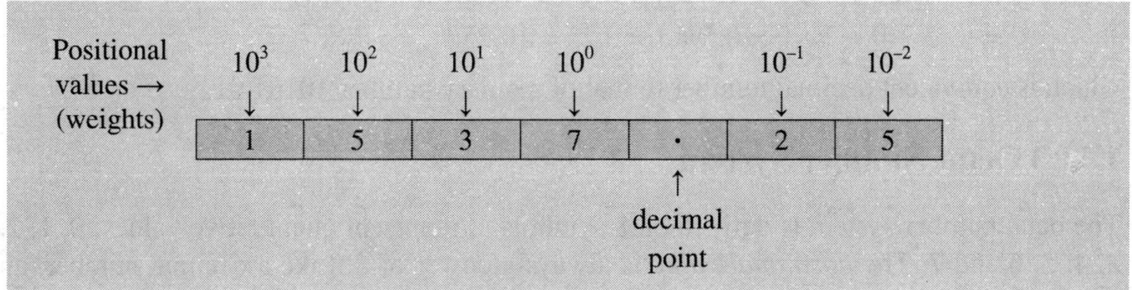

Figure 1.13: Positional values (weights) in decimal number

Therefore, the decimal number 1537.25 can be expressed as a sum of products of face values and positional values, as shown below:

$$1\times10^3 + 5\times10^2 + 3\times10^1 + 7\times10^0 + 2\times10^{-1} + 5\times10^{-2}$$
$$= 1 \times 1000 + 5 \times 100 + 3 \times 10 + 7 \times 1 + 2 \times (1/10) + 5 \times (1/100)$$
$$= 1000 + 500 + 30 + 7 + 0.2 + 0.05$$

Adding all these terms, we get back the given number.

1.11.2 Binary Number System

The binary number system is defined on digits 0 and 1 only and has a base of 2. The *positional value* is always a power of 2. Like a decimal number, a binary number can also be expressed as a sum of products of face values and positional values.

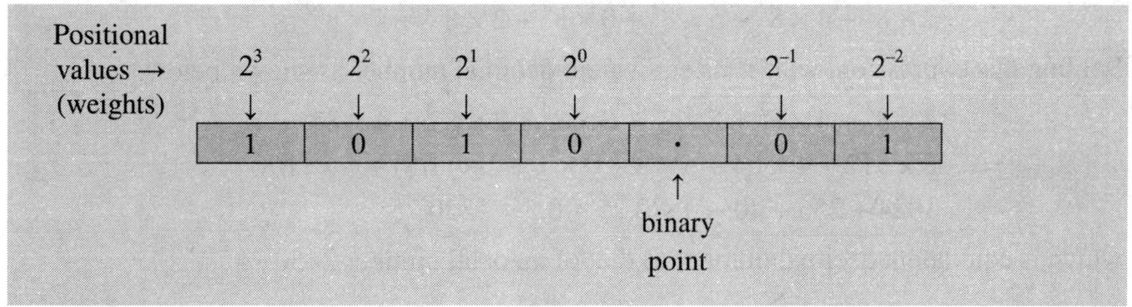

Figure 1.14: Positional values (weights) in binary number

In binary number system, each positional value (each weight) is 2 raised to the power of its position in the number.

For example, the binary number 1010.01 can be expressed as

$$1 \times 2^3 + 0 \times 2^2 + 1 \times 2^1 + 0 \times 2^0 + 0 \times 2^{-1} + 1 \times 2^{-2}$$

Solving this expression, we get its equivalent decimal number as shown below:

$$\begin{aligned} & 1 \times 2^3 + 0 \times 2^2 + 1 \times 2^1 + 0 \times 2^0 + 0 \times 2^{-1} + 1 \times 2^{-2} \\ = \; & 1 \times 8 + 0 \times 4 + 1 \times 2 + 1 \times 1 + 0 \times (1/2) + 1 \times (1/4) \\ = \; & 8 + 0 + 2 + 0 \times 0.5 + 1 \times 0.25 = 10.25 \end{aligned}$$

which is equivalent decimal number to that of a binary number 10101.01.

1.11.3 Octal Number System

The octal number system is defined on 8 symbols to represent quantitative values: 0, 1, 2, 3, 4, 5, 6 and 7. The *positional value* is always a power of 8. Like a decimal number, an octal can also be expressed as a sum of products of face values and positional values.

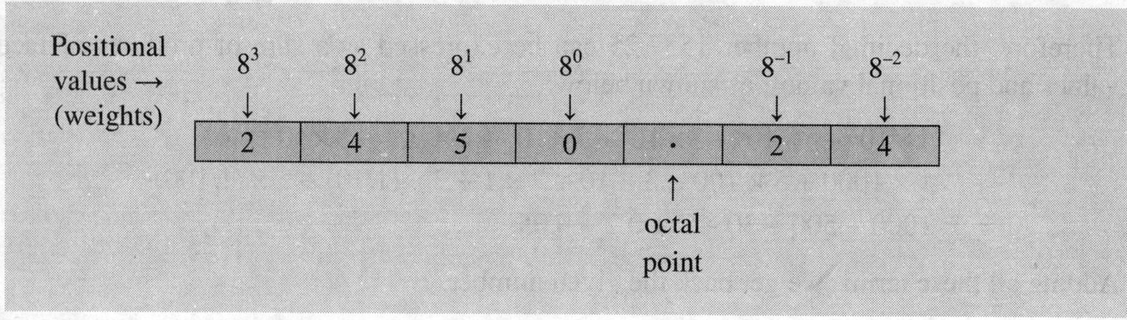

Figure 1.15: Positional values (weights) in octal number

In octal number system, each positional value (each weight) is 8 raised to the power of its position in the number.

For example, the octal number 2450.24 can be expressed as

$$2 \times 8^3 + 4 \times 8^2 + 5 \times 8^1 + 0 \times 8^0 + 2 \times 8^{-1} + 4 \times 8^{-2}$$

Solving this expression, we get its equivalent decimal number as shown below:

$$\begin{aligned} & 2 \times 8^3 + 4 \times 8^2 + 5 \times 8^1 + 0 \times 8^0 + 2 \times 8^{-1} + 4 \times 8^{-2} \\ = \; & 2 \times 512 + 4 \times 64 + 5 \times 8 + 0 \times 1 + 2 \times (1/8) + 4 \times (1/8) \\ = \; & 1024 + 256 + 40 + 0 + 0.25 + 0.5 = 1320.75 \end{aligned}$$

which is equivalent decimal number to that of an octal number 2450.24.

Since there are only 8-digits in the octal number system, therefore, only 3-bits ($2^3 = 8$) are sufficient to represent an octal digit as binary number, as shown in Table 1.5.

Table 1.5: Octal Number System

Octal Digit	Equivalent Binary Representation
0	000
1	001
2	010
3	011
4	100
5	101
6	110
7	111

1.2.4 Hexadecimal Number System

The hexadecimal number system is defined on 16 symbols to represent quantitative values: 0, 1, 2, 3, 4, 5, 6, 7, 8, 9, *A*, *B*, *C*, *D*, *E* and *F*. The letters *A*, *B*, *C*, *D*, *E* and *F* have associated with decimal values 10, 11, 12, 13, 14 and 15 respectively.

The *positional value* is always a power of 16. Like a decimal number, a hexadecimal can also be expressed as a sum of products of face values and positional values.

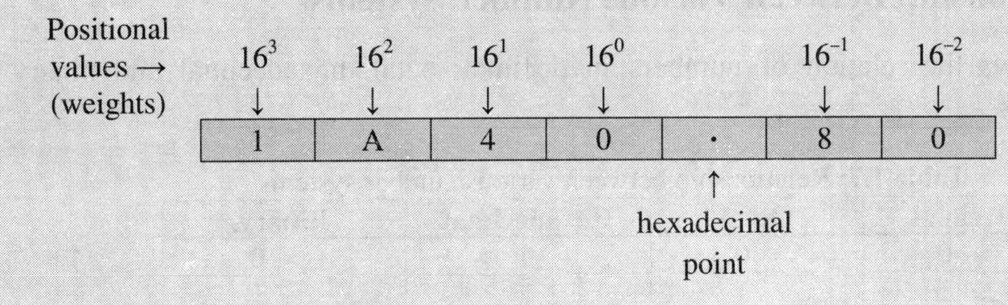

Figure 1.16: Positional values (weights) in hexadecimal number

In hexadecimal number system, each positional value (each weight) is 16 raised to the power of its position in the number.

For example, the hexadecimal number 1A40.80 can be expressed as

$$1 \times 16^3 + 10 \times 16^2 + 4 \times 16^1 + 0 \times 16^0 + 8 \times 16^{-1} + 0 \times 16^{-2}$$

Solving this expression, we get its equivalent decimal number as shown below:

$$1 \times 16^3 + 10 \times 16^2 + 4 \times 16^1 + 0 \times 16^0 + 8 \times 16^{-1} + 0 \times 16^{-2}$$

$$= 1 \times 4096 + 10 \times 64 + 4 \times 16 + 0 \times 1 + 8 \times (1/16) + 0 \times (1/256)$$

$$= 4096 + 640 + 64 + 0 + 0.25 + 0 = 4800.25$$

which is equivalent decimal number to that of an octal number 2450.24.

Since there are 16-digits in hexadecimal number system, therefore, 4-bits ($2^4 = 16$) are sufficient to represent a hexadecimal digit as binary number, as shown in Table 1.6.

Table 1.6: Hexadecimal Number System

Hexadecimal Digit	Equivalent Binary Representation
0	0000
1	0001
2	0010
3	0011
4	0100
5	0101
6	0110
7	0111
8	1000
9	1001
A	1010
B	1011
C	1100
D	1101
E	1110
F	1111

1.11.5 Relationship Between Various Number Systems

Table 1.7 shows the relation of numbers in decimal, octal, hexadecimal and binary number systems.

Table 1.7: Relationship between various number systems

Decimal	Octal	Hexadecimal	Binary
0	0	0	0
1	1	1	1
2	2	2	10
3	3	3	11
4	4	4	100
5	5	5	101
6	6	6	110
7	7	7	111
8	10	8	1000
9	11	9	1001
10	12	A	1010
11	13	B	1011
12	14	C	1100
13	15	D	1101
14	16	E	1110
15	17	F	1111

1.11.6 Number Conversions

In this section, we will learn about the way a number in a given number system can be converted to another number system.

1.11.6.1 Decimal to Other Number Systems

Suppose we want to convert a decimal number into a number in another number system with radix r, we proceed as follows.

Divide the number successively by r till the quotient is reduced to 0. The remainder after each division is recorded. The entire sequence of remainders taken in reverse order represents the equivalent number in a given number system with radix r.

1.11.6.1.1 Decimal to Binary

Example 1.4: Convert decimal number 55 to binary.

Solution: Reading the remainders from bottom to top, we have

$(55)_{10} = (110111)_2$

		Remainders
2	55	
2	27	1
2	13	1
2	6	1
2	3	0
2	1	1
	0	1 ↑

1.11.6.1.2 Decimal to Octal

Example 1.5: Convert decimal number 267 to octal.

Solution:

Reading the remainders from bottom to top, we have

$(267)_{10} = (413)_8$

		Remainders
8	267	
8	33	3
8	4	1
	0	4 ↑

1.11.6.1.3 Decimal to Hexadecimal

Example 1.6: Convert decimal number 267 to hexadecimal.

Solution:

Reading the remainders from bottom to top, we have

$(423)_{10} = (1A7)_{16}$

		Remainders
16	423	
16	26	7
16	1	10 → A
	0	1 ↑

1.11.6.2 Binary to Other Number Systems

1.11.6.2.1 Binary to Decimal

For binary to decimal conversion, we express the binary number in the form of sum of products of positional value and face value of binary digits.

Example 1.7: Convert $(1011)_2$ to decimal.

Solution:

$$(1011)_2 = 1 \times 2^3 + 0 \times 2^2 + 1 \times 2^1 + 1 \times 2^0 = 8 + 0 + 2 + 1 = 11$$

$\therefore \quad (1011)_2 = (11)_{10}$

1.11.6.2.2 Binary to Octal

For binary to octal conversion, we use the following steps:

1. Form combinations of 3-bits. Start from right-to-left, and if in the last combination bits left are less than 3, we add 0 bits on the left.
2. Now replace each combination of 3-bits by the corresponding octal digit, as given in Table 1.5.

Example 1.8: Convert $(10110110)_2$ to octal.

Solution: Start making combination of 3-bits as shown below:

10 110 110

Now, we find that the last combination is of 2-digit only, therefore, we add 0-bit on the left of the last combination as shown below:

$$\underbrace{010}\ \underbrace{110}\ \underbrace{110}$$

↓ ↓ ↓

2 6 6

$\therefore \quad (10110110)_2 = (266)_8$

1.11.6.2.3 Binary to Hexadecimal

For binary to hexadecimal conversion, we use the following steps:

1. Form combinations of 4-bits. Start from right-to-left, and if in the last combination bits left are less than 4, we add 0 bits on the left.
2. Now replace each combination of 4-bits by the corresponding hexadecimal digit, as given in Table 1.6.

Example 1.9: Convert $(10110110)_2$ to hexadecimal.

Solution: Start making combination of 4-bits as shown below:

1011 0110

1011	0110
↓	↓
B	6

∴ $(10110110)_2 = (B6)_{16}$

1.11.6.3 Octal to Other Number Systems

1.11.6.3.1 Octal to Decimal

For octal to decimal conversion, we express the octal number in the form of sum of products of octal digits and their positional weights.

Example 1.10: Convert $(345)_8$ to decimal.

Solution: $(345)_8 = (3 \times 8^2) + (4 \times 8^1) + (5 \times 8^0) = (3 \times 64) + (4 \times 8) + (5 \times 1)$
$= 192 + 32 + 5 = 299$

∴ $(345)_8 = (299)_{10}$

1.11.6.3.2 Octal to Binary

For octal to binary conversion, the procedure is very simple and straight. Replace each octal digit with corresponding 3-bit binary number as given in Table 1.5. The resultant binary number is the required number.

Example 1.11: Convert $(502)_8$ to binary.

Solution: Replace each octal digit by its equivalent 3-bit binary number as shown below:

5	0	2
↓	↓	↓
101	000	010

∴ $(502)_8 = (101000010)_2$

1.11.6.3.3 Octal to Hexadecimal

For octal to hexadecimal conversion, we use the following steps:

1. Replace each octal digit with corresponding 3-bit binary number.
2. From step 1, we get a binary number equivalent to given octal number. Now form the combinations of 4-bits, as done earlier with the binary-to-hexadecimal conversion.
3. Now replace each combination of 4-bits by the corresponding hexadecimal digit.

Example 1.12: Convert $(512)_8$ to hexadecimal.

Solution: Replace each octal digit by its equivalent 3-bit binary number as shown below:

$$\underbrace{5} \quad \underbrace{1} \quad \underbrace{2}$$
$$\downarrow \quad \downarrow \quad \downarrow$$
$$101 \quad 001 \quad 010$$

After this step, we get binary number 101001010

Now make combinations of 4-bits starting from right, as shown

1 0100 1010

Adding zeroes to the combinations with fewer bits, as shown

0001 0100 1010

Replace each 4-bit combination with its equivalent hexadecimal digits as shown

$$\underbrace{0001} \quad \underbrace{0100} \quad \underbrace{1010}$$
$$\downarrow \quad \downarrow \quad \downarrow$$
$$1 \quad 4 \quad A$$

$\therefore \quad (512)_8 = (14A)_{16}$

1.11.6.4 Hexadecimal to Other Number Systems

1.11.6.4.1 Hexadecimal to Decimal

For hexadecimal to decimal conversion, we express the hexadecimal number in the form of sum of products of hexadecimal digits and their positional weights.

Example 1.13: Convert $(321.2)_{16}$ to decimal.

Solution:

$$(321.2)_{16} = (3 \times 16^2) + (2 \times 16^1) + (1 \times 16^0) = (3 \times 256) + (2 \times 16) + (1 \times 1)$$

$$= 768 + 32 + 1 = 801.125$$

$\therefore \quad (321)_8 = (801)_{10}$

1.11.6.4.2 Hexadecimal to Binary

For hexadecimal to binary conversion, the procedure is very simple and straight. Replace each hexadecimal digit with corresponding 4-bit binary number as given in Table 1.3. The resultant binary number is the required number.

Example 1.14: Convert $(AB3)_{16}$ to binary.

Solution: Replace each hexadecimal digit by its equivalent 4-bit binary number as shown below:

```
 A      B      3
 ↓      ↓      ↓
1010   1011   0011
```

$\therefore \quad (AB3)_{16} = (101010110011)_2$

1.11.6.4.3 Hexadecimal to Octal

For hexadecimal to octal conversion, we use the following steps:

1. Replace each hexadecimal digit with corresponding 4-bit binary number.
2. From step 1, we get a binary number equivalent to given hexadecimal number. Now form the combinations of 3-bits, as done earlier with the binary-to-octal conversion.
3. Now replace each combination of 3-bits by the corresponding octal digit.

Example 1.15: Convert $(7AF)_{16}$ to octal.

Solution: Replace each hexadecimal digit by its equivalent 4-bit binary number as shown below:

```
 7      A      F
 ↓      ↓      ↓
0111   1010   1111
```

After this step, we get binary number 011110101111.

Now make combinations of 3-bits, as shown

011	110	101	111
↓	↓	↓	↓
3	6	5	7

∴ $(7AF)_{16} = (3657)_8$

1.12 FACTORS AFFECTING THE PERFORMANCE OF A COMPUTER

In the modern competitive world, quick results are the most desired. We are using computers to perform various certain tasks quickly & accurately so that we can get our work done. But there are many factors that could affect the performance of the computer system, which are discussed below:

- **Registers**
  - CPU contains of small memory areas called registers, which store data and instructions while the CPU processes them.
  - Size of the registers determines the amount of data with which the computer can work at a one time.
  - Today most PC's have 32-bit registers, mean the CPU can process four bytes data at one time. Register sizes are rapidly growing to 64 bits.

- **RAM**
  - Amount of RAM in a PC has a direct affect on the system's speed.
  - More RAM a PC has the more program instructions and data can be held in memory, which is faster than storage on disk.
  - If a PC does not have enough memory to run a program, it must move data between RAM and the hard disk frequently. This process called swapping can greatly slow a PC's performance.

- **System Clock**
  - Computer's system clock sets the pace the CPU by using a vibrating quartz crystal.
  - A single *tick* of the clock is the time required to turn a transistor off and back on. This is called a clock cycle.
  - Clock cycles are measured in Hertz(Hz), a measure of cycles per second. If a computer has a clock speed of 300 MHz, then its system clock *ticks* 300 million times every seconds.
  - Faster a PCs clock runs, the more instructions the PC can execute each second.

- **System Bus**
  - A bus is a path between the components of a computer. Data and instructions travel along these paths.
  - Data bus width determines how many bits can be transmitted between the CPU and other devices.
  - Address buss runs only between the CPU and RAM, and carries nothing but memory addresses for the CPU to use.
  - Peripheral devices are connected to the CPU by an expansion bus.

- **Cache Memory**
  - Cache memory is high-speed memory that holds the most recent data and instructions that have been loaded by the CPU.
  - Cache is located directly on the CPU or between the CPU and RAM, making it faster than normal RAM.
  - CPU-resident cache is called Level-1 (L1) cache. External cache is called Level-2 (L2) cache.
  - Amount of cache memory has a tremendous impact on the computer's speed.

1.13 SPECIFICATIONS OF A COMPUTER

The *specification* or *spec* is a list of the key components that make up the computer. It is helps the buyers to decide which PC, and which combination of features, they need.

When buying a PC, it is important to start by deciding what it is you want the PC to do. This then informs what specification you actually need. When reviewing a computer specification, the most important components to take account of are the processor, the amount of RAM and the size of the hard drive as these are central to the overall capability of the system.

In general, specifications of a computer include:

- Processor
- RAM
- Hard Drive
- Graphics processor/card
- Wireless (for laptops)
- Monitor
- Operating System
- Warranty

For example, following are the specifications of **Dell Inspiron 15 3542** Laptop:

- **Processor –** Intel core i5 (4th generation), 2.7 GHz
- **RAM –** 4 GB, DDR3, 1600 MHz
- **Hard Drive –** 1 TB, 5400 rpm
- **Graphics processor** – Intel HD Graphics 4400
- **Wireless –** WiFi, Bluetooth 4.0
- **Monitor** – 15.6", 1366 x 768, HD LED Backlit Display with Truelife
- **Operating System –** Windows 8.1 64 bit
- **Warranty –** 1 Year Accidental Damage Protection with 24x7 Phone Support -Retail

EXERCISE

1. What is a computer? What are its characteristics?
2. Differentiate between
   (*a*) hardware and software
   (*b*) data and information
3. What is/are the limitation(s) of a computer?
4. Explain the working of functional components of a computer with the help of a block diagram.
5. What are the functions of the control unit?
6. What are the functions of the memory unit?
7. When we say memory is volatile, what does it mean?
8. What are the functions of the secondary storage unit?
9. What is a machine cycle?
10. Explain booting process and its types.
11. What is role played by the input/output devices in a computer system?
12. RAM is volatile memory. What does it mean?
13. Why do we use secondary storage? Name any two secondary storage devices.
14. What are various types of software?
15. What is system software? Give its examples.
16. What is operating system? Give its examples.
17. What is application software? Give its examples.
18. Differentiate between system software and application software.
19. What do you mean my open source software?
20. What is proprietary software?
21. Differentiate between proprietary software and freeware.
22. Name the types of software where there is no end-user license agreement or terms of service agreement.

23. How are freeware and free software different?
24. What is the difference between open source software and a freeware? Write two examples of each.
25. What is the difference between office tools and domain specific tools?
26. Do the conversions as directed:
    (*a*) $(120.125)_{10} = (\ldots)_2$
    (*b*) $(1001.001)_2 = (\ldots)_{10}$
    (*c*) $(1001.75)_{10} = (\ldots)_{16}$
    (*d*) $(124.4)_8 = (\ldots)_{16}$
    (*e*) $(111110.111)_2 = (\ldots)_{16}$
    (*f*) $(1AC)_{16} = (\ldots)_8$
    (*g*) $(10110110.11)_2 = (\ldots)_8$
    (*h*) $(202.4)_8 = (\ldots)_{10}$
    (*i*) $(A00A.8)_{16} = (\ldots)_{10}$
    (*j*) $(A00A.8)_{16} = (\ldots)_8$
27. What are the factors on which the performance of a computer system depends?
28. What do you understand by specifications of a computer system?

❑ ❑ ❑

Chapter 2

Problem Solving and Program Planning

Learning Outcomes

After reading this chapter, students will be able to

- explain the importance of problem solving
- explain the importance of program planning and design
- explain thc various program design tools
- design solutions of variety of problems

2.1 INTRODUCTION

Suppose you are asked by your mathematics teacher to solve an arithmetic problem and you are not familiar with the steps involved in solving that problem. In such a situation, you will not be able to solve the problem. The same principle applies to writing the program. A programmer cannot write instructions for a computer unless the programmer knows how to solve the problem manually.

Let us imagine that you know the steps to be followed for solving the given problem but while solving the problem, you forget to apply some of the steps or you apply some of the steps in wrong sequence. Obviously, you will get a wrong answer. Similarly, while writing a program, if the programmer leaves out some of the instructions or writes the instructions in wrong sequence, the computer will give a wrong answer.

Thus, to write an effective and efficient program, it is necessary that the programmer must write each and every instruction in the proper sequence. However, the sequence of instructions, *referred to as logic*, of a program can be very complex. Therefore, in order to ensure that the program instructions are appropriate for the given problem and are in the correct sequence, program must be planned before they are written.

In this chapter, we will learn about the various tools that can be used to plan the program. Further, the use of each of these tools will be demonstrated by considering number of problems with whom you are already familiar.

2.2 PROBLEM SOLVING

As mentioned earlier, you will not be able to solve a problem if you don't know the steps involved in solving that problem. The same principle applies to writing a program. A programmer cannot write instructions to be given to the computer unless the programmer knows how to solve the problem manually.

Therefore, it is very important that before attempting to write a program, the programmers must solve the problem manually. While solving a problem, the programmer will be able to learn about the various steps to be performed and the sequence in which these steps are to be performed.

To demonstrate how the programmer should proceed, let us consider a familiar example of finding the roots of a quadratic equation of type

$ax^2 + bx + c = 0$ provided $a \neq 0$.

The roots of such an equation are given by the formula:

$$root_{1,2} = \frac{-b \pm \sqrt{b^2 - 4ac}}{2a}$$ where $b^2 - 4ac$ is known as the discriminant.

Depending on the sign of the discriminant, there are three are mutually exclusive possibilities for the roots:

Case 1: If $b^2 - 4ac < 0$, then the root are imaginary, and we can compute real part and imaginary part separately as

$$\text{real part} = -\frac{b}{2a} \quad \text{and} \quad \text{imaginary part} = \frac{\sqrt{-(b^2 - 4ac)}}{2a}$$

Once we know the real and imaginary part of the root, we can determine the actual roots as (real part ± i imaginary part).

Case 2: If $b^2 - 4ac = 0$, then the roots are real and equal, and are computed as

$$\text{each root} = -\frac{b}{2a}$$

Case 3: If $b^2 - 4ac > 0$, then the roots are real and distinct and are computed as

$$\text{root1} = \frac{-b - \sqrt{b^2 - 4ac}}{2a} \quad \text{and} \quad \text{root2} = \frac{-b + \sqrt{b^2 - 4ac}}{2a}$$

Therefore, we will proceed as

For a given set of values of a, b and c, compute expression $b^2 - 4ac$. Now compare the computed value of $b^2 - 4ac$ with zero. If it is less than zero, compute the roots as shown in case 1; if it is equal to zero, compute the roots as shown in case 2; and if it is greater than zero, compute the roots as shown in case 3.

In this book, for each problem considered, we will first see how to solve it.

2.3 PROGRAM PLANNING AND DESIGN

Planning is very important in every walk of life. Even for our routine kind of things, we have to plan. Such a plan of doing things is also as *process*. For example, we have to plan our monthly budget otherwise things can go astray and we may not enough money left in our pocket towards the end of the month.

Planning should not be in the mind only (*adhoc process*), you must write it down, *i.e.*, you must document it (*defined process*). The results of *adhoc process* are not guaranteed where as the results of *defined process* are guaranteed and predictable.

Likewise, if you know the steps to be followed for solving a given problem but you have not documented them, you may forget to apply some of the steps or you may apply some of the steps in wrong sequence. Obviously, you will get a wrong answer. Similarly, while writing a program, if the programmer leaves out some of the instructions or writes the instructions in wrong sequence, the computer will give a wrong answer.

The various tools collectively referred to as *program design tools*, that helps in planning the program are *algorithm*, *flowchart* and *pseudocode*.

In the subsequent section, we will discuss about these tools, and demonstrate their use through illustrative examples.

2.4 ALGORITHMS

An *algorithm* is defined as a finite sequence of instructions defining the solution of a particular problem, where each instruction is numbered.

However, in order to qualify as an algorithm, every sequence of instructions must satisfy the following criteria:

- ***Input***: There are zero or more values which are externally supplied.
- ***Output***: At least one value is produced.
- ***Definiteness***: Each step must be clear and unambiguous, *i.e.*, having one and only one meaning).
- ***Finiteness***: If we trace the steps of an algorithm, then for all cases, the algorithm must terminate after a finite number of steps.
- ***Effectiveness***: Each step must be sufficiently basic that it can in principle be carried out by a person using only paper and pencil. In addition, not only each step is definite, it must also be feasible.

Example 2.1: Formulate an algorithm to display the nature of roots of a quadratic equation of type

$ax^2 + bx + c = 0$ provided $a \neq 0$.

Solution: Roots of the quadratic equation are given by the formula:

$$x_{1,2} = \frac{-b \pm \sqrt{b^2 - 4ac}}{2a}$$

where the expression $b^2 - 4ac$ is known as the discriminant.

Depending on the sign of discriminant, there are three possibilities for roots:

1. If $b^2 - 4ac < 0$, then the root are imaginary.
2. If $b^2 - 4ac = 0$, then the roots are real and equal.
3. If $b^2 - 4ac > 0$, then the roots are real and distinct.

The complete algorithm will look like

Algorithm 2.1

Step 1: Input values of *a, b* and *c*

Step 2: Compute $b^2 - 4ac$ and denote its value by variable *disc*

Step 3: Is *disc* < 0? If yes then goto step 7 else goto step 4

Step 4: Is *disc* > 0? If yes then goto step 9 else goto step 5

Step 5: Output "roots are real and equal"

Step 6: Goto step 10

Step 7: Output "roots are imaginary"

Step 8: Goto step 10

Step 9: Output "roots are real and distinct"

Step 10: Stop

2.5 FLOWCHARTS

A flowchart is a pictorial representation of an algorithm. A flowchart uses different shapes to denote different types of instructions. The actual instructions are written within the shapes using clear and concise statements. These shapes are connected by directed lines to indicate the sequence in which instructions are to be executed.

Normally, an algorithm is first represented in the form of a flowchart and then the flowchart is expressed in some programming language to prepare a program. The main advantage of this approach in program writing is that while drawing a flowchart, one is not concerned with the details of the syntax of the programming language. Moreover, since a flowchart shows the sequence of instructions in a pictorial form, any error in logic can be detected more easily that in the case of a program. Once the flowchart is ready, the programmer can forget about the logic and can concentrate only on coding the instructions in each shape in terms of the syntax of the programming language. This will ensure an error-free program.

A flowchart, therefore, is a picture of the logic to be implemented by the program. It is simply a tool assisting the programmer to lay out, in a visual, two-dimensional format, ideas on how to organize a sequence of steps necessary to solve a problem by computer. It is basically the plan to be followed when the program is written. It acts like a road map for a programmer and guides him/her how to go from the starting point to the final point while writing a program.

Experienced programmers most often skip the flowchart while writing the program. However, for a beginner, it is recommended that flowchart must be drawn first in order to reduce the number of errors and omissions in the programs. Moreover, it is a good practice to have a flowchart because flowchart is very helpful during the testing of a program as well as while incorporating further modifications in the program.

Flowchart Symbols

As mentioned, flowchart uses boxes of different shapes to denote different types of instructions. The communication of the program logic through flowcharts is made easier through the use of symbols that have standardized meanings. For example, a diamond always means a decision. Only few symbols are needed to represent the necessary instructions in a flowchart. These symbols are standardized by American National Standards Institute (ANSI). Table 2.1 shows various symbols used in flowcharts along with their name and brief description.

Oval Symbol – This symbol is used to indicate the beginning (START) and ending (STOP) in the program logic flow. It is the first and the last symbol in the program logic.

Parallelogram Symbol – This symbol is used to denote the input to the program or output from the program. If there is a program instruction to input data from any input device, that may include keyboard, disk, tape, etc., that step will be indicated in the flowchart with this symbol. Similarly, if there is an instruction to output data to any output device, that may include terminal screen, printer, disk, tape, etc., that step will be indicated in the flowchart with this symbol.

Table 2.1: Various flowchart symbols and their brief description

Symbol	Name	Purpose
	Oval	*Terminal* - to mark the beginning and end of the program logic flow.
	Parallelogram	*Input/Output* - to denote the input to the program or output from the program.
	Rectangle	*Processing* - to denote the arithmetic operations and movement of data.
	Diamond	*Decision* - to denote a point where decision has to be made to branch to one of the alternatives.
	Small circle	*Connector* - To provide a logical link between segments of a flowchart.
	Directed lines	*Flow Lines* - To indicate the sequence in which instructions are to be executed.

Rectangle Symbol – This symbol is used to denote the arithmetic and data movement instructions. Thus, all arithmetic instructions such as addition, subtraction, multiplication and division are shown by this symbol. The other instructions such as moving data from one location in memory to other locations are also shown using this symbol. When more that one arithmetic and data movement instructions are to be executed consecutively, they can be placed in the same symbol and they are assumed to be executed in the order of their appearance.

Diamond Symbol – This symbol is used to denote a point where decision has to be made to branch to one of the alternatives.

It is important to note that the criteria for making the decision should be indicated clearly within the decision box. Moreover, the expression upon which each of the possible exit paths will be executed should be identified and all the possible paths should be accounted for. During execution, the appropriate path is followed depending upon the result of the decision.

Small Circle Symbol – This symbol is used to provide a logical link between segments of a flowchart. If a flowchart is very long, the flow lines start criss-crossing at many places that cause confusion and thus reduce understandability of the flowchart. Moreover, there are instances when a flowchart becomes too long to fit in a single page and the use of flow lines becomes impossible.

Thus, whenever a flowchart becomes complex enough that the number and direction of flow lines is confusing or it spreads over more than one page, it is useful to utilize the connector symbol as a substitute for flow lines. This symbol represents an entry from, or an exit to another part of the flowchart. A connector symbol is represented by a small circle and a letter or a digit is placed within the small circle to indicate the link. A pair of identically labeled connector symbols is commonly used to indicate a continued flow when the use of flow line is confusing. Therefore, two connectors with identical label serve the same function as a long flow line. That is, they show an exit to another section of the flowchart, or they indicate an entry from another section of the flowchart.

Now the question arises – *How is it possible to determine if a connector is used as an entry or an exit point*?

It is very simple – if an arrow enters into a connector, it is an exit point and the program control is transferred to an identically labeled connector that does have an outlet; if an arrow leaves a connector, it is an entry point and the control is received from an identically labeled connector. It may be noted that connectors do not represent any operation and their use in a flowchart is only for the sake of convenience and clarity.

Directed Lines – These lines are used to indicate the sequence in which instructions are to be executed. The normal sequence is from top to bottom and left to right. The crossing of these lines should be avoided whenever possible.

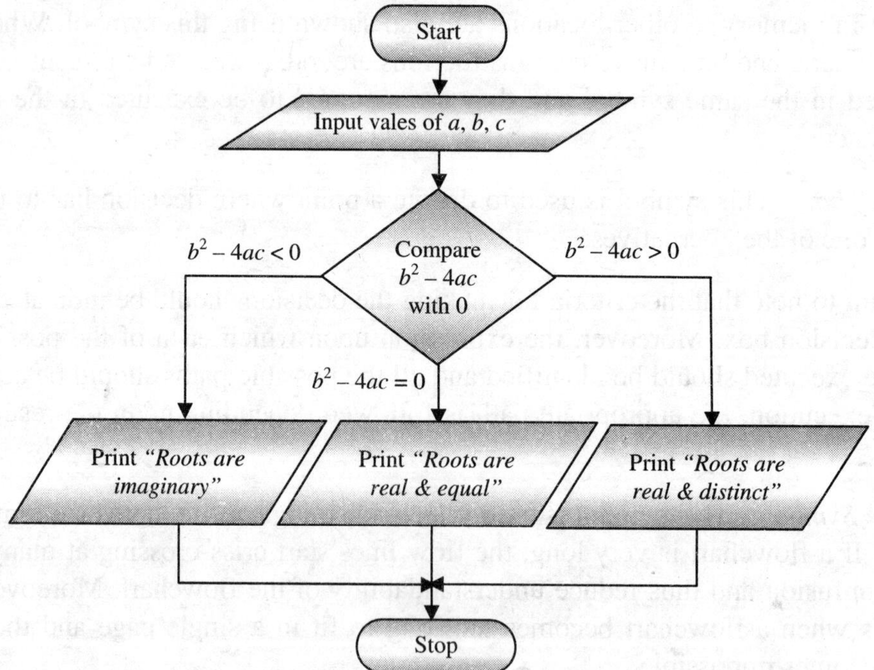

Figure 2.1: Flowchart to find the nature of roots of a quadratic equation

2.6 PSEUDOCODE

Pseudocode is another design tool that can be used for planning the program. The word "pseudo" means imitation or false and the word "code" refers to the instruction written in a programming language. Pseudocode, therefore, is an imitation of actual computer instruction. Pseudo instructions are phrases written in English like statements. Instead of using symbols to describe the logic of the program, as in flowcharts, pseudocode uses a structure that resembles computer instructions. Because, it emphasizes the design of the program, pseudocode is also called *Program Design Language* (PDL).

Pseudocode is made up of the following basic logic structures that have proved to be sufficient for writing any computer program:

1. Sequence
2. Selection (*If . . . then . . . Endif, If . . . then . . . Else . . .Endif*)
3. Iteration (*For . . . Endfor, While . . . Endwhile, Do . . . While*)

Sequence logic is used for performing instructions one after another in sequence. Thus, for sequence logic, pseudocode instructions are written in the sequence in which they are to be executed. The flow of logic is from top to bottom.

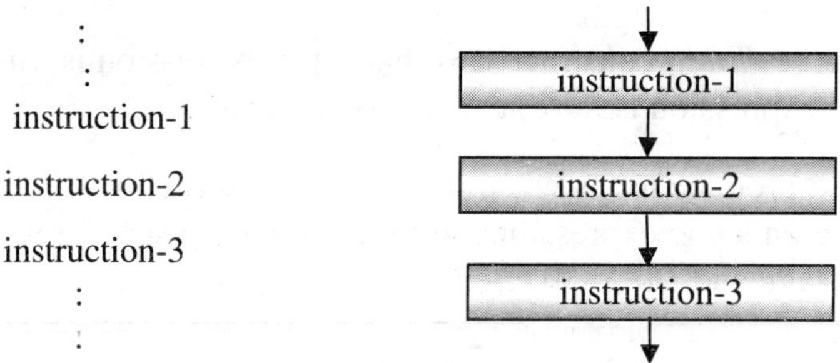

Figure 2.2: Pseudocode and flowchart for sequence structure

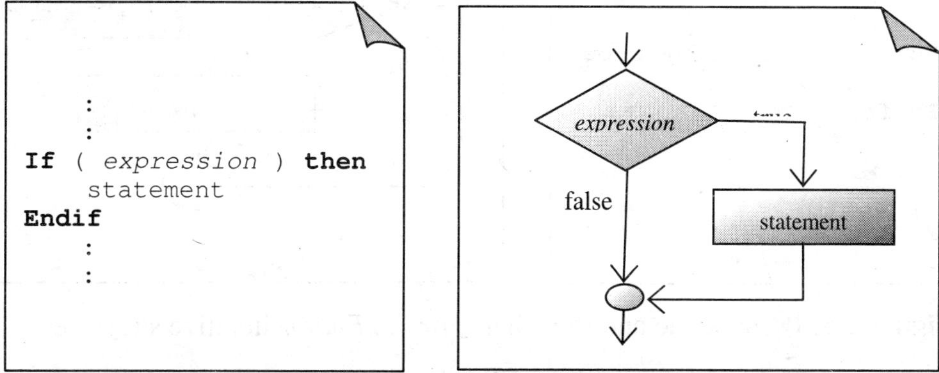

Figure 2.3: Pseudocode and flowchart for *If . . . then . . . Endif* selection structure

Selection logic, also known as decision logic, is used for making decision. It is used for selecting a proper path out of the alternative paths in the program logic. Decision logic is depicted as either *If . . . then . . . Endif* or an *If . . . then . . . Else . . . Endif* structure.

The *If . . . then . . . Endif* construct says that if the expression is true, then execute statement else (if the expression is false) skip over the statement.

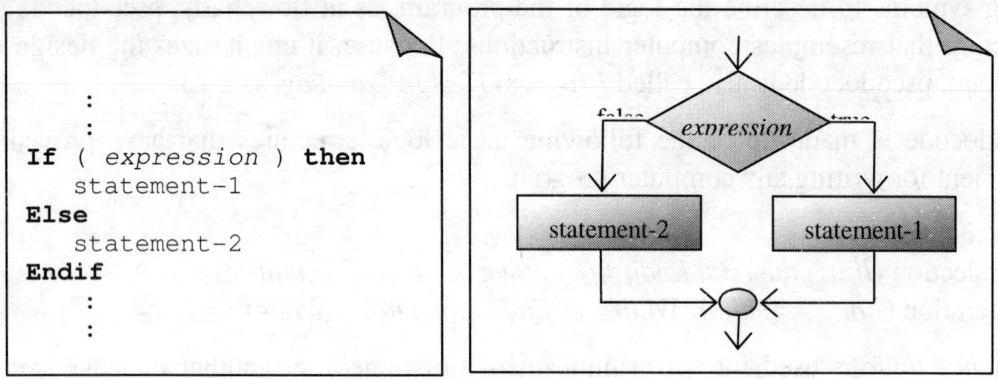

Figure 2.4: Pseudocode and flowchart for *If . . . then . . . Else . . . Endif* selection structure

The *If . . . then . . . Else . . . Endif* construct says that if the expression is true then execute statement-1, else (if the expression is false) execute statement-2.

Iterative logic is used to produce loops when one or more instructions are to be executed several times depending on some expressions. It uses three loop structures called *For . . . Endfor*, *While . . . Endwhile* and *Do . . . While*.

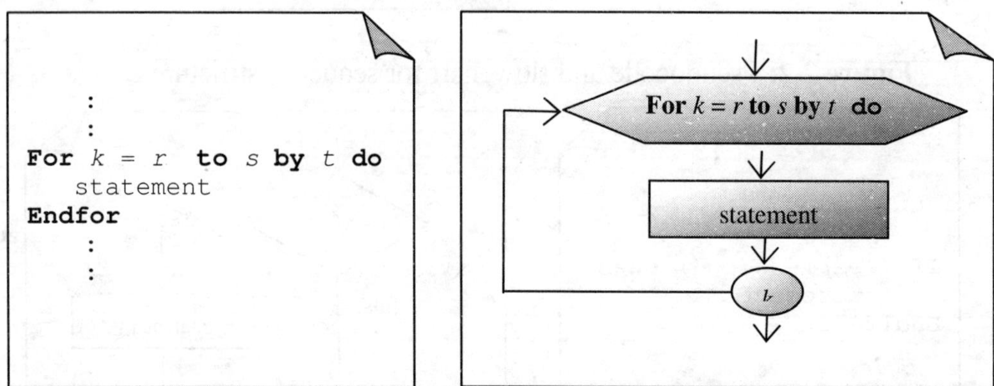

Figure 2.5: Pseudocode and flowchart *For . . . Endfor* iterative structure

It uses an index variable *k* to control the loop. Here *r* is called the initial value, *s* is called the final value and *t* is called the step size, which may be positive (increment) or negative (decrement).

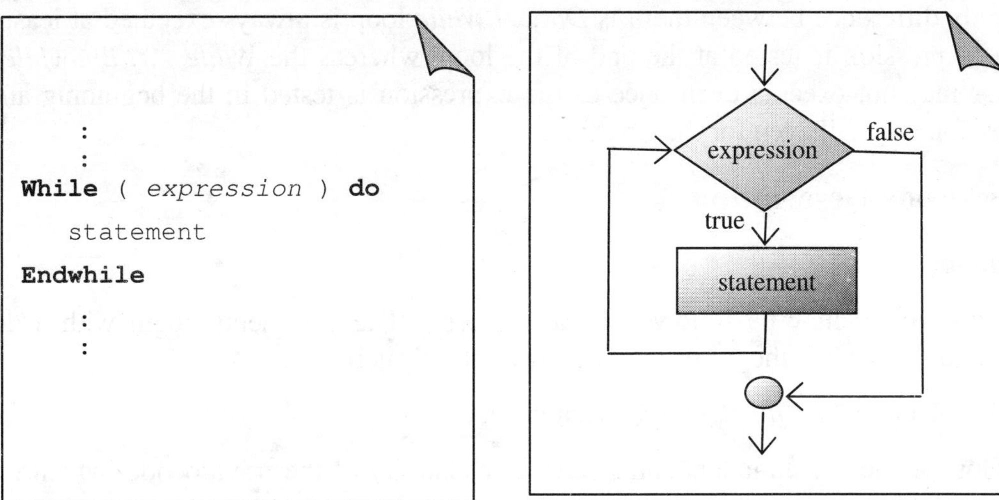

Figure 2.6: Pseudocode and flowchart for *While . . . Endwhile* iterative structure

The *For . . . Endfor* loop is used when the number of iterations is known. On the other side, *While . . . Endwhile* and *Do . . . While* loops are used when the number of iterations is not known. Both *While . . . Endwhile* and *Do . . . While* continues to execute as long as the expression is true.

The only difference between them is *Do . . . While* loop is always executed at least once as the expression is tested at the end of the loop, whereas the *While . . . Endwhile* loop may or may not execute even once as the expression is tested in the beginning and the expression may fail even for the first time.

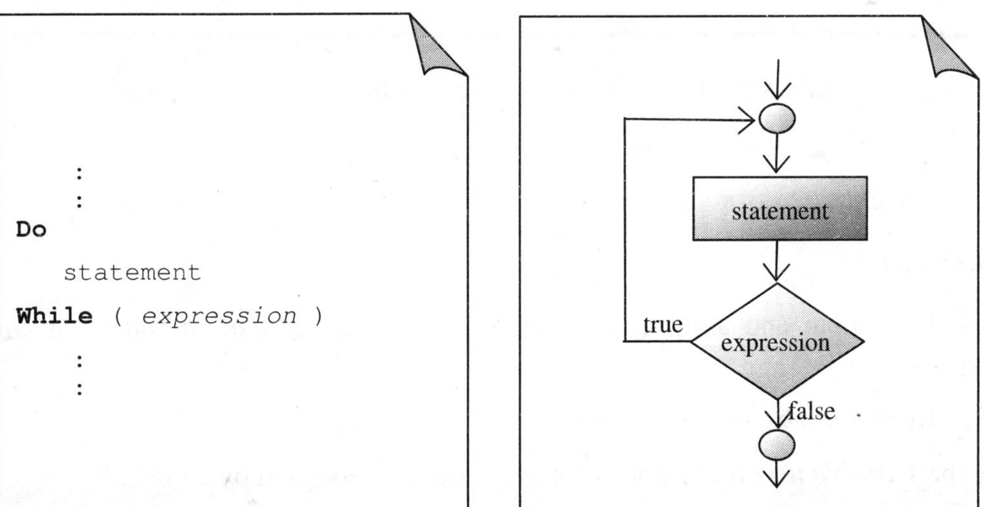

Figure 2.7: Pseudocode and flowchart for *Do . . . While* iterative structure

The only difference between them is *Do . . . While* loop is always executed at least once as the expression is tested at the end of the loop, whereas the *While . . . Endwhile* loop may or may not execute even once as the expression is tested in the beginning and the expression may fail even for the first time.

Pseudocode Description

Comments

Each instruction may be followed by a comment. The comments begin with a double slash, and the explain the purpose of the instruction, such as

Read: *n* // Enter the value of variable *n*

Appropriate use of comments enhances the readability of the pseudocode, which in turn helps in maintaining the pseudocode.

Variable Names

For variable names, we will use italicized lowercase letters such as *max*, *loc*, etc., whereas for defined constants, if any, we will use uppercase letters.

Assignment Statement

The assignment statement will use the notation as

Set $max = a_i$

to assign the value of a_i to *max*. The right hand side of the assignment statement can have a *value*, *a variable* or *an expression*.

However, if several assignment statements appear in the same line, such as

Set $k = 1$, $loc = 1$, $max = a_i$

then they are executed from left to right.

Input/Output

Data may be input and assigned to variables by means of a *read* statement with the following format

Read: *Variable list*

where the *Variable list* consists one or more variables separated by comma.

Similarly, the data held by the variables and the messages, if any, enclosed in double quotes can be output by means of a *write* statement with the following format

Print: *message* and/or *Variable list*

where the *message* and the variables in the *Variable list* are separated by comma.

Execution of Instructions

The instructions in the pseudocode are usually executed one after the other as they appear in the pseudocode. However, there may be instances when some instructions are skipped or some instructions may be repeated as a result of certain expressions.

Completion of the Algorithm

The pseudocode is completed with the execution of the last instruction. However, the pseudocode can be terminated at any intermediate state using the *exit* instruction.

Pseudocode to display the nature of roots of a quadratic equation of the type

$ax^2 + bx + c = 0$ provided $a \neq 0$.

Pseudocode 2.1

```
Begin
  Read: a, b, c
  Set disc = b^2 – 4ac
  If ( disc < 0 ) then
     Print: "roots are imaginary"
  Else
     If ( disc > 0 ) then
        Print: "roots are real and distinct"
     Else
        Print: "roots are real and equal"
     Endif
  Endif
End.
```

If you look at the algorithm, flowchart and pseudocode for program design to find the nature of the roots of the quadratic equation, you will find that the logic in flowchart is easy to understand whereas the pseudocode is much closer to the actual program.

In this chapter, we will consider flowchart and pseudocode to express the design of the program.

ILLUSTRATIVE EXAMPLES

Example 2.2: Draw flowchart and write pseudocode to find the largest of three numbers, say *a*, *b*, *c*.

Solution: We first compare *a* with *b*. If *a* is greater than *b* then we compare *a* with *c*. If *a* is greater than *c*, then *a* is taken as the largest number otherwise we take c as the largest number.

However, if *a* is not greater than *b*, we compare *b* with *c*. If *b* is greater than *c* then *b* is taken as the largest number otherwise we take *c* as the largest number.

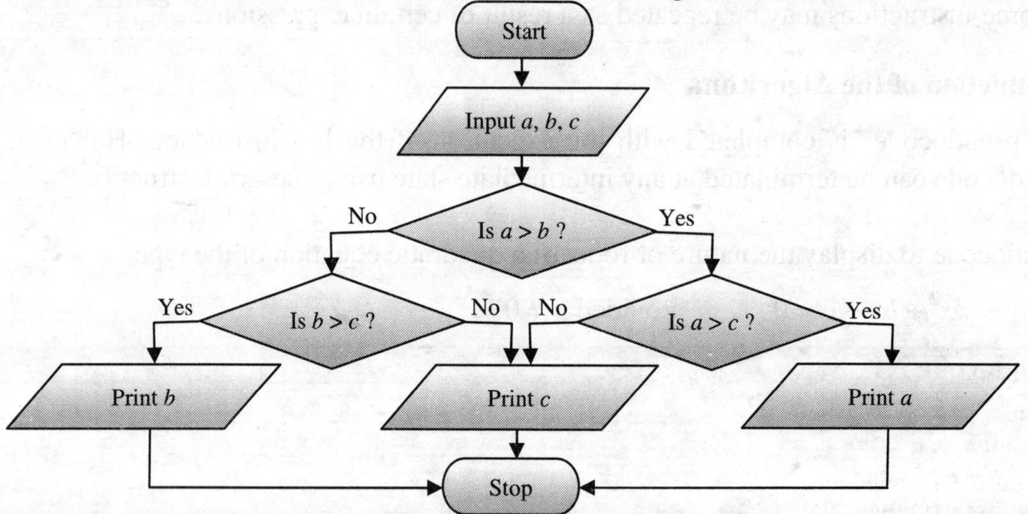

Pseudocode 2.2

```
Begin
   Read: a, b, c
   If ( a > b ) then
        If ( a > c ) then
             Print: a
        Else
             Print: c
        Endif
   Else
        If ( b > c ) then
             Print: b
        Else
             Print: c
        Endif
   Endif
End.
```

Figure 2.8: Flowchart & pseudocode to find largest of three numbers

Example 2.3: Draw flowchart and write pseudocode to find the largest of four numbers, say *a*, *b*, *c*, *d*.

Solution: For four or more numbers, the solution of Example 2.2 will become very cumbersome. Therefore, we will follow an alternate approach.

We take first number as the largest number and store it in a variable, say *max*. Then, we compare the second number with the *max*, and if second number is greater than max, then second number is taken as the largest by storing it in *max*.

Same process is repeated for remaining numbers and the final value of *max* given the largest number among given four numbers.

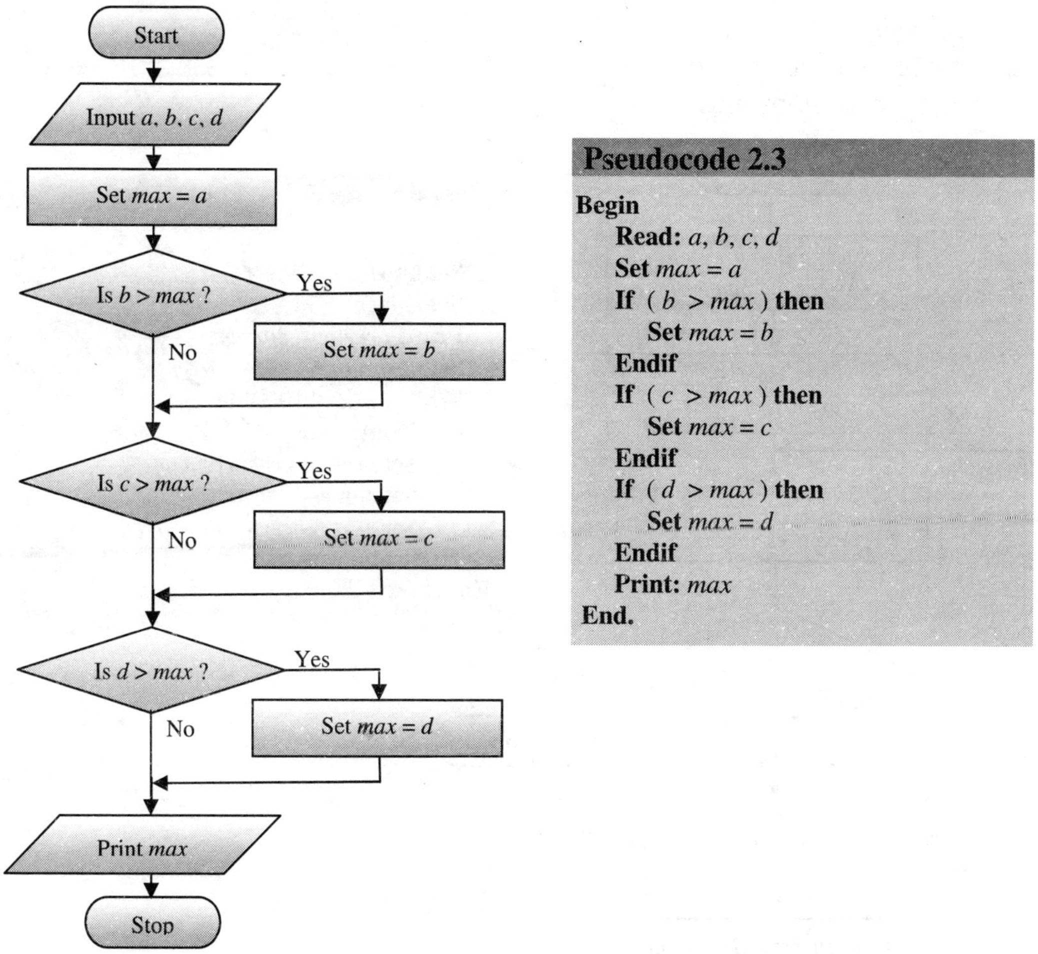

Figure 2.9: Flowchart & pseudocode to find largest of four numbers

Example 2.4: Fibonacci sequence is given below:

0 1 1 2 3 5 8 13 21 . . .

Draw a flowchart and write pseudocode to generate the above sequence up to the term less than or equal to 100.

Solution: Observe that, leaving first two terms, each term is obtained as the sum of the immediately preceding two terms.

If we use variable *previous* for previous term, *current* for current term, *next* for next term, and setting *previous* and *current* to values 0 and 1 respectively, i.e., first two terms of the sequence, then the entire sequence can be generated by using the recurrence relation

next = *previous* + *current*
replace *previous* by *current*
replace *current* by *next*

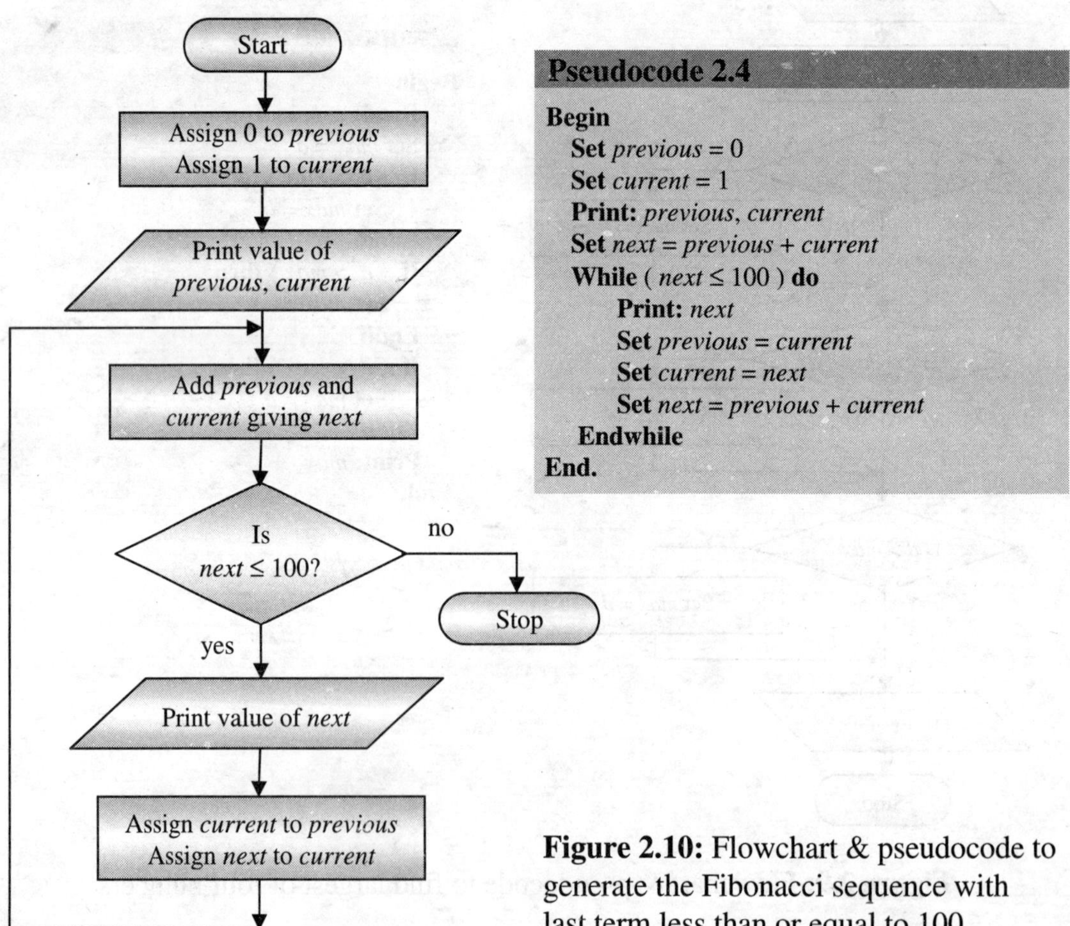

Pseudocode 2.4

```
Begin
   Set previous = 0
   Set current = 1
   Print: previous, current
   Set next = previous + current
   While ( next ≤ 100 ) do
        Print: next
        Set previous = current
        Set current = next
        Set next = previous + current
    Endwhile
End.
```

Figure 2.10: Flowchart & pseudocode to generate the Fibonacci sequence with last term less than or equal to 100

Example 2.5: Draw a flowchart and write pseudocode to print first *n* terms of the Fibonacci sequence.

For example, if input value for *n* is 8, the output should be

0 1 1 2 3 5 8 13

Solution:

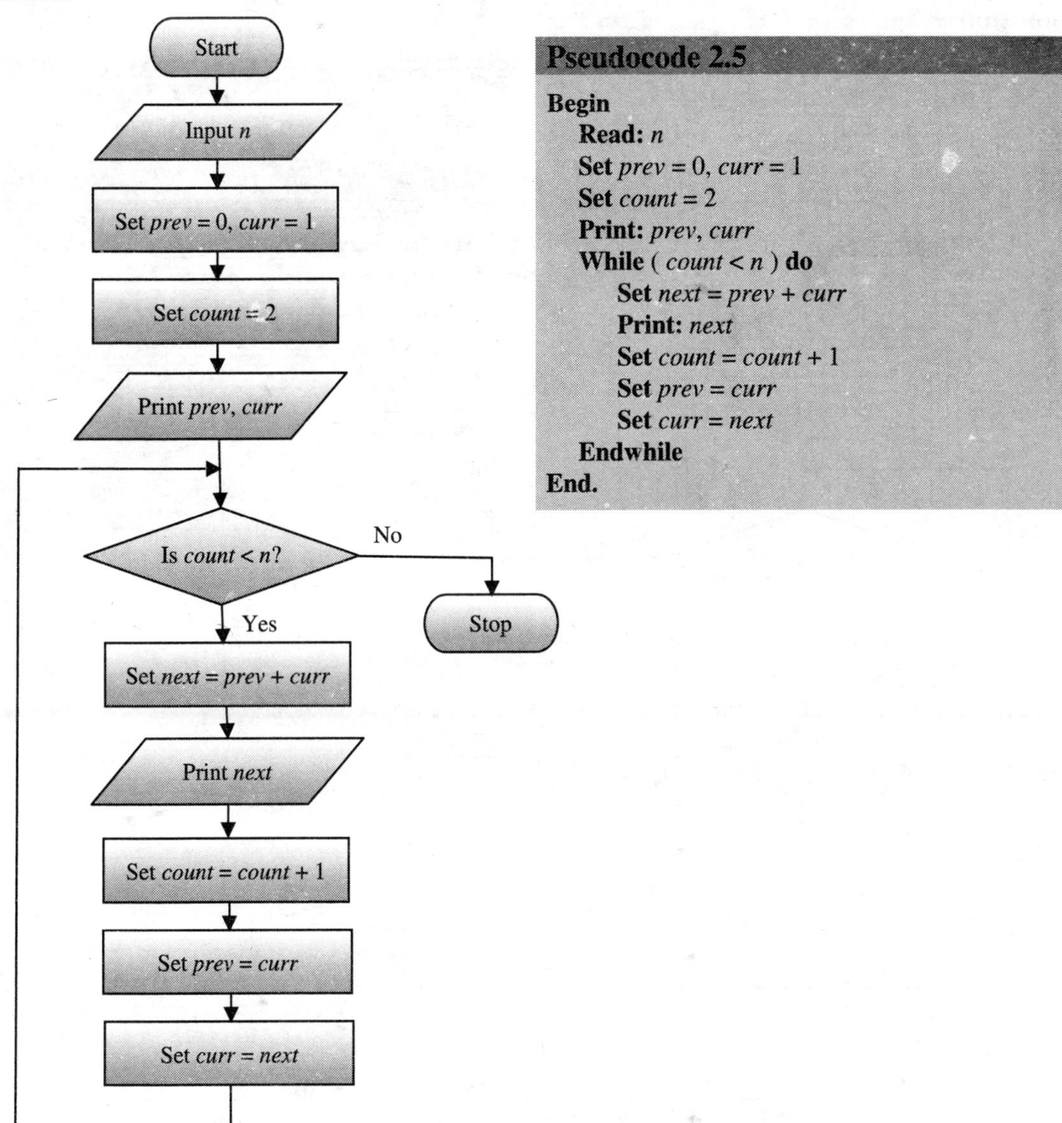

Pseudocode 2.5

```
Begin
   Read: n
   Set prev = 0, curr = 1
   Set count = 2
   Print: prev, curr
   While ( count < n ) do
       Set next = prev + curr
       Print: next
       Set count = count + 1
       Set prev = curr
       Set curr = next
   Endwhile
End.
```

Figure 2.11: Flowchart & pseudocode to print first *n* terms of the Fibonacci sequence

Example 2.6: Draw a flowchart and write pseudocode to print *n*th term of the Fibonacci sequence.

For example, if input value for *n* is 8, the output should be

13

Solution:

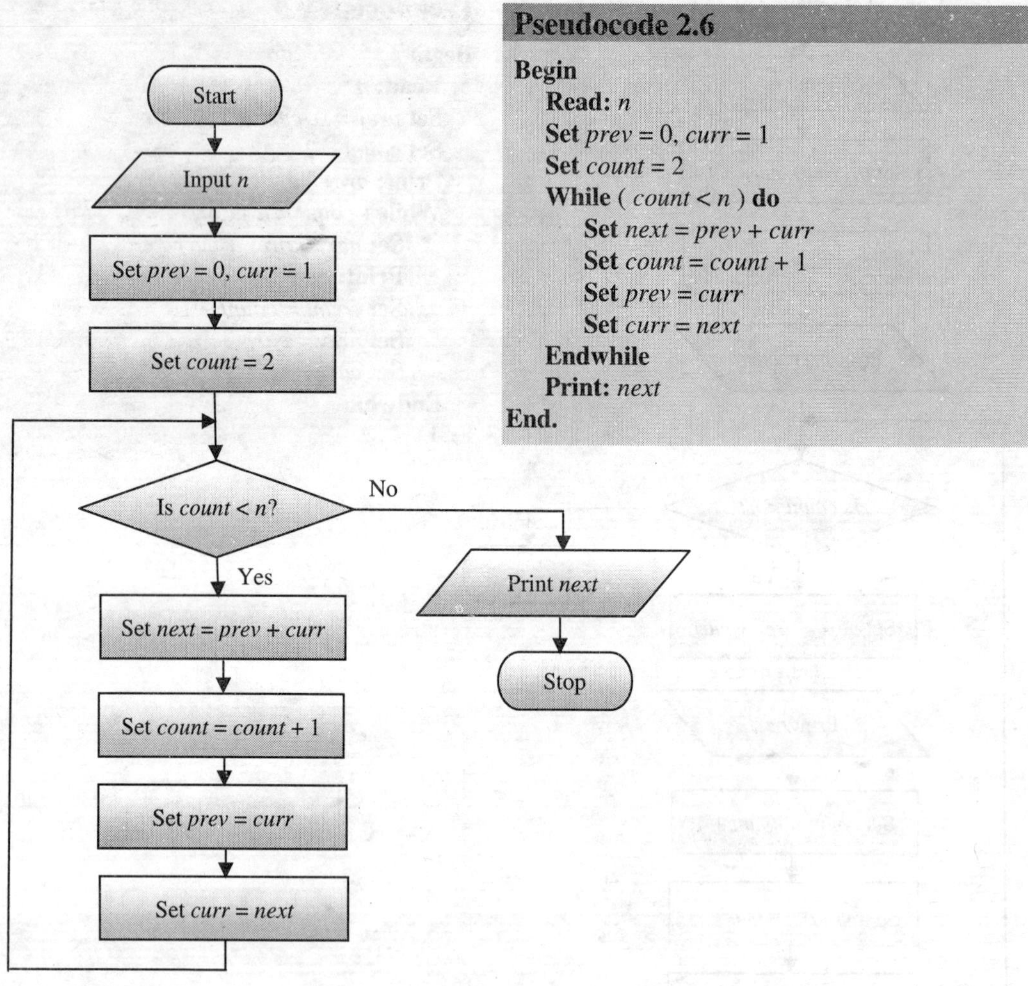

Pseudocode 2.6

```
Begin
   Read: n
   Set prev = 0, curr = 1
   Set count = 2
   While ( count < n ) do
      Set next = prev + curr
      Set count = count + 1
      Set prev = curr
      Set curr = next
   Endwhile
   Print: next
End.
```

Figure 2.12: Flowchart & pseudocode to print *n*th term of the Fibonacci sequence

Example 2.7: Draw a flowchart to find whether the given natural number n is a prime number or not.

Solution: A natural number is said to be prime if it divisible by 1 and itself only, *i.e.*, it cannot be factorized. In addition, to this definition, an even number except 2 is not a prime number. Therefore, our test criteria becomes

1. If n is greater than 2 and is even then n is not a prime number.
2. If test at step 1 fails, then we try to divide number n by factors $k = 3, 5, 7, \ldots \sqrt{n}$. Therefore, if n is divisible by any value of k, number n is not a prime number.
3. If test at step 2 also fails, then n is a prime number.

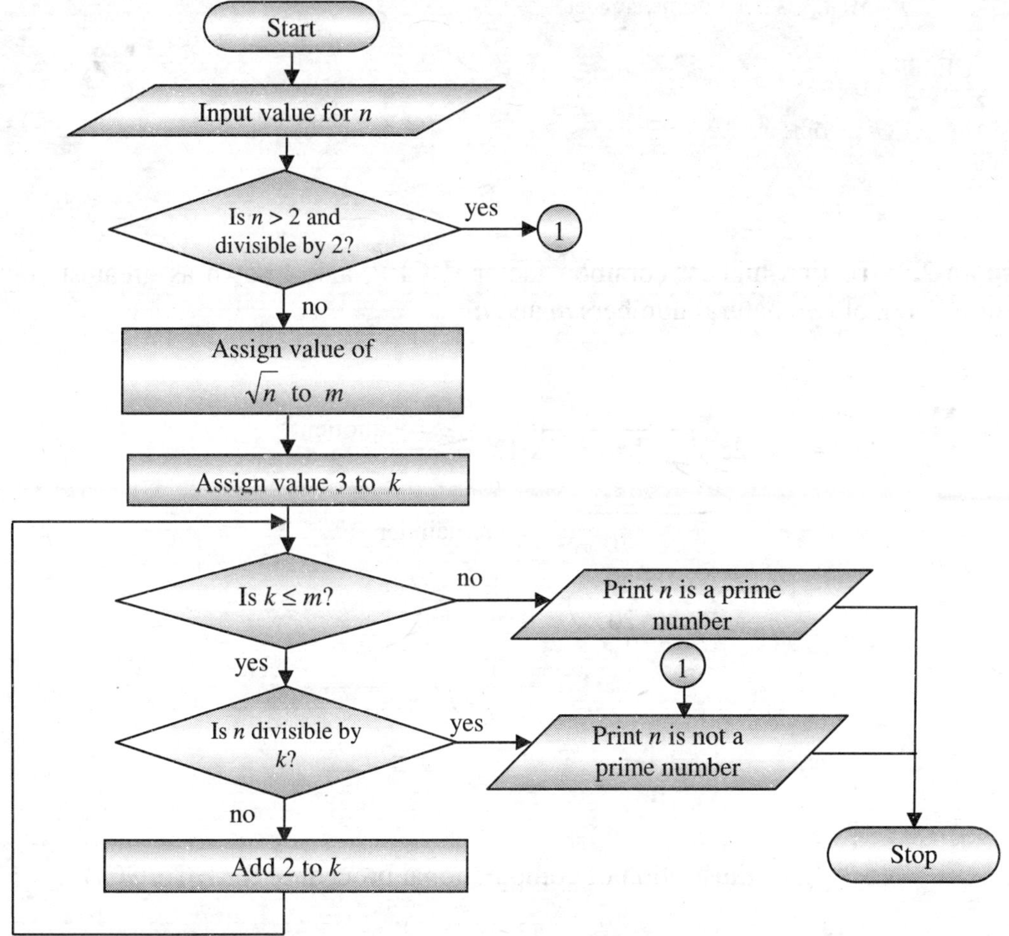

Figure 2.13: Flowchart to check whether number n is prime or not

Following is the pseudocode to find whether the given positive number n is a prime number or not.

Pseudocode 2.7

```
Begin
   Read: n
   If ( n > 2 and n mod 2 == 0 ) then
      Print: n, " is not a prime number"
      Exit
   Else
      Set m = √n
      For k = 3 to m by 2 do
         If ( n mod k == 0 ) then
            Print: n, " is not a prime number"
            Exit
         Endif
      Endfor
      Print: n, " is a prime number"
   Endif
End.
```

Example 2.8: To find highest common factor (HCF)/, also known as greatest common divisor (GCD), of two natural numbers m and n.

Solution:

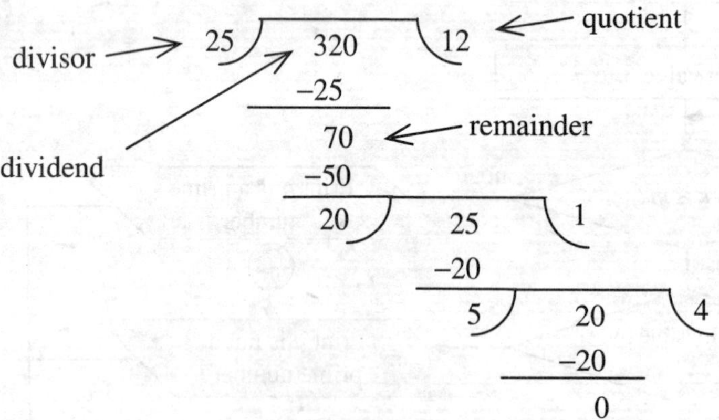

Figure 2.14: Illustration of computational procedure for HCF/GCD

Figure 2.14 demonstrates the long/continued division method to find the HCF/GCD of two natural numbers. You must have observed that in successive divisions, the divisor of the previous division becomes dividend, remainder of becomes divisor, and division is again carried out. This process is carried out till the reminder becomes zero, and the current divisor is taken as HCF/GCD of the given natural numbers.

This process can be implemented by using the following steps

1. Perform division.
2. If remainder is zero, then stop and take the divisor as HCF/GCD.
3. Replace dividend by divisor.
4. Replace divisor by remainder.
5. Repeat from step 1.

Start

Input m, n

Divide m by n and let the remainder be r

Is $r = 0$?

yes

Print "HCF/GCD = ", n

Stop

no

Replace m by n and n by r

Pseudocode 2.8

```
Begin
   Read: m, n
   Set r = m mod n
   While ( r ≠ 0 ) do
         Set m = n
         Set n = r
         Set r = m mod n
   Endwhile
   Print: "HCF/GCD = ", n
End.
```

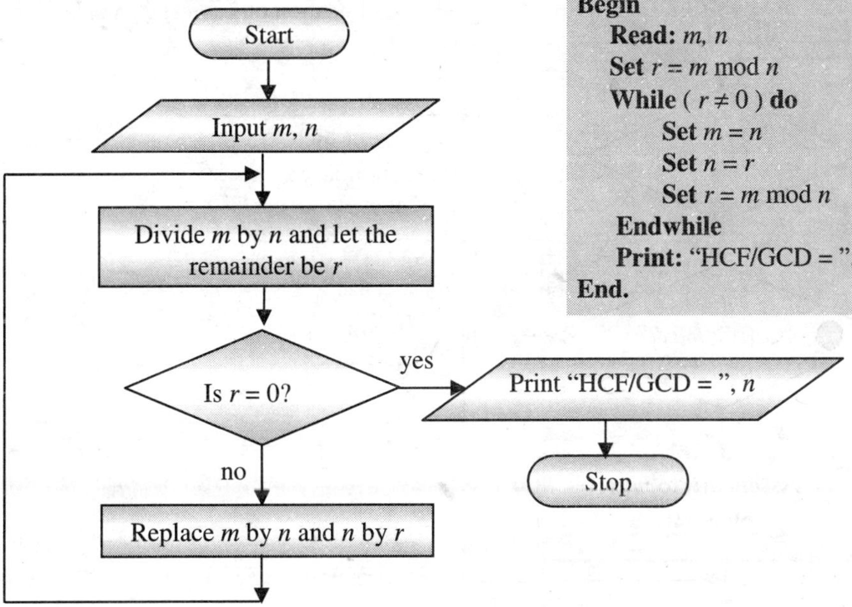

Figure 2.15: Flowchart & pseudocode to compute HCF of two numbers

Example 2.9: Suppose *a* is a list of *n* real numbers, and we want to find the location, say *loc*, of the largest element of the list.

Solution: *This problem can be solved as follows*: We begin with *loc* = 1, and *max* = *a*[1]. We then compare *max* with each successive element *a*[*k*] of the list. If *a*[*k*] is greater than *max*, we then update *loc* and *max* as *loc* = *k* and *max* = *a*[*k*], respectively. The final value of *loc* gives the location of the largest element of the list.

Pseudocode 2.9

```
Begin
  Read: n
  For i = 1 to n by 1 do
     Read: a[i]
  Endfor
  Set loc = 1
  Set max = a[1]
  For i = 2 to n by 1 do
     If ( a[i] > max ) then
        Set max = a[i]
        Set loc = i
     Endif
  Endfor
  Print: "Location of Largest element = ", loc
End.
```

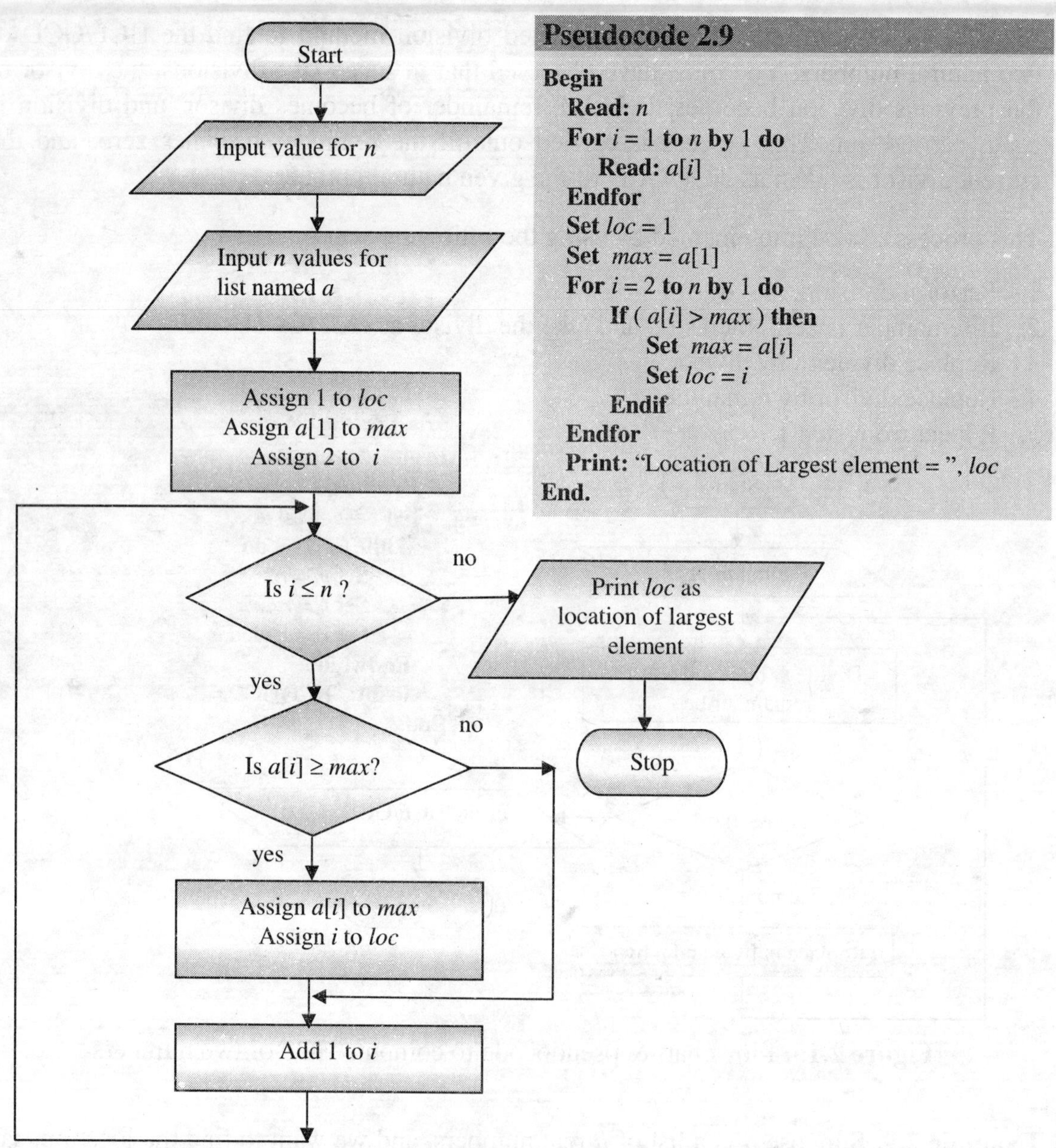

Figure 2.16: Flowchart to find the location of the largest element from a list of numbers

Example 2.10: Income tax for individuals is computed on slab rates as follows:

Income	Tax Payable
Upto Rs. 1,00,000/-	Nil
From Rs. 1,00,001/- to Rs. 2,00,000/-	10% of the excess over Rs. 1,00,000/-
From Rs. 2,00,001/- to Rs. 3,00,000/-	20% of the excess over Rs. 2,00,000/-
Above Rs. 3,00,000/-	30% of the excess over Rs. 3,00,000/-

Draw a flowchart or write a pseudocode that reads the income and prints the income tax due.

Solution: Note the following:

- If the income is upto Rs. 1,00,000/- only, then the tax is *nil*.
- If the income is in the range of Rs. 1,00,001/- to Rs. 2,00,000/-, then the tax is 10% of the amount in excess of Rs. 1,00,000/-.
- If the income is in the range of Rs. 2,00,001/- to Rs. 3,00,000/-, then the tax is 10% of Rs. 1,00,000/- (Rs. 10,000/-) for a slab of Rs. 1,00,001 to Rs. 2,00,000/- plus 20% of the amount in excess of Rs. 2,00,000/-.
- If the income is more than Rs. 3,00,000/-, then the tax is 10% of Rs. 1,00,000/- (Rs. 10,000/-) for a slab of Rs. 1,00,001 to Rs. 2,00,000/- plus 20% of Rs. 1,00,000/- (Rs. 20,000/-) for a slab of Rs. 2,00,001 to Rs. 3,00,000/- plus 30% of the amount in excess of Rs. 3,00,000/-.

Let us have hands on by computing the tax if the given income is

(*a*) Rs. 2,50,000/-
(*b*) Rs. 1,75,000/-
(*c*) Rs. 3,20,000/-

(*a*) *Tax Computations for income of Rs. 2,50,000/-*

For first Rs. 1,00,000/- Tax is Nil
For next Rs. 1,00,000/- Tax is 10,000/- (@ 10%)
For next Rs. 50,000/- Tax is 10,000/- (@ 20%)

Total Tax due Rs. 20,000/-

(*b*) *Tax Computations for income of Rs. 1,75,000/-*

For first Rs.1,00,000/- Tax is Nil
For next Rs. 75,000/- Tax is 7,500/- (@ 10%)

Total Tax due Rs. 7,500/-

(*c*) *Tax Computations for income of Rs. 3,20,000/-*

For first Rs. 1,00,000/-	Tax is Nil
For next Rs. 1,00,000/-	Tax is 10,000/- (@ 10%)
For next Rs. 1,00,000/-	Tax is 20,000/- (@ 20%)
For next Rs. 20,000/-	Tax is 6,000/- (@ 30%)
Total Tax due	Rs. 36,000/-

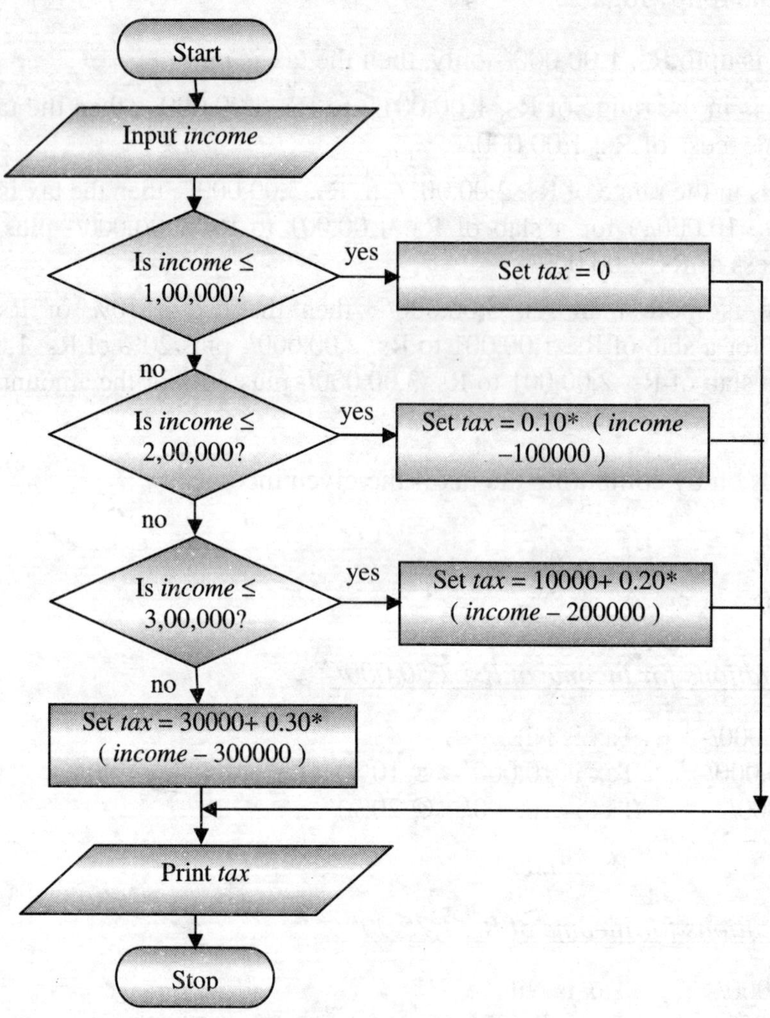

Figure 2.17: Flowchart to compute tax

Pseudocode 2.10

```
Begin
    Read: income
    If (income < 100000) then
        Set tax = 0
    Else if (income < 200000) then
        Set tax = 0.01 × (income – 100000)
    Else if (income < 300000) then
        Set tax = 10000 + 0.02 × (income – 200000)
    Else
        Set tax = 30000 + 0.03 × (income – 300000)
    Endif
    Print: tax
End.
```

Example 2.11: Given measure of three line segments as *a*, *b*, and *c*. Draw a flowchart and also write pseudocode to check whether the line segments can be used to make a triangle.

Solution: By the Triangle Inequality, in order for three line segments to make a triangle, the sum of the lengths of any two line segments must exceed the length of the third line segment.

Pseudocode 2.11

```
Begin
    Read: a, b, c
    If ( a + b > c ) then
        If ( b + c > a ) then
            If ( c + a > b ) then
                Print: "Yes"
            Else
                Print: "No"
            Endif
        Else
            Print: "No"
        Endif
    Else
        Print: "No"
    Endif
End.
```

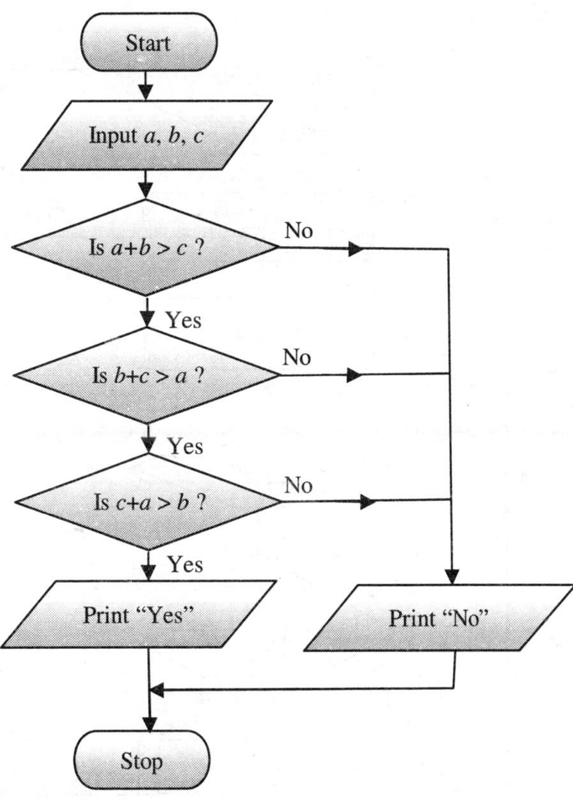

Figure 10.18: Flowchart & pseudocode to check whether a triangle can be formed from given three line segments

Example 2.12: Draw a flowchart to print first *n* prime numbers.

Solution:

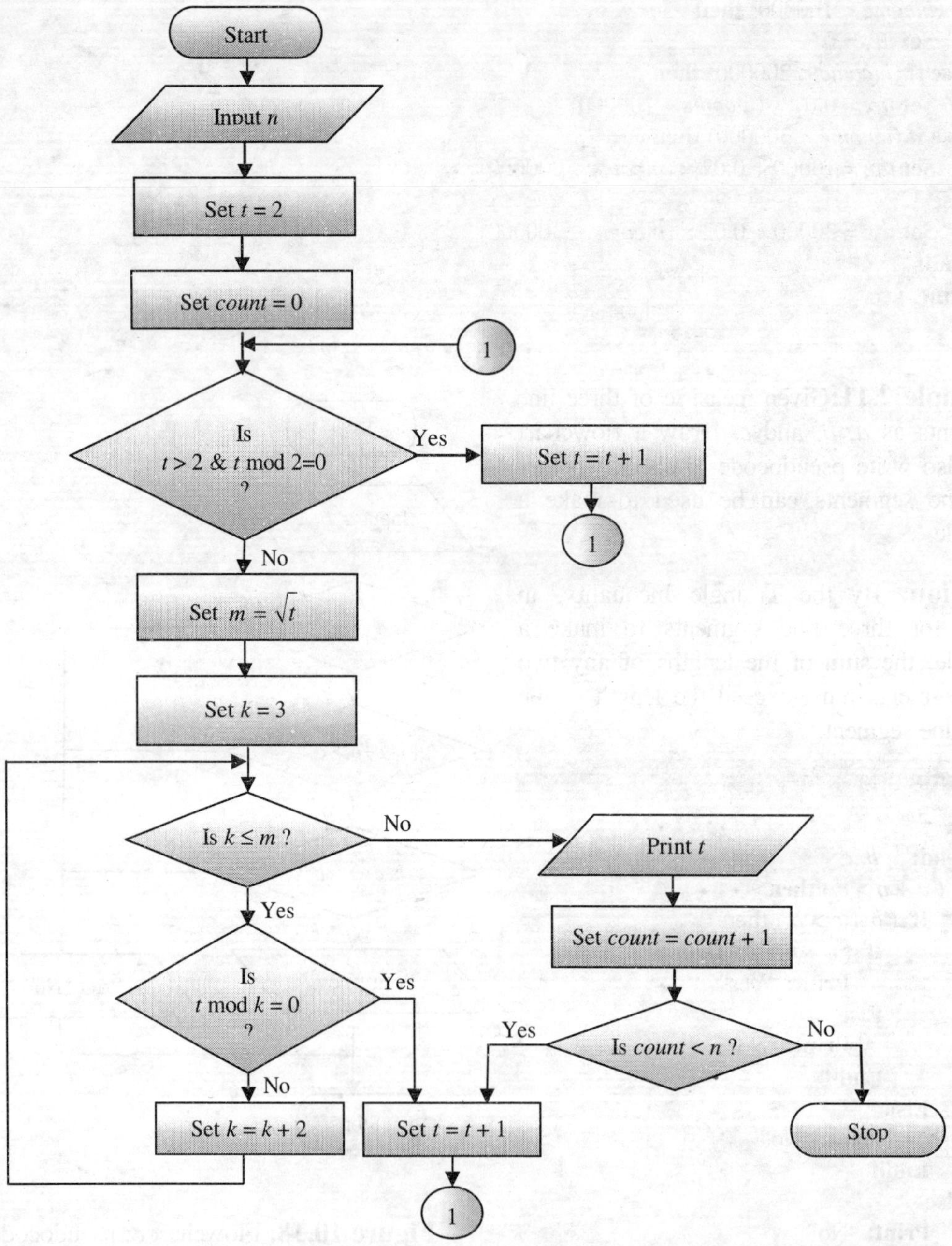

Figure 2.19: Flowchart to print first *n* prime numbers

Example 2.13: Draw a flowchart to print *n*th prime number.

Solution:

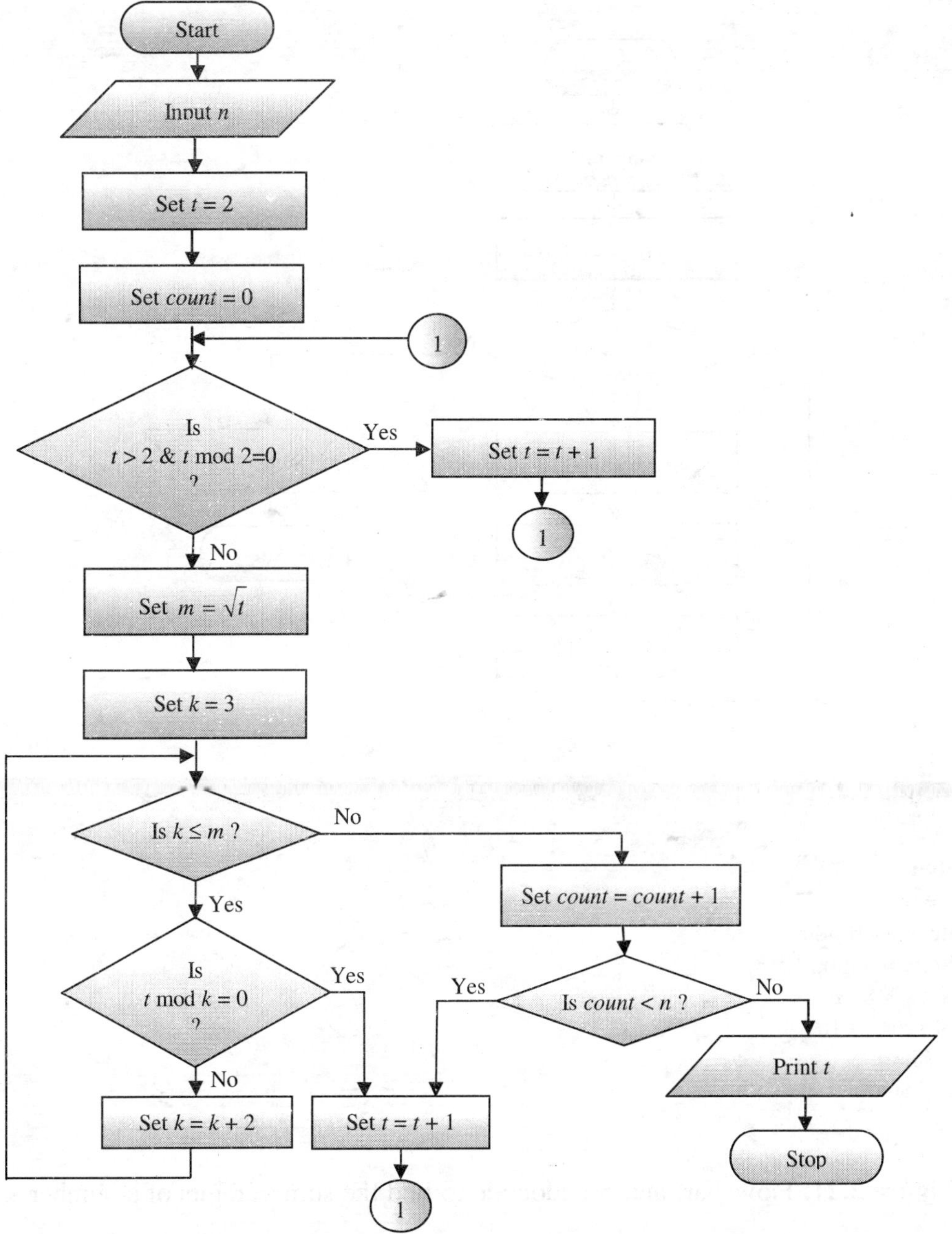

Figure 2.20: Flowchart to print *n*th prime number

Example 2.14: Draw a flowchart and write pseudocode to find the sum of digits of a number *n*.

Solution:

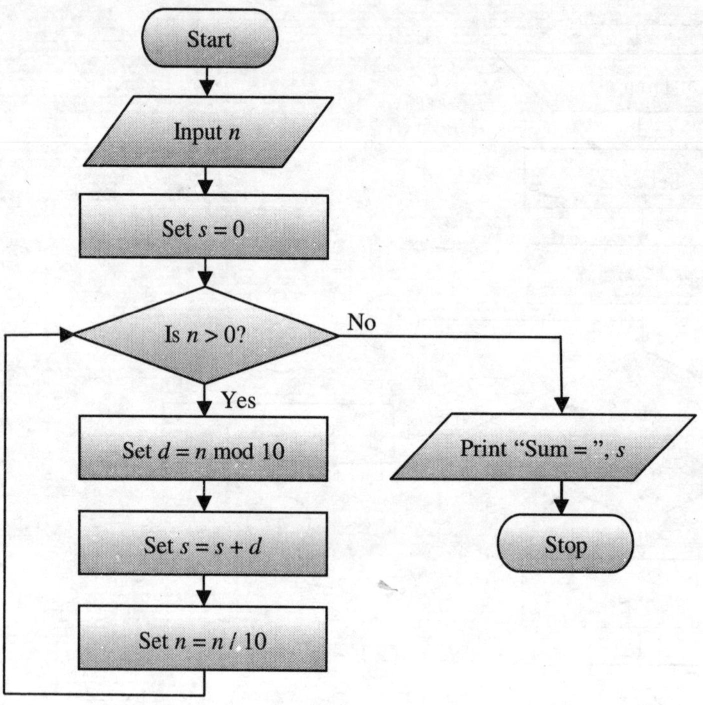

Pseudocode 2.12

```
Begin
   Read: n
   Set s = 0
   While ( n > 0 ) do
      Set d = n mod 10
      Set s = s + d
      Set n = n / 10
   Endwhile
   Print: "Sum = ", s
End.
```

Figure 2.21: Flowchart and pseudocode to find the sum of digits of a number

Example 2.15: Draw a flowchart and write pseudocode to find the largest digit of a number *n*.

Solution:

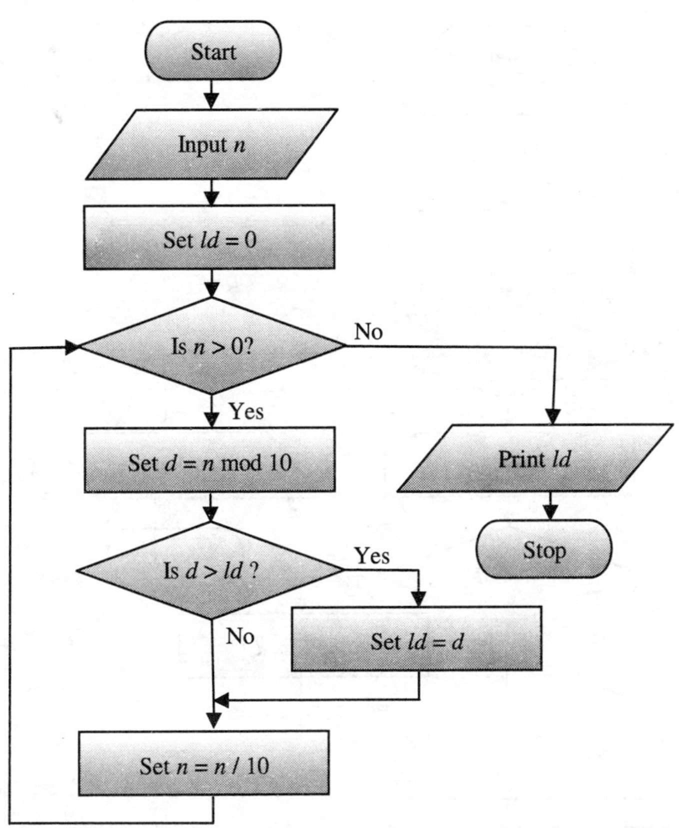

Pseudocode 2.13

```
Begin
    Read: n
    Set ld = 0
    While ( n > 0 ) do
        Set d = n mod 10
        If ( d > ld ) then
            Set ld = d
        Endif
    Endwhile
    Print: "Largest digit = ", ld
End.
```

Figure 2.22: Flowchart and pseudocode to find the largest digit of a number

Example 2.16: Draw a flowchart and write pseudocode to find the largest digit of a number *n*.

Solution:

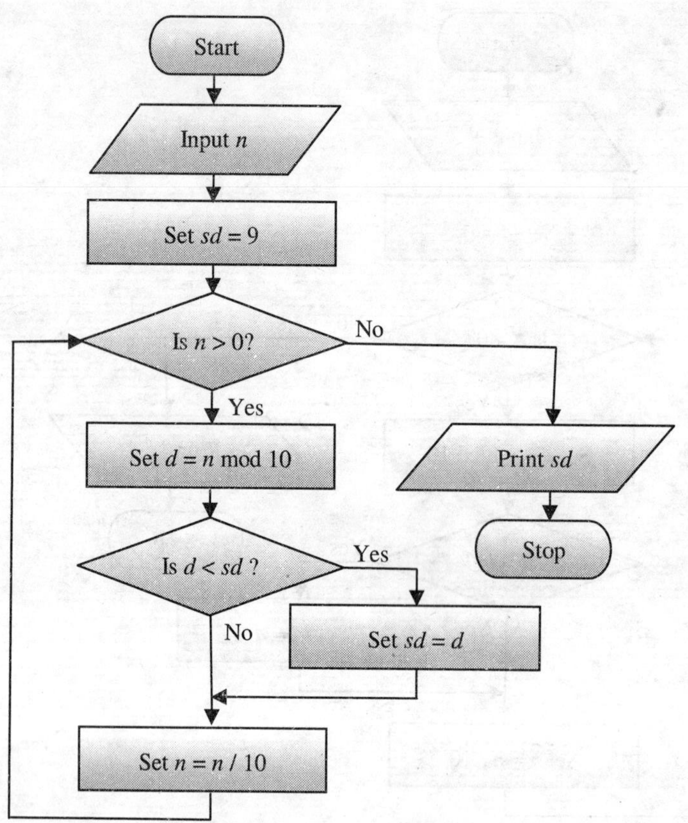

Pseudocode 2.14

```
Begin
    Read: n
    Set sd = 9
    While ( n > 0 ) do
        Set d = n mod 10
        If ( d < sd ) then
            Set sd = d
        Endif
    Endwhile
    Print: "Smallest digit = ", sd
End.
```

Figure 2.23: Flowchart and pseudocode to find the smallest digit of a number

Example 2.17: Draw a flowchart and write pseudocode to check whether the given number *n* is palindrome or not.

Solution:

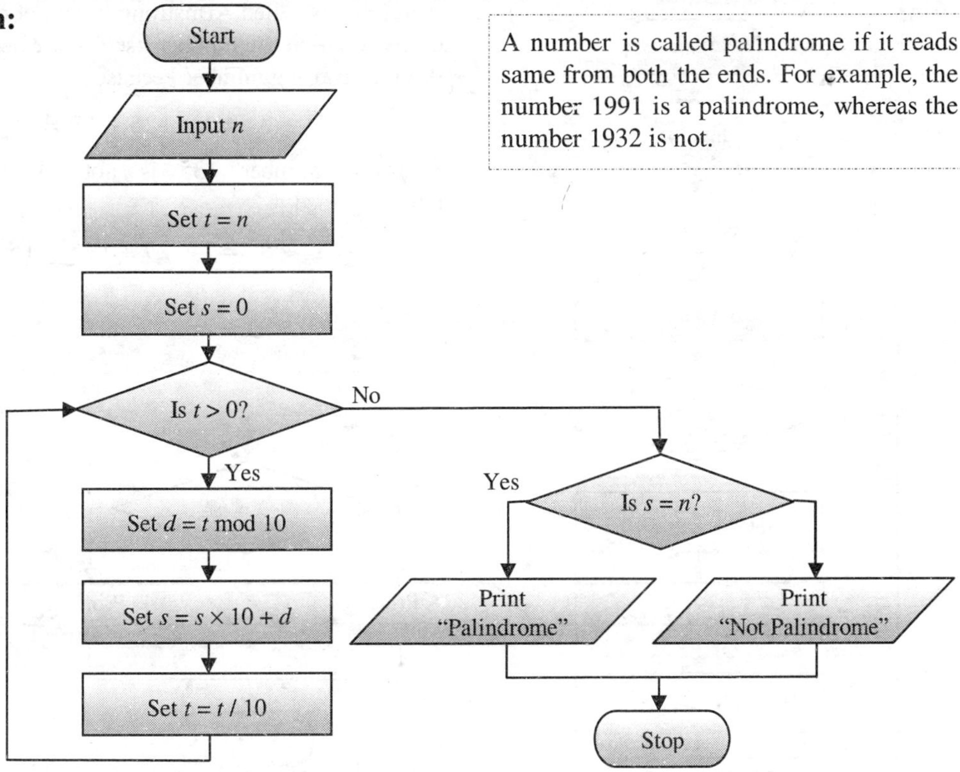

A number is called palindrome if it reads same from both the ends. For example, the number 1991 is a palindrome, whereas the number 1932 is not.

Start

Input *n*

Set $t = n$

Set $s = 0$

Is $t > 0$?

No

Yes

Set $d = t \bmod 10$

Set $s = s \times 10 + d$

Set $t = t / 10$

Yes

Is $s = n$?

Print "Palindrome"

Print "Not Palindrome"

Stop

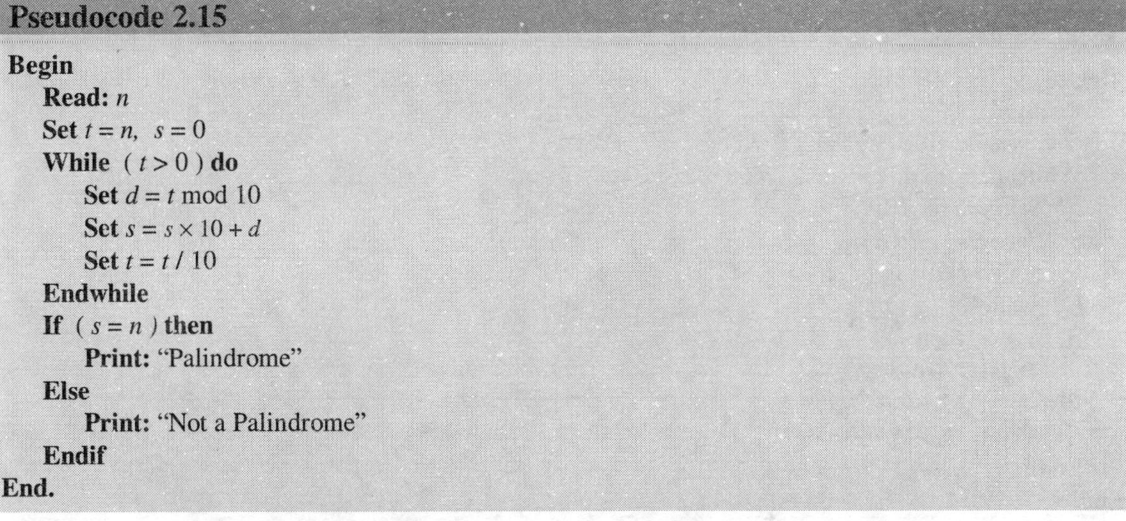

Pseudocode 2.15

```
Begin
   Read: n
   Set t = n,  s = 0
   While ( t > 0 ) do
      Set d = t mod 10
      Set s = s × 10 + d
      Set t = t / 10
   Endwhile
   If ( s = n ) then
      Print: "Palindrome"
   Else
      Print: "Not a Palindrome"
   Endif
End.
```

Figure 2.24: Flowchart & pseudocode to check whether the given number *n* is palindrome or not

Example 2.18: Draw a flowchart and write pseudocode to check whether the given number n is Armstrong number or not.

Solution:

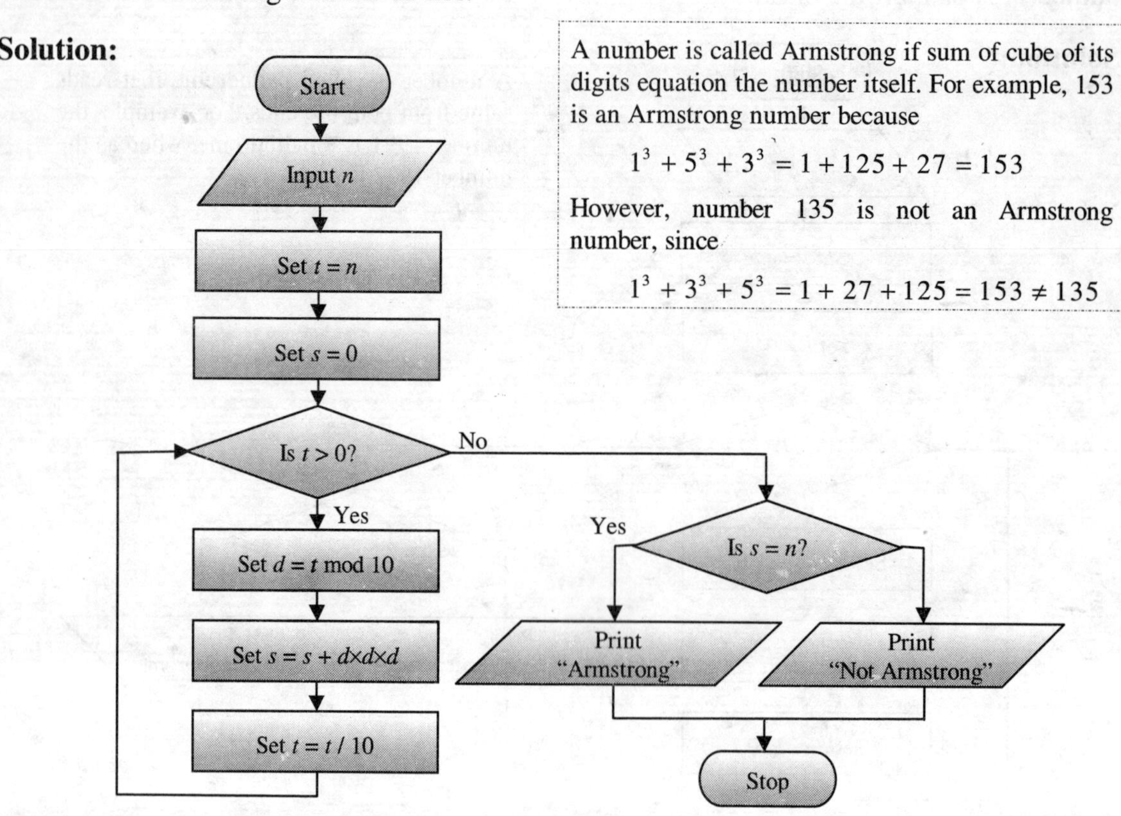

Pseudocode 2.16

```
Begin
   Read: n
   Set t = n,  s = 0
   While ( t > 0 ) do
       Set d = t mod 10
       Set s = s + d × d × d
       Set t = t / 10
   Endwhile
   If ( s = n ) then
        Print: "Armstrong"
   Else
        Print: "Not Armstrong"
   Endif
End.
```

Figure 2.25: Flowchart and pseudocode to check whether the given number n is Armstrong number or not

Example 2.19: A notion of *well-ordered* number is introduced here. Any number whose digits from left to right are in increasing order (with no digit repeated) is called a *well-ordered* number. For example, number 138 is called *well-ordered* because the digits in the number (1, 3, 8) increase from left to right, *i.e.*, 1 < 3 < 8. On the other hand, the number 365 is not *well-ordered* because 6 is larger than 5.

Draw a flowchart and also write pseudocode to check whether the given number *n* is *well-ordered* not.

Solution:

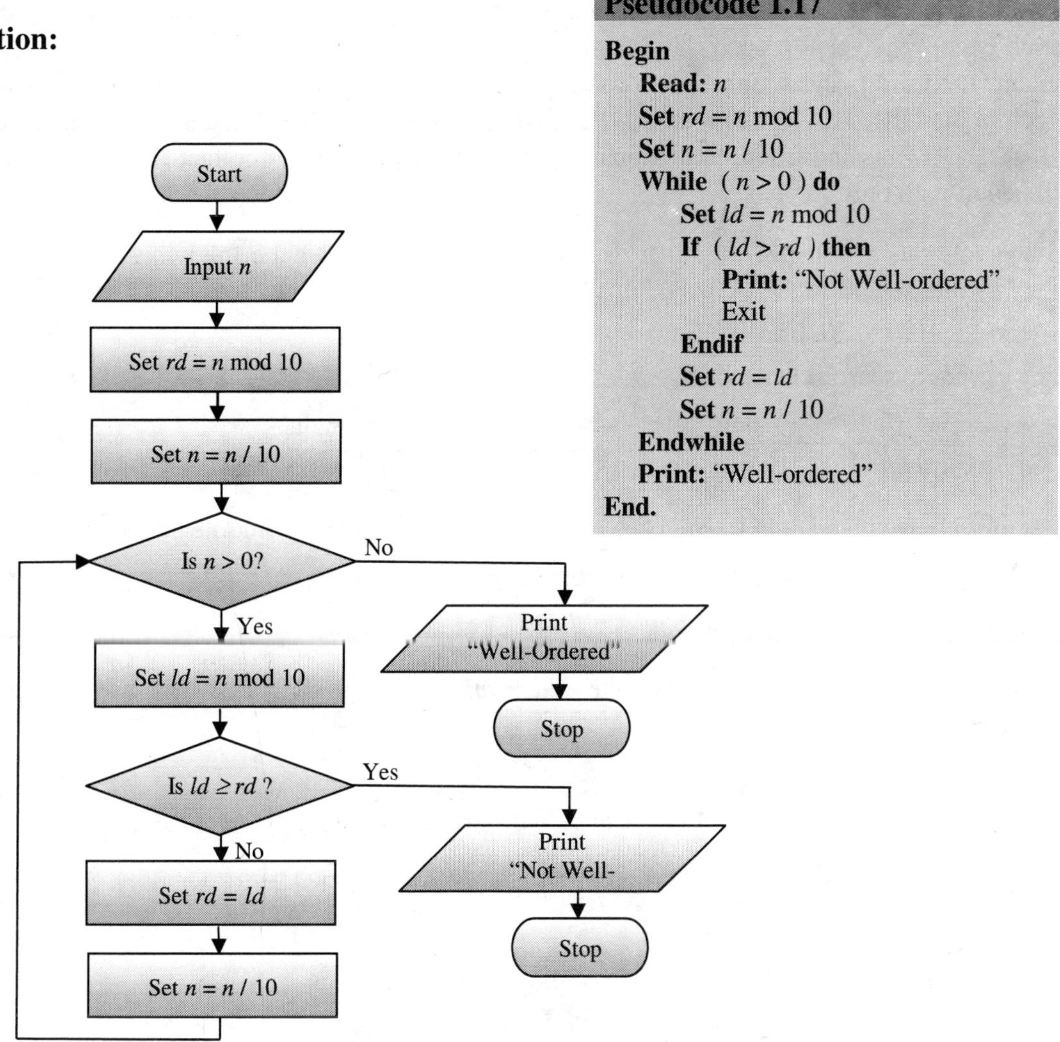

Pseudocode 1.17

```
Begin
    Read: n
    Set rd = n mod 10
    Set n = n / 10
    While ( n > 0 ) do
        Set ld = n mod 10
        If ( ld > rd ) then
            Print: "Not Well-ordered"
            Exit
        Endif
        Set rd = ld
        Set n = n / 10
    Endwhile
    Print: "Well-ordered"
End.
```

Figure 2.26: Flowchart and pseudocode to check whether the given number *n* is well-ordered or not

Example 2.20: Write a pseudocode to search a given element in a list, named *a* having *n* elements, which is maintained in ascending order.

Solution: Searching is a process of finding the location of given element in the list. The search is said to be successful if the given element is found, *i.e.*, the element does exist in the list; otherwise unsuccessful.

As the list is sorted, the best searching algorithm, called *binary search,* is used to find the location of the given element. We do use this approach in our daily life. For example, suppose we want to find the meaning of the term *modem* in a computer dictionary. Obviously, we don't search page by page. We open the dictionary in the middle (roughly) to determine which half contains the term being sought. Then for the subsequent search one half is discarded, and we search in the other half. This process is continued till either we have located the required term or that term is missing from the dictionary, which will indicated by the fact that at the end we will be left with only one page.

To illustrate the working of the binary search technique, consider the following sorted array *a* with 7 elements

3, 10, 15, 20, 35, 40, 60

and we want to search element 15.

	a[0]	*a*[1]	*a*[2]	*a*[3]	*a*[4]	*a*[5]	*a*[6]
Given Array *a*	3	10	15	20	35	40	60

To start with, we take $beg = 0$, $end = 6$, and compute location of the middle element as

$$mid = (beg + end) / 2 = (0 + 6) / 2 = 3$$

Since *a*[*mid*], *i.e.*, $a[3] \neq 15$, and $beg < end$. We start next iteration.

As $a[mid] = 20 > 15$, therefore, we take $end = mid - 1 = 3 - 1 = 2$, where as *beg* remains unchanged. Thus

$$mid = (beg + end) / 2 = (0 + 2) / 2 = 1$$

Since *a*[*mid*], *i.e.*, $a[1] \neq 15$, and $beg < end$. We start next iteration.

As $a[mid] = 10 < 15$, therefore, we take $beg = mid + 1 = 1 + 1 = 2$, where as *end* remains unchanged.

Since $beg = end$, again compute location of the middle element as

$$mid = (beg + end) / 2 = (2 + 2) / 2 = 2$$

Since *a*[*mid*], *i.e.*, $a[2] = 15$, the search terminates on success.

In case, value of variable *beg* becomes greater than variable *end*, it indicates that search is unsuccessful.

Pseudocode 2.18

```
Begin
    Read: n
    For i = 0 to (n–1) by 1 do          // input elements of sorted list
        Read: a[i]
    Endfor
    Read: item
    Set beg = 0
    Set end = n–1
    Set mid = ( beg + end ) / 2
    While ( ( beg ≤ end ) and ( a[mid] ≠ item ) ) do
        If ( item < a[mid] ) then
            Set end = mid-1
        Else
            Set beg = mid+1
        Endif
        Set mid = ( beg + end ) / 2
    Endwhile
    If ( beg > end ) then
        Print: "Search unsuccessful"
    Else
        Print: item, " found at location ", mid
    Endif
End.
```

Example 2.21: Write a pseudocode to sort a list, named *a* having *n* elements, in ascending order.

Solution: Sorting is the process of arranging the elements in some logical order. This logical order may be ascending or descending in case of numeric values or dictionary order in case of alphanumeric values. We do, almost, use this process daily in our routine activities. For example, we do arrange our class notes in order of date so we can refer them quickly later on. Like this there are many situations, which you can enumerate.

Bubble sort is one of most popular technique for this purpose. It requires $(n–1)$ pass to sort an array. In each pass, every element $a[i]$ is compared with $a[i+1]$, for $i = 0$ to $(n–k)$, where k is the pass number and if they are out of order, *i.e.*, if $a[i] > a[i+1]$, they are swapped (interchanged).

This will cause the largest element to move or bubble up. Thus, after the end of first pass the largest element in the array will be placed in nth position, and on each successive pass, the next largest element is placed at position $(n–1)$, $(n–2)$, ..., 2, respectively. For more clarity, carefully examine the following steps.

Pass 1.

Step 1. If $a[0] > a[1]$ then swap $a[0]$ and $a[1]$.
Step 2. If $a[1] > a[2]$ then swap $a[1]$ and $a[2]$.
:
Step n–1. If $a[n-2] > a[n-1]$ then swap $a[n-2]$ and $a[n-1]$.

Pass 2.

Step 1. If $a[0] > a[1]$ then swap $a[0]$ and $a[1]$.
Step 2. If $a[1] > a[2]$ then swap $a[1]$ and $a[2]$.
:
Step n–2. If $a[n-3] > a[n-2]$ then swap $a[n-3]$ and $a[n-2]$.

:

Pass *k*.

Step 1. If $a[0] > a[1]$ then swap $a[0]$ and $a[1]$.
Step 2. If $a[1] > a[2]$ then swap $a[1]$ and $a[2]$.
:
Step n–k. If $a[n-k+1] > a[n-k]$ then swap $a[n-k+1]$ and $a[n-k]$.

:

Pass *n*–1.

Step 1. If $a[0] > a[1]$ then swap $a[0]$ and $a[1]$.

Note that, the output of a given pass becomes the input for the next pass.

To illustrate the working of bubble sort method, consider the sorting of the following array in ascending order.

12 40 3 2 15

Note that, the output of a given pass becomes the input for the next pass.

Given array can be shown as

$a[0]$	$a[1]$	$a[2]$	$a[3]$	$a[4]$
12	40	3	2	15

Pass-1:

$a[0]$	$a[1]$	$a[2]$	$a[3]$	$a[4]$
(12)	(40)	3	2	15

(*a*) Compare $a[0]$ and $a[1]$. Since $a[0] < a[1]$ $(12 < 40)$, no action is taken.

$a[0]$	$a[1]$	$a[2]$	$a[3]$	$a[4]$
12	(40)	(3)	2	15

(*b*) Compare $a[1]$ and $a[2]$. Since $a[1] > a[2]$ $(40 > 3)$, swap these elements.

$a[0]$	$a[1]$	$a[2]$	$a[3]$	$a[4]$
12	3	(40)	(2)	15

(*c*) Compare $a[2]$ and $a[3]$. Since $a[2] > a[3]$ $(40 > 2)$, swap these elements.

$a[0]$	$a[1]$	$a[2]$	$a[3]$	$a[4]$
12	3	2	(40)	(15)

(*d*) Compare $a[3]$ and $a[4]$. Since $a[3] > a[4]$ $(40 > 15)$, swap these elements.

$a[0]$	$a[1]$	$a[2]$	$a[3]$	$a[4]$
12	3	2	15	40

Thus, after first pass, the largest element, 40, has moved to its final position. However, the rest of the numbers, which may have changed their positions, are yet to be sorted. The shaded part shows the elements that have been sorted.

For the remaining passes, only comparisons and swaps, if required, are shown.

Pass-2:

$a[0]$	$a[1]$	$a[2]$	$a[3]$	$a[4]$	Action
(12)	(3)	2	15	40	Swap
3	(12)	(2)	15	40	Swap
3	2	(12)	(15)	40	No swap
3	2	12	15	40	

After second pass, the second largest element, 15, has moved to its final position.

Pass-3:

$a[0]$	$a[1]$	$a[2]$	$a[3]$	$a[4]$	Action
(3)	(2)	12	15	40	Swap
2	(3)	(12)	15	40	No swap
2	3	12	15	40	

After third pass, the third largest element, 12, has moved to its final position.

Pass-4:

a[0]	a[1]	a[2]	a[3]	a[4]	Action
2	3	12	15	40	No swap
2	3	12	15	40	

After fourth pass, the fourth largest element, 3, has moved to its final position. Thus, the array is completely sorted in 4 passes.

Pseudocode 2.19

```
Begin
   Read: n
   For i = 0 to (n–1) by 1 do
       Read: a[i]
   Endfor
   For k = 1 to (n–1) by 1 do
       For j = 0 to (n–k–1) by 1 do
           If a[j] > a[j+1] then
               Set temp = a[j]          // variable temp is used to facilitate
               Set a[j] = a[j+1]        // interchange of values
               Set a[j+1] = temp
           Endif
       Endfor
   Endfor
   For i = 0 to (n–1) by 1 do
       Print: a[i]
   Endfor
End.
```

REVIEW EXERCISE

1. Why is it advisable to plan the program before writing it?
2. What is an algorithm? What are the essential characteristics for a sequence of instructions to qualify as an algorithm?
3. What is a flowchart? What are the various symbols used in flowcharts?
4. What is a pseudocode? Why is it so called?
5. Among algorithm, flowchart and pseudocode, which makes the logic of the underlying program clearer and easily understandable?
6. Given a chance to plan a program, will you prefer to choose algorithm, flowchart or a pseudocode? Give reasons for your choice.
7. Draw a flowchart to find largest of the five numbers *a*, *b*, *c*, *d*, and *e*.

8. Write a pseudocode to display first 20 prime numbers.
9. Write a pseudocode to find the sum of first 100 odd numbers.
10. Suppose an amount p is deposited in a commercial bank, which pays compound interest at the rate of $r\%$ annually, for n years. Write a pseudocode that prints the amount in account after each year.
11. Write a pseudocode to find LCM and HCF of two natural number m and n.
12. Write a pseudocode to find factorial of a natural number n.
13. Write a program to print first n pairs of twin prime numbers. Note that two consecutive prime numbers are said to twin prime numbers if they differ by 2.
14. The monthly telephone bill is to be computed as follows:

    Minimum Rs. 200 for upto 100 calls
    plus Rs. 0.60 per call for next 50 calls
    plus Rs. 0.50 per call for next 50 calls
    plus Rs. 0.40 per call for any call beyond 200 calls.

    The input contains name of the customer and number of calls made and the desired output is the name and telephone bill to be paid by the customer. Draw a flowchart for logical steps needed to accomplish the above task.
15. State pollution control board has the following classification policy:

Pollution Index	Classification
< 30	Pleasant
30 – 60	Unpleasant
> 60	Hazardous

    Draw a flowchart that accepts pollution index and prints the appropriate classification.
16. A department store places an order with a company for n pieces of miners, m pieces of toasters, and p number of fans. The cost of items are as follows:

Item Description	Price per Unit (in Rupees)
Miners	1,500
Toaster	200
Fan	450

    The discount allowed for various items are 5% for miners, 15% for fan, and 10% for toaster. The company charge 10% as sales tax on all items on net value after deducting the discount. Write pseudocode that reads m, n, and p and computes the amount to be paid by the store.

❑ ❑ ❑

Chapter 3

Overview of C Language

Learning Outcomes

After reading this chapter, students will be able to explain

- history and evolution of C language
- why C language is called middle level language
- key characteristics of C language
- potential areas where C language is predominantly used
- general structure of C program
- steps required in building a C program
- usage of various C Compilers

3.1 INTRODUCTION

Invented and first implemented by Dennis Ritchie, a system programmer at Bell Laboratories, on a DEC PDP-11 computer system running the UNIX operating system. Later on, the UNIX operating system was rewritten almost entirely in C language. The development of C language was the result of the development process that started with an older language known as Basic Combined Programming Language (BCPL), developed by Martin Richards. The BCPL language influenced a language called B language, developed by Ken Thompson, which lead to the development of C language in the 1970s. It was a mere coincidence that the letters B and C are in alphabetic order.

For many years the de facto standard for C language was the one supplied with the UNIX operating system and described in book entitled "*The C Programming Language*" by Brain Kernighan and Dennis Ritchie (Prentice-Hall, 1978). The C language was formally standardized in December 1989, when the American National Standards Institute (ANSI) standard for C was adopted.

The creation of C language was considered many to have marked the beginning of the modern age of programming languages. It was designed, developed, and implemented by the working programmers, reflecting the way they approached the job of programming. Its features were honed, tested, thought about, and rethought by the people who actually used the language. The result was a language that programmers liked to use. Indeed, the C language quickly attracted many followers who had a near-religious zeal for it. As such it found wide and rapid acceptance in the programmer community.

In short, the C language is a language designed by and for programmers.

3.2 ANATOMY OF COMPUTER LANGUAGES

To see how C language compares with other programming languages, let us have a look on the programming languages in general. The programming language can be divided into two broad categories, depending on their level of interaction with the underlying hardware of the computer:

- Low-level Languages
- High-level Languages

3.2.1 Low-level Languages

In this category, we have machine language and assembly language. These languages permit the efficient use of the computer.

But the problems with these languages are:

- These languages are hardware dependent, *i.e.*, programs written using these languages is not portable, i.e., cannot be used on other computers.
- Programming using these languages is not an easy job. One must have thorough knowledge of the architecture of the computer.

3.2.2 High-level Languages

In this category, we have FORTRAN, BASIC, PASCAL, COBOL, PL/1, etc. These languages are designed for better programming efficiency, *i.e.*, faster program development. These languages have following advantages:

- The syntax for writing program instructions is very much like English statements. This enables the readers to learn high-level languages (HLLs) quickly. In addition, the programs written in HLLs can be easily understood, which facilitates its maintenance.
- The programs written in HLLs are not hardware dependent. This means that program written for one machine can be transferred to another machine with minimal changes or none at all.

The C language stands between these two categories. That is why it is often called *middle-level* language, since it was designed to have best of both the worlds, *i.e.*, a good programming efficiency as well as good machine efficiency.

- To achieve programming efficiency, the C language has all the elements as of any other modern high-level language.
- To achieve machine efficiency, the C language has requisite features to access any hardware component of the system, to operate at register level, and to interface with high-speed assembly language routines.

3.3 CHARACTERISTICS OF C LANGUAGE

The C language was and will remain as one of the popular programming languages. Some of the key characteristics that add to its popularity are:

- It is a general purpose programming language, and therefore, can be used to solve vide variety of problems.
- It is supports structuring programming, and therefore, programs developed in it can easily be understood.

- It is free form, *i.e.*, no rigid format for writing programs. Any instruction can start from anywhere and end anywhere. In addition, a single instruction can span (can be written in) many lines.
- It is case sensitive. It distinguishes between lowercase (small) and uppercase (capital) alphabets.
- It was the first language to provide rich set of operators.
- It allows you, through pointers, to access any storage location and the devices connected to your computer through your program.
- It allows you to manipulate internal processor registers, and thus it is very useful for low-level programming.
- It is portable, that means any program written in it can be used on any computer without or with little modification(s).
- It allows you to develop your own library of functions that can be linked to your any program like standard library functions.

3.4 APPLICATION AREAS OF C LANGUAGE

The strength of C language makes it a natural choice for

- Systems programming that include writing software for language translators, device drivers, editors, linkers, loaders, etc.
- Network software to implement different communication protocols.
- Graphics programming that include writing software for graphical user interfaces (GUIs), scientific visualization, and presentation graphics, etc.
- Embedded systems where C routines are interfaced high speed assembly language routines and the resulting code is stored in ROM chip which is a part of the embedded system.

3.5 GENERAL STRUCTURE OF A C PROGRAM

The usual order of statements written in a C program is as shown in Figure 3.1.

Section 1 is optional. If present it contains the description about the program. This description usually contains the information about the task being accomplished by the program. In addition, it can also contain the name of the author (programmer), the date on which it was written, the date on which it was last modified, etc.

Section 2 contains the preprocessor directives. The frequently used preprocessor directives are *include* and *define*. These directives tell the preprocessor how to prepare the program for compilation. The *include* directive tells which header files are to be included in the program and the *define* directive is usually used to associate an identifier with a literal (constant) that is to be used at many places in the program.

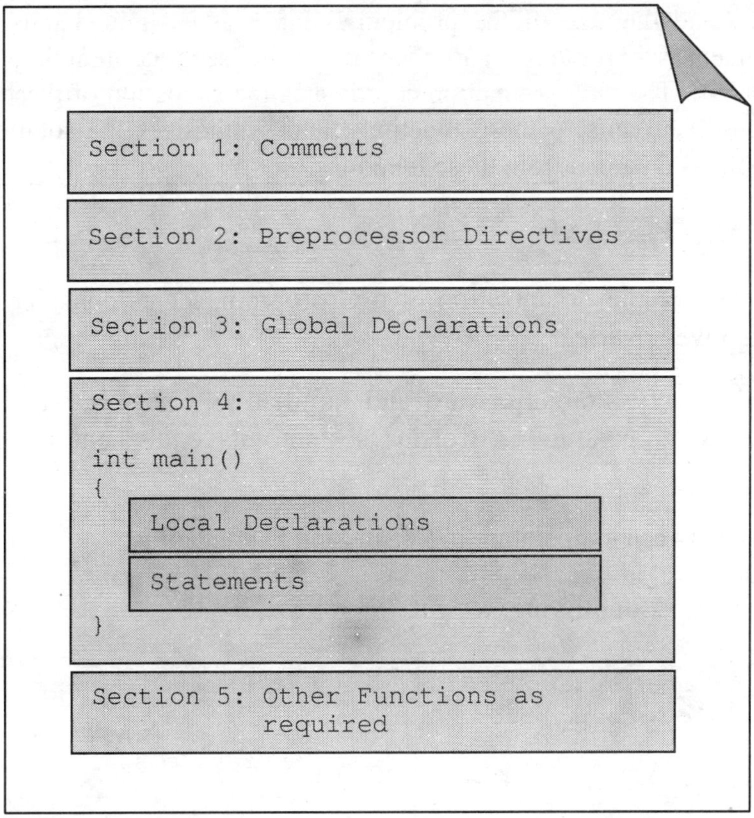

```
Section 1: Comments

Section 2: Preprocessor Directives

Section 3: Global Declarations

Section 4:

int main()
{
    Local Declarations
    Statements
}

Section 5: Other Functions as
           required
```

Figure 3.1: General Structure of a C Program

Section 3 is optional. If present, it contains the global declarations. These declarations usually include the declaration of the data items (variables) which are to be shared between many functions in the program. In addition, these declarations can also include the decorations of functions (prototypes), except the *main()* function, to be used in the program.

Section 4 contains the *main()* function. The execution of the program always begins with the execution of the *main()* function. The *main()* function can call any number of other functions, and those called function can further call other functions. The first section in the *main()* function, as well as other functions, contains local declarations. These declarations are local in the sense that they pertain to the requirements of that function only. The second section in the *main()* function, as well as in other functions, contains the statements that defines the actions to be performed by the function.

Section 5 is also optional. If present, it contains the other functions.

If the problem to be solved is simple and small in size, then only the *main()* function is sufficient to accomplish the task. However, if the problem is complex and the size of the problem is large, it is divided into small and independent subproblems, and then we write separate functions for each subproblem. The *main()* function coordinates the execution of these functions by appropriate calls to these functions, and synthesizes the solutions of the subproblems obtained from these functions.

3.6 SAMPLE C PROGRAM

In order to have a feel of the organization of a C program, let us consider the simple example C program, given overleaf.

Problem considered is very straightforward and familiar to you. We want to write a program that converts temperature in Celsius scale to its equivalent temperature in Fahrenheit scale.

We know the relation between temperature in Celsius and Fahrenheit is:

$$\frac{C}{100} = \frac{F-32}{180} \qquad \text{Simplifying, we get} \quad F = 1.8 \times C + 32$$

Listing 3.1

```
/*
 * Program to convert temperature from Centigrade scale
 * to Fahrenheit scale
 */
#include <stdio.h>
int main(void)
{
   float fahrenheit, centigrade;
   printf( "\nEnter temperature in Celsius scale: ");
   scanf( "%f", &centigrade );
   fahrenheit = 1.8 * centigrade + 32;
   printf( "\nEquivalent temperature in Fahrenheit scale");
   printf( " = %.2f\n",fahrenheit);
   return 0;
}
```

Here the line numbers are added for ready reference.

Test Run

```
Enter temperature in Celsius scale: 30
Equivalent temperature in Fahrenheit scale = 86.00
```

Dissection of the Program

If you are familiar with other programming languages, this program may look strange to you. One reason for this strangeness is that the entire program is written in lowercase letters. In most other languages, usually uppercase letters are used, though they permit the use of lowercase letters. So you have to become accustomed to lowercase as quickly as possible. Also, it is important to remember that C language is case sensitive, *i.e.*, it differentiates between lowercase and uppercase letters. Lowercase letters are characteristic of the C language.

A C program is written in a free format. By free format, we mean that an instruction can start anywhere in a line and can end anywhere. Even an instruction can span many lines. Spacing is of no consequence in C language, it is used to enhance the readability of the program.

Lines 1-4 demonstrate the style used in C language to add comments in a program. This is done by starting the comments with two characters '/*' and ending with the characters '*/'. Between these pair of characters, called *delimiters*, any characters may be included in either uppercase or lowercase.

In other words, a comment appears as follows:

```
/*  You type here what you want   */
```

In addition to this style, comments can also be added after the instruction on the same line as shown below:

```
#include<stdio.h>      /* include stdio.h file */
```

Comments may not be essential in a simple program, but in complex programs they are lifesaver. For any program, no matter how well it is written, the day will come when we have to make amendments to incorporate the changing conditions. The chances are quite high that the original author of the program may have left the organization. Even the original author may not remember the logic of a program that he wrote several months or years ago.

In order to reduce the burden on the person, who will make amendments to the program, ample comments should be included at the time when the program is being written. Too often we postpone the task of adding comments until the program is finally Okayed, but this practice is not very useful.

Some programmers feel that C language is such a concise and cryptic language that every line of code should be accompanied by a comment. But too many comments actually reduce the readability of the program. Still, it is better to have too many comments than few.

Line 5 demonstrates the way pre-processor directives are used. These directives instruct the compiler process the source code in a specific way. In this case, *include* directive, also called *file inclusion* directive, and instructs the compiler to include the specified file at this point. In this case, the specified file is a header file (a file with extension *.h*) that is part of the C compiler.

The other often-used pre-processor directive is *#define* whose syntax is

```
#define MAX 100
```

This directive, called *macro expansion* directive, tells the compiler to replace every occurrence of MAX with value 100 during pre-processing. There are many other directives that we will discuss later.

Line 6 specifies a function named *main*. This is a special name that is recognized by the system. It points to the precise place in the program where execution begins. Every C program must have a *main* function. We cannot have more than one main function in a program.

Every C function has a return type associated with it. Where return type is one of the data types that determine the type of the value returned by the function. If a function is not supposed to return any value, its return type is specified as *void*. Since main function is the function from where execution begins, it will usually be declared with type *int*. A pair of parentheses follows the word main. The matter, if any, contained in these parentheses specifies the formal arguments of the function. If there are no arguments, we write *void*.

The *main()* Function

The *main()* function is a part of every C program. The C language permits following forms of *main()* function:

```
o  main()
o  main(void)
o  void main()
o  int main()
o  void main(void)
o  int main(void)
```

The empty pair of parenthesis indicates that the *main()* function takes no arguments. This may be explicitly indicated by using the keyword *void* inside the parenthesis. We may also specify the keyword *void* or *int* before the word main. When the keyword *void* is used, it indicates that the function *main()* does not return any value to the operating system. When the keyword *int* is used, it indicates that the function *main()* will return an integer value to the operating system. Value 0 is returned to the operating system to indicate that the program terminated on completion of its intended task whereas the value 1 is used to indicate that the program terminated prematurely (could not complete its intended task).

Line 7 contains character '{', called left brace or left curly bracket. There is also matching right brace '}' in line 14, which appears at the end of the main function. These pair of matching braces encloses the body of the function.

Line 8 declares two variables, *fahrenheit* and *centigrade* of type *float*, which can represent and store two real numbers (float) in computer memory. Variable *centigrade* is used to hold the value, representing temperature in Celsius scale, entered by the user during program execution. Variable *fahrenheit* is to used to hold the computed value, representing temperature in Fahrenheit scale, equivalent to given temperature in Celsius scale. Note that the names of the variables used are self-explanatory. It is always recommended to use meaningful names. This will help to increase the readability of the program.

Line 9 uses library function *printf()* that displays "*Enter temperature in Celsius scale:* " on the computer screen. We call these kinds of messages as *user prompts* as they guide the user to enter the desired input. In this case, the value to be entered is temperature in Celsius scale.

You must have observed that the entire sequence of characters enclosed in double quotes is not printed. Then, *what is the role of characters* '\n' *in the beginning*? These two characters together represent the *new line* character. Although *new line* character is a combination of two characters, *i.e.*, '\' and 'n', they are translated by the C compiler into a single character. The *new line* character instructs the compiler to advance to the next line before printing subsequent information.

Line 10 uses another library function *scanf()* that accepts the user input and stores in variable *centigrade* (in fact, in a memory location that is reserved for the *centigrade*). The first argument is a *format string* that is enclosed in double quotes and tells the system how the entered values are to be interpreted. Following the format string, there will be a list of one or more variables; each variable prefixed with character '&', called *address of operator*.

Line 11 is an assignment statement that computes the temperature in Fahrenheit scale equivalent to given temperature in Celsius scale, and assigns to variable *fahrenheit*, *i.e.*, stores in variable *fahrenheit*.

Lines 12-13 uses library function *printf()* that outputs the stored value in variable *fahrenheit* along with the message *"Equivalent Temperature in fahrenheit = "*. The use of these kinds of messages is not mandatory, but is very useful as they make the output easy to interpret.

Line 14 uses *return* statement, which on execution terminates the execution of the program and return the control back to the operating system. In addition, it also returns value 0 to the operating system indicating that the program terminated on success. In case of failure, value 1 is returned. This retuned value can be used by the operating system for deciding further line of action.

Line 15 marks the end of the *main* function as well as end of the program as this is the only function in the program.

- If there is more than one function in a file, each will have its own body enclosed in pair of braces.
- Variables can only be declared in the beginning of a block before any executable statement.

3.7 PROGRAM DEVELOPMENT LIFE CYCLE

Program development life cycle (PDLC) is a systematic way of developing quality programs (and hence software). It provides an organized plan for breaking down the task of program development into manageable chunks, each of which must be successfully completed before moving on to the next phase. The program development process is divided into following phases:

- Defining the problem
- Designing the program
- Building the program
- Testing the program
- Documenting the program
- Deploying and maintaining the program

3.7.1 Defining the Problem

The first step in developing a program is to define the problem. In major software projects, this is a job for system analyst, who provides the results of their work to programmers in the form of a *program specification*. The program specification precisely defines the input data, the processing that should take place, the format of the output reports and the user interface. Depending on the size of the job, the program development might be handled by an individual or a team.

3.7.2 Designing the Program

An architect doesn't design a home just by going to the site with a bunch of boards and bricks. After determining the needs of the house's owner, the next step in designing a house is an architectural drawing – a plan drawn up on the paper that can be discussed and reviewed until everything is right. Likewise, programmers create program designs.

Like an architect's drawing, the program design specifies the components that make the program work.

Program design begins by focusing on the main goal that the program is trying to achieve and then breaking the program into manageable components, each of which contributes to this goal. This approach of program design is called *top-down program design* or *modular programming*. The first step involves identifying the *main routine* (*main()* function), which is the program's major activity. From there, programmers try to break down the various components of the main routine into smaller units called *modules*, until each module is highly focused. Experience shows that this is the best way to ensure program quality. If an error appears in a program designed this way, it is relatively easy to identify the module that must be causing the error.

For each module, the programmer draws a conceptual plan using an appropriate program design tool to visualize how the module will do its assigned job.

The various familiar program design tools are:

- **Structure Chart** – A *structure chart*, also called *hierarchy chart*, show the top-down design of a program. Each box in the chart indicates a task that the program must accomplish. The top module, called the *main module* or *control module*, oversees the transfer of control to other modules.

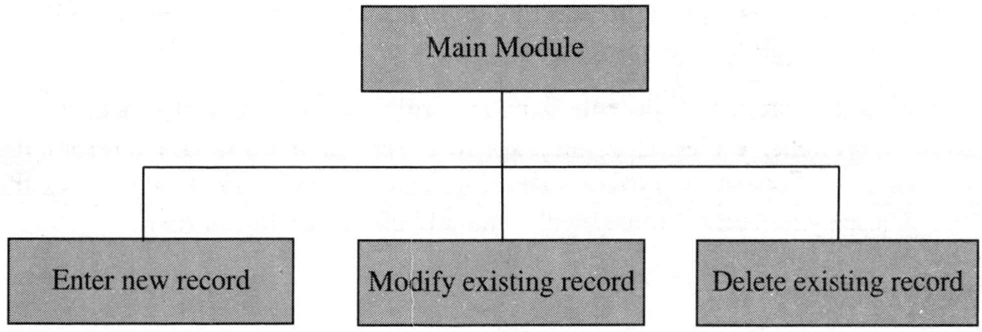

Figure 3.2: Structure Chart for a Data Entry Program

- **Algorithm** – An *algorithm* is a step-by-step description of how to arrive at a solution. Algorithms are not restricted to computers. In fact, we use them every day.

- **Flowchart** – A *flowchart* is a diagram that shows the logic of the program. Programmers create flowcharts either by hand using a flowchart template or on the computer. Each flowcharting symbol has a meaning. For example, parallelogram indicates an input or output, rectangle indicates a process, and diamond indicates a condition.

- **Pseudocode** – A *pseudocode* is another tool to describe the way to arrive at a solution. In some sense, it is considered as a form of an algorithm. The only difference is that instead of expressing the action to be carried in form of steps, they are expressed in program language like constructs.

We have already discussed in detail about various program design tools such as *algorithms*, *flowcharts* and *pseudocodes*, in *Chapter 2*. The use of these tools was also demonstrated by taking familiar illustrative examples.

3.7.3 Building the Program

Once the design of the program is ready, the next step is to convert the program design into a computer program. During this conversion each of the steps of the program design is coded as one or more C language instructions. Here also it is recommended that one should write the program first on a piece of paper before keying in the computer.

Creating and Editing Programs

Once the program is ready on paper, we key in computer memory using a *text editor*. A text editor helps us to enter the character data into computer memory, allows editing (changing) the data in computer memory, and save the data from memory in a disk file on secondary memory with extension "*.c*".

This stored file is known as *source file*, and its contents are known as *source code*. This source file will be the input for the compiler.

The programmer must carefully follow the language rules, which precisely specify how to express certain operations. Violation of language rules results in grammatical errors, more precisely known as *syntax errors*. Program development tool - *Compiler*, will check these syntax errors. These syntax errors must be eliminated before moving further.

Compiling Programs

The source code in the source file, stored on the disk, must to be translated into machine language. This job is done by the *compiler*. The C compiler actually is a combination of two separate programs – the *preprocessor* and the *translator*.

The *preprocessor* reads the source code and prepares it for translation. While reading the source code, it scans the code for preprocessor directives and processes them accordingly. For example, when it encounters the *include* directive, it substitutes that directive with the contents of the specified header file (such as *stdio.h*), and when it encounters the *define* directive, it substitutes the identifier with the specified literal (constant). The output from the preprocessor is an intermediate file, known as *translation unit*.

The translator reads the translation unit instruction-by-instruction and checks them for their grammatical accuracy. If there is a syntax error in it, it flags an error message – called *diagnostic message* on the screen. These diagnostics messages help the programmer to identify the cause of the syntax error and the place where it is present.

Therefore, if there is even a single syntax error, the translation process, known as *compilation*, is terminated. In that case, open the source file using the text editor, and make the necessary corrections and repeat the compilation.

However, if there are no syntax errors in the translation unit, the translator rereads the instruction from the beginning, translates them into machine language, and writes them onto a disk file. The translated version of the source code is known as *object code*, and is stored in the disk file with extension "*.obj".

Linking Programs

Once the source code is translated into object code, though it is in machine language, still it is not in executable form. The reason being is that it may be referring to other functions which are not included in the object code. These referred functions can be library functions (pre-written functions supplied with the compiler in the form of libraries) or user-defined functions stored in other source files and which are compiled separately. All these functions also need to be included in the object code to get a final machine code which is in the executable form, known as *executable code*. This task is performed by the *linker*.

A *linker* is a development tool that extracts the referred library functions from the system libraries, reads the object codes of the user-defined functions from the object files and integrates them with the current object code to produce the final executable code, which is stored in disk file with extension "*.exe". This executable code is the final form of the program that is ready for execution.

Executing Programs

Once the program is linked, it is ready for execution. To execute a program we give an operating system command, such as *run*, to load the program into computer memory and execute it. Getting the program into memory is the function of an operating system program known as *loader*. The loader locates the executable program in the secondary storage, reads it and brings into the computer memory. Once the program is loaded, the operating system transfers the control to the program and the program begins its execution.

The complete process of *designing*, *creating*, *compiling*, *linking* and *executing* the program, referred to as building the program, is depicted in Figure 3.3.

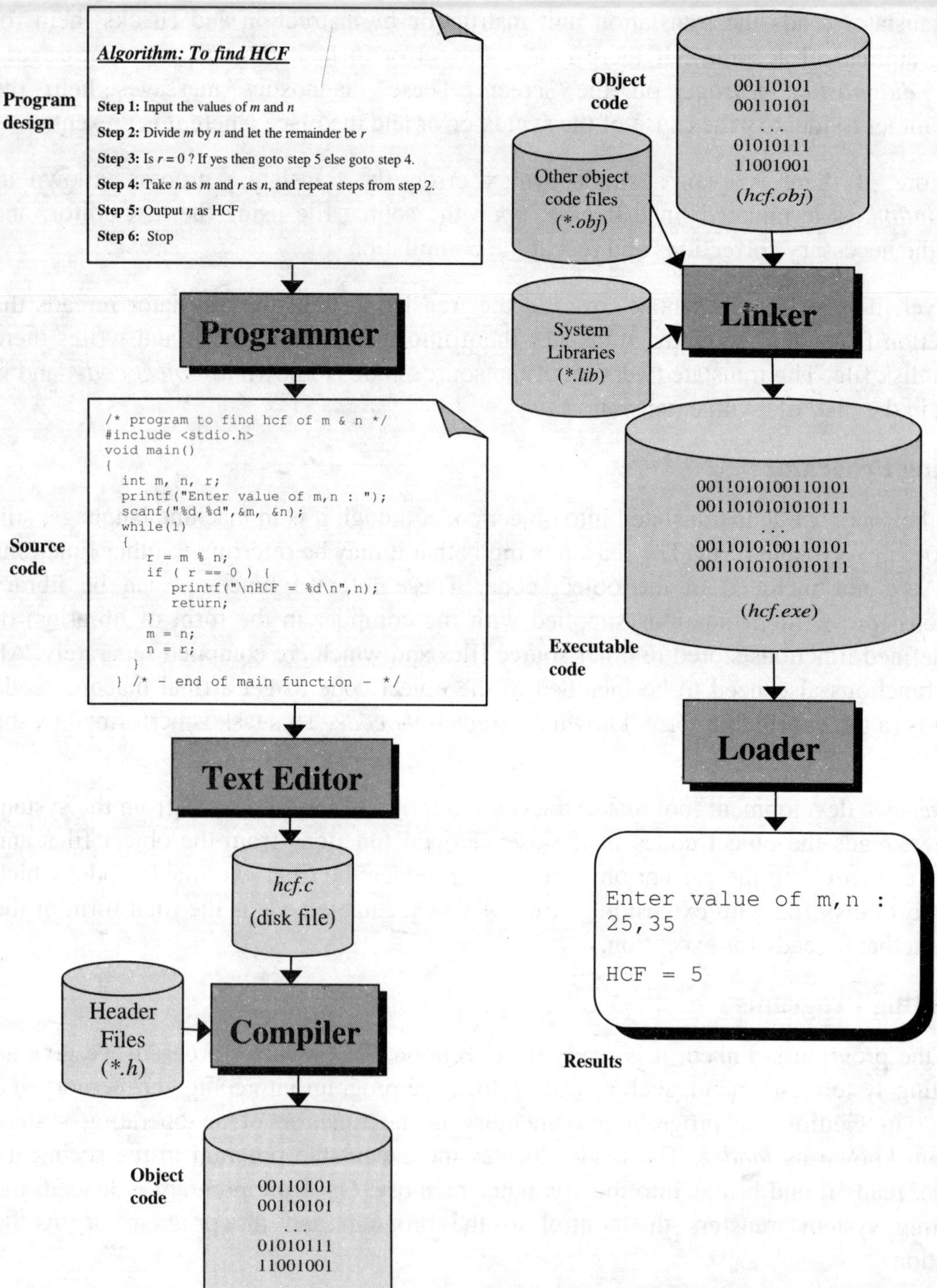

Figure 3.3: Steps in building C program

3.7.4 Testing the Program

Even when the program is executing, the output of the program may not be correct. This will be because of logical errors in the program. A *logical error* is a mistake that the programmer made while designing the solution to the problem. For example, a programmer tells the computer to calculate the net pay by adding deductions to the gross salary instead of subtracting. A program development tool, such as compiler, cannot detect these errors. Therefore, the programmer must find and correct logical errors by carefully examining the program output for a set of data for which results are already known. Such type of data is known as *test data*.

Syntax errors and logical errors collectively are known as *bugs*. The process of identifying and eliminating these errors is known as *debugging*.

3.7.5 Documenting the Program

After testing, the program development is almost complete. The *structure charts*, *algorithms*, *pseudocodes*, and *flowcharts* developed during the design phase become documentation for others who are associated with the program. In addition, more documentation needs to be done as the programs are being coded such as list of variable names and definitions, description of files that the program need to work with, and format of output that the program produces. All of this documentation needs to be saved and placed together for future reference.

This phase ends by writing a manual that provides an overview of the program's functionality, tutorials for the beginner, in-depth explanations of major program features, reference documentation of all program commands and a thorough description of the error messages generated by the program. This manual is given to the user when the program is installed.

3.7.6 Deploying and Maintaining the Program

In the final phase, the program is deployed (installed) at the user's site. Here also, the program is kept under watch till the user gives green signal to it. Users may discover errors that were not caught in the testing phase, no matter how exhaustively the program was tested.

Even after the development project is complete, it needs to be maintained and evaluated regularly. In program maintenance, the programming team fixes program errors that users discover during its day-to-day use. In periodic evaluations, the team asks whether the program is fulfilling its objectives. This evaluation may lead to modifications to update the program or to add more features for the user. It may even lead to a decision to abandon the current system and develop a new one, and so the program development life cycle begins afresh.

3.8 USING VARIOUS C COMPILERS

3.8.1 Using C Compiler on Windows Platform

If you are using Window based platform, then Turbo C/C++ Compiler is the user's preferred choice.

Turbo C/C++ Compiler provides an Integrated Development Environment (IDE)., where the necessary operations for program development are available in a unified screen display with menus and windows. It provides an ideal environment for learning C language.

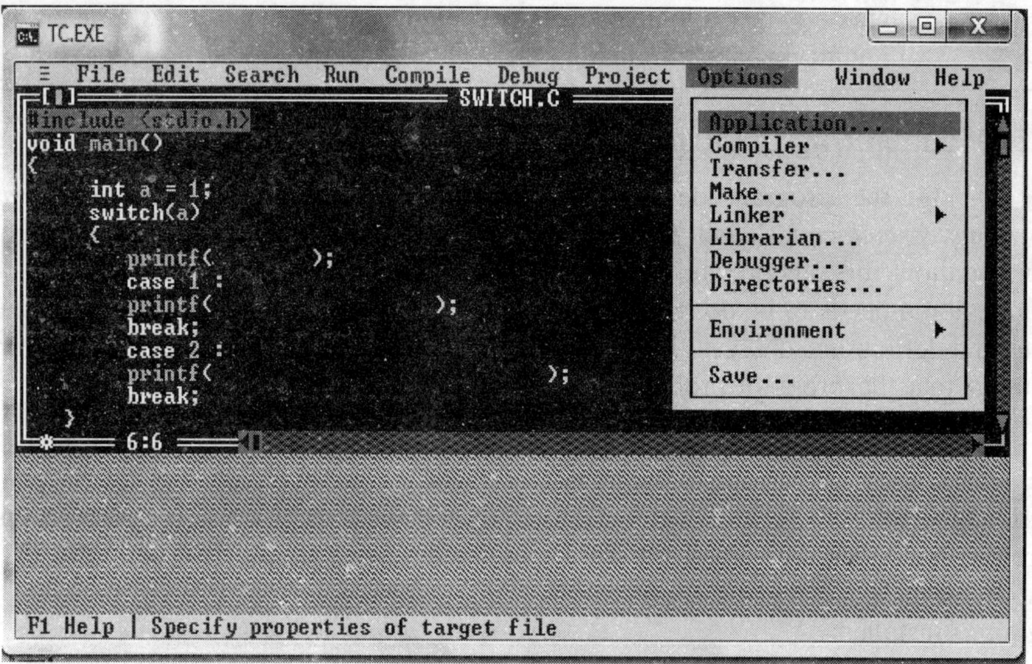

Figure 3.2: IDE of Turbo C++, which can be used create C programs as well as C++ programs

3.8.2 Using C Compiler on Unix Platform

To compile a C program, type the command as

```
cc myprog.c
```

The program will be compiled and if there is no syntax error, it subsequently will be linked with system libraries and the executable code is stored in file *a.out*.

To run the program, enter the command as

```
./a.out
```

However, if you want to save the executable file on the user specified file, type the command as

```
cc -o myprog myprog.c
```

Now to run the program, enter the command as

```
./myprog
```

3.8.3 Using C Compiler on Linux Platform

To compile a C program, type the command as

```
gcc myprog.c
```

The program will be compiled and if there is no syntax error, it subsequently will be linked with system libraries and the executable code is stored in file *a.out*.

To run the program, enter the command as

```
./a.out
```

However, if you want to save the executable file on the user specified file, type the command as

```
gcc -o myprog myprog.c
```

Now to run the program, enter the command as

```
./myprog
```

3.9 SOME KEY POINTS TO REMEMBER

List of some key points about which every C programmer must be aware:

- The C language is a free form language, *i.e.*, any instructions can start from anywhere and end anywhere. Even a single instruction can span many lines.
- Pre-processor directive can be used anywhere in the C program but the recommended place is in the beginning of the source file.
- Each instructions end with character semicolon ';'.
- The C language is a case sensitive, *i.e.*, it will treat identifiers abc, Abc, aBc, ABC, abC, etc. as different identifiers.
- The C program is usually written in lower case. Uppercase letters are usually used for symbolic names and predefined names.
- Comments can be placed anywhere.
- Execution of program always begins with first executable statement in *main* function.
- Variable can only be declared in the beginning of the function, *i.e.*, prior to any executable statement.
- Entire calculations involving real numbers are done in double precision mode, *i.e.*, with 12 decimal digits precision.

- A C program can contain any number of functions. It basically depends on the size and complexity of the problem under consideration.

MULTIPLE CHOICE QUESTIONS

1. If a program written in any language on a given platform can be transported to another platform without any change or with minimal changes, then the language is said to ______ .
   (*a*) Understandable (*b*) Readable
   (*c*) Portable (*d*) Maintainable
2. Who developed C language?
   (*a*) Von Neuman (*b*) Dennis Ritchie
   (*c*) Peter Norton (*d*) Ken Thompson
3. Which of the following character is used to terminate an instruction in a C program?
   (*a*) , (comma) (*b*) : (colon)
   (*c*) ; (semicolon) (*d*) . (period)
4. How many *main* functions can be used in a C program?
   (*a*) 0 (*b*) 1
   (*c*) 2 (*d*) Any number
5. Which of the following cannot be a first letter of an identifier?
   (*a*) digit (*b*) underscore
   (*c*) alphabet (*d*) All of above
6. Which of the following are used to enclose the body of a function?
   (*a*) [ ] (*b*) { }
   (*c*) ( ) (*d*) None
7. The C language can be used on which of the following platforms?
   (*a*) Unix (*b*) Windows
   (*c*) Linux (*d*) All of above
8. The C language is a ____________ language.
   (*a*) High (*b*) Low
   (*c*) Middle (*d*) Symbolic
9. The total number of phases in the development of a program is ___.
   (*a*) 6 (*b*) 7
   (*c*) 5 (*d*) 9
10. Which of the following errors does the compiler catch?
   (*a*) Logical errors (*b*) Runtime errors
   (*c*) Syntax errors (*d*) Al of above
11. The acronym ANSI stands for ____________.
   (*a*) American National Standards International
   (*b*) American National Software Incorporation
   (*c*) American National Standards Institute
   (*d*) American National Standards Instructions

12. Find the odd term
    (*a*) Source code
    (*b*) Object code
    (*c*) Executable code
    (*d*) ASCII code
13. Testing of a program is done to ensure that
    (*a*) it should not contain any syntax error
    (*b*) it performs its intended task
    (*c*) it compiles successfully
    (*d*) it should not run infinitely
14. The term *bug* refers to ________.
    (*a*) Syntax error
    (*b*) Logical error
    (*c*) Runtime error
    (*d*) All of above
15. The acronym ASCII stands for ________________.
    (*a*) American Standard Code for International Information
    (*b*) American System for Compiler Information International
    (*c*) American System for Code Information International
    (*d*) American Standard Code for Information Interchange
16. Which of the following statements is not true about C language?
    (*a*) Every instruction is terminated by semicolon
    (*b*) It is case insensitive
    (*c*) Comments can be placed any where in the program code
    (*d*) It has features of both high-level and low-level languages
17. The term *debugging* refers to the process of ________.
    (*a*) translating the source code to object code
    (*b*) linking the object code with the system libraries
    (*c*) identifying the bugs and fixing them
    (*d*) None of above
18. The term *portability* means that ________________.
    (*a*) Program code is understandable
    (*b*) Program code is maintainable
    (*c*) Program code is bug free
    (*d*) Program code can be transported without/with minimal changes to another computer
19. The C programs are converted into machine language using ________.
    (*a*) Assembler
    (*b*) Interpreter
    (*c*) Compiler
    (*d*) Operating system
20. Extension of file containing machine code produced by the compiler is ______.
    (*a*) *.com
    (*b*) *.exe
    (*c*) *.c
    (*d*) *.obj

REVIEW EXERCISE

1. Name the predecessor of the C language.
2. Name the person who developed C language.
3. When and where the C language was developed?
4. Why C language is called middle-level language?
5. List the application areas of C language.

6. Describe the general structure of a C program.
7. What is PDLC? Describe various phases of PDLC.
8. Describe the following - *text editor*, *compiler*, *linker*, *loader*.
9. What symbol or character is used to terminate every C statement?

❑ ❑ ❑

Chapter 4

Data Types, Variables & Constants

Learning Outcomes

After reading this chapter, students will be able to

- explain the concept of character set
- explain what does tokens, identifiers and keywords mean
- explain the rules for naming an identifier
- explain the concept of data types and various data types supported by C language
- explain the built-in data types
- explain about literals, constants, symbolic constants and variables
- describe the declaration and initialization of variables
- explain the various type modifiers
- explain the concept of storage class and its types
- explain the concept of type definition

4.1 INTRODUCTION

A program must be able to store its input somewhere so that it can work with the data and process it into a form that is meaningful to the user. All the data processed by a program is stored either as a *variable* or as a *constant*. Each variable in C language has some characteristics associated with it, called its *storage class*.

The ability to divide the data into different types is one of the most important features of the modern programming languages. It enables us to work with relatively complex data objects instead of elementary data objects that a computer can understand and manipulate.

This chapter attempts to describe these semantic units of the C language.

4.2 CHARACTER SET

In a most basic sense, a C program is a sequence of characters. When these characters are submitted to the compiler, they are interpreted in various contexts as characters, identifiers, constants, and statements. The characters used in a C source program belong to the American Standard Code for Information Interchange (ASCII pronounced as *ass-kee*) set.

The characters in C language are grouped into the following categories:

- Letters
- Digits
- Special characters
- White space characters

Table 4.1: Character set of C Language

Letters	**Digits**	
Uppercase : A–Z Lowercase : a–z	All decimal digit : 0–9	
	Special Characters	
, comma	. period	: colon
; semicolon	~ tilde	! exclamation mark
# hash/number sign	% percentage sign	^ caret
& ampersand	* asterisk	– hyphen/minus sign
+ plus sign	= equal sign	| vertical bar
′ single quote	″ double quote	/ slash
\ backslash	_ underscore	$ dollar sign

[left bracket	] right bracket	(left parenthesis
) right parenthesis	{ left brace	} right brace
< left angle bracket or less than sign	> right angle bracket or greater than sign	? question mark
		@ at the rate of sign
	White Space Characters	
Blank space	Horizontal Tab	Carriage return
Formfeed	Newline	

All C compilers ignore white space characters. These characters are basically used to enhance the readability and understandability of a C program.

4.3 TOKENS

A *token* is a smallest entity that has a meaning in itself. A C program is a sequence of tokens.

Tokens in C language can be classified into following categories:

- Keywords
- Identifiers
- Literals
- Punctuators
- Operators

4.3.1 Keywords

Every word in C is either classified as a *keyword* or an *identifier*.

Keywords are basically those words that have predefined and fixed meaning and these meanings cannot be changed. These keywords serve as basic building blocks for forming program instructions (statements). All keywords are written in lowercase. There incorrect usage results in a syntax error.

For example, the following statement

```
float const pie = 2.1412;
```

consist of six tokens: 'float', 'const', 'pie', '=', '2.1412', and ';'.

The list of keywords of C language, as per American National Standard Institute (ANSI) standard, is given in Table 4.2.

Table 4.2: Keywords in C

asm	default	for	short	union
auto	do	goto	signed	unsigned
break	double	if	sizeof	void
case	else	int	static	volatile
char	enum	long	struct	while
Const	extern	register	switch	
continue	float	return	typedef	

4.3.2 Identifiers

An *identifier* is basically a name in a program. Identifiers can be used to denote variables, arrays and functions. These user-defined names consist of sequence of characters, where each character can be a letter, a digit or an underscore '_', and cannot start with a digit. Both uppercase and lowercase letters are permitted, however, lowercase are commonly used. The underscore is usually used to link the two or more words to create meaning full names.

Since C language is case sensitive (it distinguishes between lowercase and uppercase), therefore, the identifiers *price* and *Price* will be treated as two different identifiers.

RULES FOR IDENTIFIERS

1. The first character must be an alphabet or an underscore.
2. Must consist of letters, digits or underscore only.
3. Cannot use a keyword.
4. Cannot contain a white space.

Examples of valid identifiers:

myFile	roll_no	date_of_birth	_chk
file10	N1TO10	area	AbCdE

Examples of invalid identifiers:

my-File	contains character other than letter, digit and underscore
pin.code	contains character other than letter, digit and underscore
12file	starts with a digit

4.3.3 Literals

A *literal* (often referred to as constants) refer to fixed value that do not change during the execution of a program.

Integer Literal

Integer literals are numbers without a decimal point. An integer number consists of a sequence of digits preceded by an optional sign, plus (+) or minus (–). Integers can be of any length, it is only limited by the memory available.

An integer literal can be expressed in *decimal*, *binary*, *octal* or *hexadecimal* notation.

- An integer without leading 0 and consisting of sequence of decimal digits (0–7) is treated as decimal integer. Few examples

```
7550        -1245       +1245       21467835
```

- An integer with leading 0b/0B and followed by sequence of binary digits (0,1) is treated as binary integer. Few examples

```
0b10101     0B111000        -0b11001100
```

- An integer with leading 0 and followed by sequence of octal digits (0–7) is treated as octal integer. Few examples

```
0165        0o3150      -0O2500
```

- An integer with leading 0*x*/0*X* and followed by sequence of octal digits (0–7, *A*–F/ *a*–*f*) is treated as hexadecimal integer. Few examples

```
0x575       0X75AF      -0xABCD
```

Real Literals

A *real literal* are numbers having fractional parts. The real literals can be specified in two forms – *fractional form* (*fixed-point form*) and *exponent form* (*scientific notation*).

A real literal in fractional form consists of signed or unsigned digits including a decimal point.

Examples of valid real literals in fractional form:

```
-125.50     +100.5      1.      .454
```

Examples of valid real literals in fractional form:

```
        155     (Decimal point missing)
  12,500.25     (Comma not allowed)
```

A real literal in exponent form consists of two parts: *mantissa* and *exponent*. For example, in literal 5.25E+2 (which is equivalent to 5.25×10^2), mantissa part is 5.25 and exponent part is +2. Here E+2 represent 10^2, and letter *E* (or *e*) stands for *exponent*.

Examples of valid real literals in exponent form:

```
  1.25E12      .0001E+2      5.E-3      100.00E0      5.5E+10
```

Examples of invalid real literals in exponent form:

```
  +145.80E     (No digit specified for exponent)
  12.35E5.5    (Exponent cannot be a fraction)
  12,75E-5     (Comma allowed)
```

Single Character Literals

A character literal consists of a single character enclosed in single quotes (') such as 'A'. Internally, each character is represented by an integral value that represents the ASCII code of the character. For example, character 'A' will be represented by integral value 65, character 'B' by 66, etc.

Table 4.3: Common Escape Sequences

Escape Sequence	Represents	Effect
\n	Newline	the subsequent output starts from new line.
\t	Tab	moves over to the next eight-space-wide field.
\b	Backspace	moves the cursor one space left.
\r	carriage return	carriage return.
\f	formfeed	advances to the top of the next page on printer.
\a	beep	produces one audio tone of the speaker
\'	single quote	insert character ' in the output
\"	double quote	insert character " in the output
\\	backslash	insert character \ in the output
\0	null	terminates a string

An escape sequence actually represents a single character.

The C language allows certain characters that cannot be entered/typed directly from the keyboard such as *backspace*, *tab*, *carriage return*, *newline*, etc. These characters are represented by a sequence that begin with character \ (back slash), and are known as *escape sequences*.

String (multi-character) Literals

A string literal is a sequence of characters enclosed in double quotes. The characters can be from any character in the character set.

Valid examples of string literals are:

```
"hello"    "Well done . . ."   "123.24"    "X"        "10+5"
```

Note that single character literal such as 'A' is not equivalent to single character string literal as "A".

Escape sequences can also appear in a string literal. For example printing string

```
"\"String literals are double-quoted.\""
```

will give the following result

```
"String literals are double-quoted."
```

4.3.4 Punctuators

Punctuators are also called separators. Various punctuators used in C++ are described in following table.

Table 4.4: List of punctuators and their description

Punctuators	Description
Brackets [ ]	Used to enclose array subscripts.
Parentheses ( )	Used to enclose arguments in function declaration as well as function definition, parameters in function calls and expressions.
Braces { }	*Used to enclose the block of statements. Also used to enclose list of elements while initializing arrays.*
Comma ,	Used to separate arguments in function definition, and parameters in function call.
Semicolon ;	Used to terminate a statement.
Equal to sign =	Used to initialize a variable and as an assignment operator.
Pound sing #	Used for preprocessor directives.

4.3.5 Operators

Operators are the verbs of C++ language that let the user perform computations on values. You can think of operators as verbs and of operands as the subjects and object of those verbs. C++ language's rich set of operators is one of its distinguished features.

These operators can be enumerated in following categories:

- Arithmetic Operators (+, -, *, /, %)
- Relational Operators (<, <=, >, >=, ==, !=)
- Logical Operators (!, &&, ||)
- Bit-wise Operators (~, >>, <<, &, |)
- Special Operators
  - Increment & Decrement Operators (++ - -)
  - The *sizeof* Operator
  - The *addressof* Operator (&)
  - Indirection Operator (*)
  - Ternary/Conditional Operator (? :)
  - Comma Operator (,)

4.4 CONCEPT OF DATA TYPE

A *data type* is an interpretation applied to a string of bits. Formally, *data type* is defined as a finite set of values along with well-defined set of rules for operations that can be performed on these values.

Data types in C are of two types:

- **Built-in Data Types**
  - Built-in data types are the most basic data types in C.
  - The term built-in means that they are pre-defined in C and can be used directly in a program.
  - *char*, *int*, *float* and *double* are built-in data types.
  - Apart from these, we also have *void* data type.
- **Derived Data Types**
  - Also called user-defined data types, are those which are first declared and defined by the user before use.
  - They are composed of built-in data types.
  - *Array*, *structure*, *union*, *enumeration*, *pointers,* etc., are derived data types

In this chapter, we will discuss about built-in data types only.

4.5 BUILT-IN DATA TYPES

The various kinds of built-in data types, also known as fundamental data types and basic data types, supported by C under this category are summarized in Table 4.5 along with description and their memory requirements.

Table 4.5: Built-in data types

Name	Description
Integer	
int	Integer numbers
Real/Floating-point	
float	Single precision floating-point numbers
double	Double precision floating-point numbers
Character	
char	A Single Character

4.5.1 The *int* Data Type

The *int* data type is used for integers. Integers are numbers with no decimal point. For example: 124, –75, 1200, 0, etc.

An integer number consists of a sequence of digits preceded by an optional sign, plus (+) or minus (–). A decimal point and comma is not allowed in an integer value.

The integer data type consists of a subset of integers. An integer value is an exact representation of the corresponding integer. Table 4.6 shows the different types of integer numbers, their memory requirements, and the range of values for each type.

Table 4.6: Type of Integer Numbers

Type	Memory Requirements (in bytes)		Range of Values
	16-bit Compiler (such as Turbo C/C++)	**32-bit Compiler** (such as Dev- C/C++)	
short	2	2	for 2-byte integer –32768 to +32767
int	2	4	
long	4	4	for 4-byte integer –2147483648 to +2147483647

An integer value consists of a sequence of digits preceded by an optional sign, plus (+) or minus (–). A decimal point and comma is not allowed in an integer value.

The following are valid integer values of type *int*:

```
123        +123        0        32765
3445       7523       -2000     -3578
```

The following are invalid integer values of type *int* because of the reasons as given:

```
 .123     (Decimal point not allowed)
3,452     (Comma not allowed)
45234     (Value larger than permissible range)
```

4.5.2 The *float* and *double* Data Types

The *float* data type is used for numbers that have a decimal point which are commonly called floating-point numbers. For example: 75.50, 12.45, –95.6, 100.25, etc.

The *double* data type is also used for floating-point numbers. The only difference between *float* and *double* data types is that the range and precision of the *double* data type is greater than that of *float* data type.

In other words, *double* data type is used for large floating-point numbers, the ones that do not fit in the scope of the *float* data type.

The real/floating-point data type consists of a subset of the real numbers. A real value is an approximation of the desired real number correct to certain decimal positions. A real value of type *float* is accurate to six decimal positions, a real value of type of *double* is accurate to twelve decimal positions, and a real value of type of *long double* is even accurate to more than twelve decimal positions.

Table 4.7 shows the different types of real numbers, their memory requirements, and the range of values for each type.

Table 4.7: Type of Real Numbers

Type	Bytes	Range
float	4	3.4×10^{-38} to $3.4\times10^{+38}$
double	8	1.7×10^{-308} to $1.7\times10^{+308}$

Real values are numbers having fractional parts. The real values can be specified in two forms – *fractional form* (*fixed-point form*) and *exponent form* (*scientific notation*). A real value in fractional form consists of signed or unsigned digits including a decimal point.

The rule for writing a real value in fractional form is:

> *"A real value in fractional form can have zero or more digit before a decimal point followed by zero or more digits after the decimal point. It may have either + or – sign preceding it. A value without sign is assumed to be positive."*

Examples of valid real values in fractional form:

```
    -123.345        +125.50.456        1234.
```

Examples of valid real values in fractional form:

```
          15        (Decimal point missing)
    12,34.23        (Comma not allowed)
```

A real value in exponent form consists of two parts: *mantissa* and *exponent*. For example, in value 2.5E–1 (which is equivalent to 2.5×10^{-1}), mantissa part is 2.5 and exponent part is –1. Here E–1 represents 10^{-1}, and letter *E* (or *e*) stands for *exponent*.

The rule for writing a real value in exponent form is given below:

> *"A real value in exponent form has two parts – mantissa and en exponent. The mantissa can either be an integer value or a real value in fractional form. The mantissa is followed by a letter E (or e) and the exponent. The exponent must be an integer value."*

Examples of valid real values in exponent form:

```
    +1.25E12     .0001E+2     1.E-2     100.00E0     5E+10
```

Examples of invalid real values in exponent form:

```
    +145.80E     (No digit specified for exponent)
    12.35E5.5    (Exponent cannot be a fraction)
    12,235E-5    (No comma allowed)
```

4.5.3 The *char* Data Type

The *char* data type is used for characters. It can be used for any valid character in C and not just alphabets.

Here is a special thing about *char* data type: characters are stored in memory using their ASCII codes which are numeric, *i.e.*, integers. Thus, *char* data type is basically the same as the *int* data type.

4.5.4 The *void* Data Type

The word *void* means empty, therefore, we can say *void* data type has no value.

It is useful in many situations. One of such situation — it is used as return type for functions that do not return a value.

4.6 LITERALS/CONSTANTS

A constant (also known as simply literal or literal constant) refers to fixed values that do not change during the execution of a program.

Integer Constants

An integer constant refers to a sequence of digits. Such a sequence without a leading sign is assumed as positive integer constant.

Following are valid integer constants of type *int*:

```
123        +123        0        32765        3445
```

An integer constant by default is treated as an integer value of type *int*, and the compiler reserves 2-bytes of memory for its storage.

However, we can suffix letter 'L' with the integer value to denote that integer value is of type *long*, and for this value compiler reserves 4-bytes of memory for its storage.

The following are valid integer constants of type *long*:

```
120L        +122L        0L        -2000L        -112L
```

Real Constants

Real constants are numbers having fractional parts. The real constants can be specified in two forms – *fractional form* (*fixed-point form*) and *exponent form* (*scientific notation*).

Real constant in fractional form consists of signed/unsigned digits including a decimal point.

Examples of valid real constants in fractional form:

```
-125.50        +100.5        1.        .454
```

A real constant in exponent form consists of two parts: *mantissa* and *exponent*. For example, in constant 5.25E+2 (which is equivalent to 5.25×10^2), mantissa part is 5.25 and exponent part is +2. Here E+2 represent 10^2, and letter *E* (or *e*) stands for *exponent*.

Examples of valid real constants in exponent form:

```
1.25E12        .0001E+2        5.E-3        100.00E0        5.5E+10
```

Single Character Constants

A character constant consists of a single character enclosed in single quotes (') such as 'A'.

String Constants

A string constant is a sequence of characters enclosed in double quotes. The characters can be from any character in the character set.

Valid examples of string constants are:

```
"hello"    "Well done . . ."    "123.24"    "X"    "10+5"
```

Note that single character constant such as 'A' is not equivalent to single character string constant as "A".

Escape sequences can also appear in a string constant. For example printing string

```
"\"String constants are double-quoted.\""
```

will give the following result

```
"String constants are double-quoted."
```

4.7 SYMBOLIC CONSTANTS

A symbolic constant is a name that substitutes a literal constant. As an example, assume that we are computing the area of circle with given radius using the following statement:

```
area = 3.14159 * radius * radius;
```

In the above statement, numeric constant 3.14159 represents the value of the mathematical constant π, and can be replaced by symbolic constant named *PI*. The task is accomplished as follows:

```
#define PI 3.14159
area = PI * radius * radius;
```

In the above statements, *#define* pre-processor directive associates the name PI with the numeric constant 3.14159. Note that no semicolon is used after the *#define* preprocessor directive, and also to other directives as well.

During the compiling, the preprocessor will replace each occurrence of name PI with numeric constant 3.14159.

The use of symbolic constant has following advantages:

- The use of symbolic constants enhances the readability of the program, as the appropriately chosen names pronounce much as their counter parts in the real world (actual use).

- Suppose the same numeric constant is used at many places in the program, and you want to change its value, you can do it in following ways:
  - At every place, change the value manually in the program, which is cumbersome and time consuming. Also there may be chances that at some place values may be left unchanged.
  - Use symbolic constant, *i.e.*, associate a name with the constant, and then at each place of the value use symbolic constant. Now, if you want to change the value of the constant, you have to do only at one place. During compilation, every occurrence of the symbolic constant will be replaced with the new value.

4.8 VARIABLES

A variable provides us with named storage that we can write to, retrieve, and manipulate throughout the course of the program. In other words, variables are memory locations in the computer's memory that holds data. Contents of a variable may vary — hence the name *variable*. Variables hold different kind of data and the same variable might hold different values during the execution of a program.

Further, each variable in C language is associated with a specific type, which determines the size and layout its associated memory, the range of values that can be stored within that memory, and the set of operations that can be applied to it.

Unlike a constant, a variable is *addressable* and its address can be accessed using *addressof* (&) operator.

There are two values associated with a variable:

1. Its data value that is stored at some memory address. This sometimes referred to as variable's *rvalue* (pronounced as "are-value"). We may think *rvalue* as meaning *read value*. Both a literal constant and a variable can serve as *rvalue*.
2. Its address value that is its address in memory within which its data value is stored. This sometimes referred to as variable's *lvalue* (pronounced as "ell-value"). We may think *lvalue* as meaning *location value*. Only a variable can serve as *lvalue*.

Consider the following statement

```
y = y + 3.5;
```

Here variable *y* appears on both sides of an assignment operator. The right-hand instance is read. Its associated data value is fetched from its address in memory. The left-hand instance is written. The result of the addition operation is stored at its location value and its previous value is overwritten. Therefore, variable *y* on the left-hand side serves as *lvalue* and on the right-hand side it serves as *rvalue*.

4.8.1 Declaring Variables

In C language, each variable in a program must be declared and defined. Declaration is used to name a variable while definition is used to create (allocate memory) a variable. The definition of variable assumes that the declaration of the variable has been done or is being done at the same time.

When we create variables, the declaration gives them a symbolic name and the definition reserves memory for them. Once variables are defined, they are used to hold the data that are required by the program for its operation.

To create a variable, we first specify the data type and then its name. If two or more variables of same data type are to be created, their names are separated using comma. The syntax for declaring and defining a variable is

```
type v1, v2, . . ., vn;
```

where *v1*, *v2*, ..., *vn* are the names of the variables separated by comma.

A variable's type can be any of the fundamental data types or user-defined data types. The one exception is the *void* type, *i.e.*, a variable cannot be of type *void*.

Following are some examples of variable declarations and definition:

```
int count, m, n;
float value, sum;
char ch;
double deviation;
```

First statement declares variables *count*, *m* and *n* of type *int*. Compiler reserves 2-bytes for each of these variables.

Second statement declares variable *value* and *sum* of type *float*. Compiler reserves 4-bytes for each of these variables.

Third statement declares variable *ch* of type *char*. Compiler reserves 1-bytes for it.

Fourth statement declares variable *deviation* of type *double*. Compiler reserves 8-bytes for it.

4.8.2 Initializing Variables

It is important to note that during the declaration and definition, a variable is normally not initialized. However, in C language, it is possible to give an initial value to a variable during its declaration and definition. To do that, a variable is followed by character '=' and then the value to be given to the variable. Consider the following statement

```
int sum = 0;
```

It declares variable *sum* of type *int*, reserves 2-bytes of memory for it, and places integer value 0 in those 2-bytes, *i.e.*, initializes variable *sum* to a value 0.

As declarations and definitions of variables of same type can be combined in one statement, likewise initialization of variables of same type can also be done. The following statement illustrates this:

```
int sum = 0, count = 0;
```

4.8.3 The *const* Access Modifier

When the *const* access modifier is used along with the variable, it allows the value to be assigned once and that too using initialization. Afterwards, the value of this variable cannot be modified, *i.e.*, variable behaves as a constant. Consider the following declaration statement

```
const float pie = 3.1428;
```

It stores real constant in a variable *pie* (π – a useful constant in mathematics).
Any further assignment to this variable *pie* is not permitted, *i.e.*, any assignment statement of type

```
pie = value/variable/expression/function-call;
```

will result in a syntax error as the compiler will not allow to do that.

4.8.4 The *unsigned* Type Modifiers

The *unsigned* type modifier can be used only with *int* and *char* fundamental data types. When it is used, a variable can hold only positive values.

In following cases, variable *x* can hold any value in the range 0 to 65535, and variable *y* can take any value in the range 0 to 255.

```
unsigned int x;              unsigned char y;
```

4.9 STORAGE CLASSES

Every variable in C language possesses a characteristic called its *storage class.* The storage class defines two main characteristics of the variable, apart from the place of storage and the initial value. These are

- Lifetime
- Visibility or scope

Lifetime — *Lifetime* of a variable is the length of time it retains a particular value.

Visibility — *Visibility* or *scope* of a variable refers to those parts of a program that will be able to recognize it.

At first instance, a question arises: *why are storage classes necessary*? The answer is that by using variables with the appropriate lifetime and visibility, we can write programs that use memory more efficiently, run faster, and are less prone to programming errors. The correct use of storage class is especially important in large programs.

The storage class precedes the variable's declaration and tells the compiler regarding *where the variable should be stored, what should be its initial value, what will be its scope*, and for *how much time it will be available for use*.

The C language supports following four storage classes:

- auto
- static
- register
- extern

If the storage class in the variable declaration is omitted, it is assumed of type *auto*, *i.e.*, storage class is *auto* by default. Let us examine these storage classes one by one.

For example, consider the following declaration statement

```
static unsigned int x;
```

It tells the compiler that the storage class for variable *x* is static, which signifies the following things:

- Variable *x* will be stored in computer main memory.
- Since variable *x* is not initialized, it will automatically be initialized with value is zero.
- Variable *x* will be visible to the block in which it is declared.
- The variable *x*, if declared in a function other than *main()* function, will retain its value during different function calls.

We will learn more about storage classes in *Chapter 9*.

4.10 TYPE DEFINITION

A type definition, *typedef*, gives a name to a data type by creating a new type that can be used anywhere a type is permitted. Type definition allows you to create short names for types, which in turn can save your time while coding the program, and feature can be very handy particularly during competitive programming.

Syntax for type definition is

```
typedef type IDENTIFIER;
```

where *type* is either built-in data type or user-defined data type, and *IDENTIFIER*, usually in uppercase, is the new and convenient name for the *type*. The *typedef* keyword tells the compiler to recognize the *IDENTIFIER* as synonymous of *type*.

Suppose, we are dealing with only whole number, typically, we will use following declaration

```
int x;
```

With this declaration, variable *x* can hold any value in the range -32768 to 32767 (on 16-bit compiler).

Since in our program we will be dealing with only whole numbers (you can say positive integers only), the following declaration will a better choice.

```
unsigned int x;
```

With this declaration, variable *x* can hold any value in the range 0 to 65535 (on 16-bit compiler).

Now imagine a situation, where we may be required to declare variables to hold such numbers in large number of sections of the program. We have to repeat the above code, which may consume our valuable time. We can save our valuable time by using the following type definition, which we need to type once only.

```
typedef unsigned int UINT;
```

Afterwards, we need to use following declaration

```
UINT x;
```

which is very short in comparison to the earlier one.

Type definition is very handy for user-defined data types. We will learn about this aspect in *Chapter 11*.

MULTIPLE CHOICE QUESTIONS

1. Which of the following is not a valid identifier in C language?

   (*a*) EMP_CODE　　(*b*) rollno123
   (*c*) a1b2c3d4　　(*d*) 3a2b1c

2. Which of the following is not a valid data type in C language?
   (*a*) Char (*b*) float
   (*c*) long (*d*) double
3. Which of the following declarations is not correct?
   (*a*) float a; (*b*) int b = 4;
   (*c*) unsigned char c; (*d*) unsigned float d;
4. Which of the following is not a valid integer constant of 2-byte *int*?
   (*a*) 3750 (*b*) 32800
   (*c*) –32767 (*d*) 0
5. A variable of 2-byte *int* in C can take a value in the range
   (*a*) 0 to 32767 (*b*) 0 to 65535
   (*c*) –32768 to 32767 (*d*) –32767 to 32768
6. Which of the following is not a reserved word in C language?
   (*a*) for (*b*) goto
   (*c*) doo (*d*) switch
7. Which of the following is not a valid storage class?
   (*a*) automatic (*b*) static
   (*c*) register (*d*) extern
8. How much memory is required to store a value of type *double*?
   (*a*) 4 bytes (*b*) 6 bytes
   (*c*) 8 bytes (*d*) 10 bytes
9. Which of the following modifier is used to declare a variable as constant?
   (*a*) unsigned (*b*) short
   (*c*) signed (*d*) const
10. Which of the following doesn't denote a primitive data value in C?
    (*a*) "a" (*b*) ‘k’
    (*c*) 35.25 (*d*) 014
11. The storage class controls
    (*a*) life time of a variable (*b*) scope of a variable
    (*c*) Both (*a*) & (*b*) (*d*) None of above
12. A character variable can store ________ character(s).
    (*a*) 1 (*b*) 256
    (*c*) 8 (*d*) 255
13. Storage class of a variable determines
    (*a*) Which section of the program can access a variable?
    (*b*) For how long variable exists.
    (*c*) Its initial value if it is not explicitly initialized.
    (*d*) All of above

14. Which of the following storage classes is a default storage class?
    (*a*) register (*b*) extern
    (*c*) static (*d*) None
15. Which of the following is not a storage class?
    (*a*) register (*b*) external
    (*c*) auto (*d*) static
16. Scope of the variable refers to
    (*a*) Part of a program in which the variable can be recognized.
    (*b*) Duration for which the variable retains a given value during execution.
    (*c*) Value of the variable.
    (*d*) Data type of the variable.
17. Lifetime of the variable refers to
    (*a*) Part of a program in which the variable can be recognized.
    (*b*) Value of the variable.
    (*c*) Duration for which the variable retains a given value during execution.
    (*d*) Data type of the variable.

EXERCISE

1. What is the role of white space characters?
2. What is a token? What are various types of tokens? Give example of each.
3. What is a keyword? Give few examples.
4. What is an identifier? What are the rules used to name an identifier?
5. Which of the following are invalid identifiers and why?

```
roll-no   12_items  temp_1   roll no
Height    float     a1234    SUM
```

6. What is a literal? What are different ways to express an integer literal?
7. What is an escape sequence? Explain the usage of various escape sequences.
8. Do you see any difference between 'A' & "A"?
9. What is a data type? Name the built-data types.
10. Which of the following are valid values of 2-byte *int*?

```
3000      25,000    -32769    15.
0b1011    0389      0X123     00150
```

11. What is symbolic constant? Give example in support of your answer.
12. What is a variable? How variables are declared and initialized? Can we initialize more than one variable in a single statement?
13. Is it possible to make a variable to behave like a constant? If yes, how?
14. What do you understand by the terms *lifetime* and *scope* of a variable?
15. What is a storage class? Name the storage classes supported by C.
16. What is the use of *unsigned* keyword?
17. Explain the usefulness of type definition.

❑ ❑ ❑

Chapter 5

Operators and Expressions

Learning Outcomes

After reading this chapter, students will be able to

- explain various operators and their usage
- explain various type of expressions and their evaluation
- explain type conversion in assignment and expressions
- explain assignment statement and its forms
- explain the concept of precedence and associativity among operators
- explain the library functions & their utility

5.1 INTRODUCTION

Operators are the verbs of a language that let the user perform computations on values. You can think of operators as verbs and of operands as the subjects and object of those verbs. C language's rich set of operators is one of its distinguished features.

An *expression* is a formula consisting of one or more operands and zero or more operators linked together to compute a value. The computed values are to be stored in a variable for future use. This is done using an assignment statement.

In a computer program, variety of computations is to be performed as per requirements of the problem in hand. Yet there are some routine type computations that may be part of almost every program. Therefore as a convenience, every modern programming language provides a support for these routine type computations in the form of library functions.

This chapter attempts to describe these semantic units of the C language.

5.2 OPERATORS

As mentioned earlier, operators are the verbs of a language that let the user perform computations on values. You can think of operators as verbs and of operands as the subjects and object of those verbs.

The C language's rich set of operators is one of its distinguished features. These operators can be enumerated in following categories:

- Arithmetic Operators
- Relational Operators
- Logical Operators
- Bitwise Operators
- Special Operators
    - Increment & Decrement Operators
    - The *sizeof* Operator
    - The *addressof* Operator
    - Indirection Operator
    - Conditional Operator
    - Comma Operator

5.2.1 Arithmetic Operators

Different arithmetic operators supported by C language are listed in Table 5.1. All of the operators may be used as binary operators, which appear between their operands. The plus (+) and minus (–) may also be used as unary, and precede their operand.

Table 5.1: Arithmetic Operators

Symbol Used	Operation Performed
+	Addition
–	Subtraction
*	Multiplication
/	Division
%	Modulus

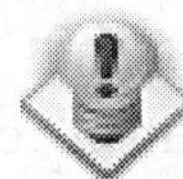

1. Modulus (%) operators, that give remainder, work only on operands of type *int, long int.*
2. There is no operator for exponentiation operation in C language.

Integer Arithmetic

When both the operands are integers, the operations will be called *integer arithmetic* and will always yield an integer value. Consider the following statements

```
int x = 13, b = 5;
```

then we have the following results:

```
x - y = 8
x + y = 18
x * y = 65
x / y = 2 (decimal part truncated)
x % y = 3 (remainder of division)
```

In the integer division, if both the operands are of same sign, the result of the division is truncated towards 0. If one of them is negative, the direction of truncation is implementation dependent. On Turbo C/C++ compilers, it is towards 0.

Therefore,

`3/4, -3/-4, -3/4` and `3/-4`

all yields value 0 on Turbo C/C++ compiler.

Similarly in modulus operation, the sign of the result is always the sign of the first operand. Therefore,

```
-13/5 = -3          -13/-5 = -3          13/-5 = 3
```

Real Arithmetic

When both the operands are real numbers, the operations will be called *real arithmetic* and will always yield a real value. The real operand may be in *fractional form* or *exponent form.* Since real numbers (floating-point values) are rounded to permissible number of significant digits, result of arithmetic operations is an approximate value of correct result.

That is

```
 6.0 /  7.0  =  0.857143
-1.0 /  3.0  = -0.333333
 3.0 / -2.0  = -1.500000
```

The modulus operation cannot be used with real operands.

Mixed-mode Arithmetic

When one of the operands is real and the other is integer, the operations will be called *mixed-mode arithmetic* and will always yield a real value. The integer operand is converted into real operand and then real arithmetic is performed, resulting in a real value.

That is

```
 6 / 7.0  →  6.0 / 7.0  = 0.857143
-1.0 / 3  → -1.0 / 3.0  = -0.333333
```

Note that

```
15 / 10.0 = 1.5
15.0 / 10 = 1.5
```

where as

```
15 / 10 = 1
```

5.2.2 Relational Operators

These operators are used to compare the values of the operands that must be compatible, in order to facilitate decision-making. If the comparison succeeds, they return a value 1 (equivalent to Boolean value *true*) and if comparison fails then returns value 0 (equivalent to Boolean value *false*).

Table 5.2: Relational Operators

Symbol Used	Operation Performed
<	Less than
<=	Less than or equal to
>	Greater than
>=	Greater than or equal to
==	Equal to
!=	Not equal to

The C language supports all six relational operators. These relational operators are listed in Table 5.2, along with their descriptions.

Note that in C language, the non-zero value is treated as equivalent to Boolean true and zero value as equivalent to Boolean false. Further note that, there is no support for Boolean data as such in C language in old versions of C language. The new version of C language, known as C99 standard, supports the Boolean type.

Table 5.3 lists some example relational expressions and the result of their evaluation.

Table 5.3: Relational Expressions

Relation Expressions	Result of operation
`20 > 10`	1
`15 != 12`	1
`15 == 13`	0
`20 >= 35`	0
`10 <= 19`	1
`13 >  20`	0

Expressions such as

```
100 >= 59        15 < 25
```

are known as relational expressions.

However, if the relational expression is of following type

```
a + b == x + y
```

then firstly arithmetic expressions on both the sides of the relational operators are evaluated then the comparison is performed, *i.e.*, arithmetic operators have a higher precedence that relational operators.

5.2.3 Logical Operators

These operators are used to form compound conditions by joining two or more simple conditions formed using relational operators. Different logical operators supported by C language are listed in Table 5.4.

Table 5.4: Logical Operators

Symbol Used	Operation Performed
!	Logical NOT
&&	Logical AND
||	Logical OR

NOT Operator

The NOT operator (!) is a unary operator. It inverts the value, *i.e.*, it changes a *true* value to *false* and false value to *true*.

Table 5.5: Rules for Logical Operators

NOT Operation

x	*!x*
false	true
true	false

AND Operation

x	*y*	*x&&y*
false	false	false
false	true	false
true	false	false
true	true	true

OR Operation

x	*y*	*x* ‖ *y*
false	false	false
false	true	true
true	false	true
true	true	true

AND Operator

The AND operator (&&) is a binary operator. Four distinct combinations of values of its operands are possible. The result is *true* only when both the operands are *true*, in all other cases, the result is *false*.

OR Operator

The OR operator (||) is also a binary operator. Four distinct combinations of values of its operands are possible. The result is *false* only when both the operands are *false*, in all other cases, the result is *true*.

Evaluating Logical Expressions

There are two methods to evaluate a logical expression. In the first method, the expression must be completely evaluated before the result is determined. This means that AND expression must be completely evaluated even when the first operand is *false* and it is therefore known that result will be *false*. Likewise, an OR expression must be completely evaluated even when first operand is *true* and it is therefore known that result will be *true*.

In the second method, also known as *short-circuit evaluation*, the expression need not be evaluated completely. As soon as it is known that what the final result will be, the rest of the expression is not evaluated. Therefore, in evaluating the AND expression, if the first operand is *false*, the second operand is not evaluated since it is apparent that the result will be *false*. Likewise, in evaluating the OR expression, if the first operand is *true*, the second operand is not evaluated since it is apparent that the result will be *true*.

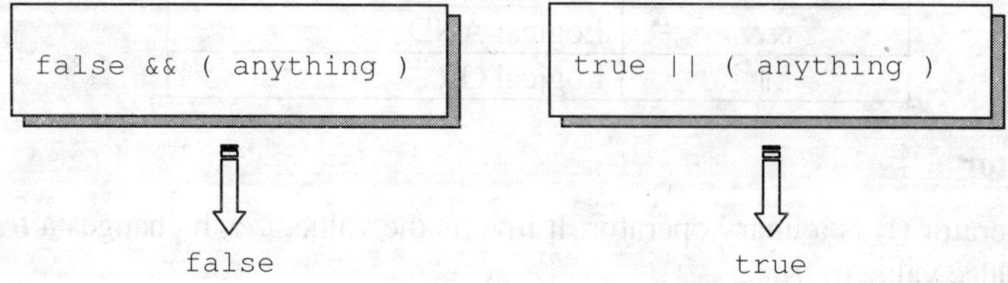

Figure 5.1: Short-circuit evaluation of AND/OR Operation

5.2.4 Bitwise Operators

These operators operate at bit-level, and allow us to manipulate the individual bits. Specifically, these operators find their use in low-level programming.

Different bit-wise operators supported by C language are listed in Table 5.6.

Table 5.6: Bitwise Operators

Symbol Used	Operation Performed
>>	Shift-right
<<	Shift-left
~	1's Complement
&	Bitwise AND
^	Bitwise Exclusive OR
|	Bitwise Inclusive OR

To illustrate working of bitwise operators, consider the following declarations statement

```
int a = 20, b = 94, c, d;
```

The value of *a* will be internally stored as 0000000000010100, and of *b* as 0000000001011110, *i.e.*, as two byte binary numbers.

5.2.4.1 Logical Bitwise Operators

The logical operators look at data to be manipulated as individual bits. Following four operators are used to manipulate the bits: bitwise AND (&), bitwise inclusive OR (|), bitwise exclusive OR (^), and one's complement (~). The first three are binary operators while the fourth one is unary operator.

5.2.4.1.1 Bitwise AND Operator

The bitwise AND operator (&) is a binary operator, and it performs the bitwise AND operation on the corresponding bits of the operands. The rules for bitwise AND operation are given in Table 5.7.

Table 5.7: Bitwise AND Operator

bit1	bit2	bit1 & bit2
0	0	0
0	1	0
1	0	0
1	1	1

```
Operand 1        ----    0000000000010100
Operand 2        ----    0001000010001110
Bitwise AND      ----    0000000000000100
```

This can be accomplished using the statement of type

```
c = a & b;
```

Using the value of *a* and *b*, as per type declaration statement, find out the value of *c*.

5.2.4.1.2 Bitwise Exclusive OR Operator

The bitwise exclusive OR operator (^) is a binary operator, and it performs the bitwise exclusive OR operation on the corresponding bits of the operands. The rules for bitwise exclusive OR operation are given in Table 5.8.

Table 5.8: Bitwise Exclusive OR Operator

bit1	bit2	bit1 ^ bit2
0	0	0
0	1	1
1	0	1
1	1	0

```
Operand 1                 ----   0000000000010100
Operand 2                 ----   0001000010001110
Bitwise Exclusive OR      ----   0001000010011010
```

This can be accomplished using the statement of type

```
c = a ^ b;
```

Using the value of *a* and *b*, as per type declaration statement, find out the value of *c*.

5.2.4.1.3 Bitwise Inclusive OR Operator

The bitwise inclusive OR operator (|) is a binary operator, and it performs the bitwise inclusive OR operation on the corresponding bits of the operands. The rules for bitwise inclusive OR operation are given in Table 5.9.

Table 5.9: Bitwise Inclusive OR Operator

bit1	bit2	bit1 | bit2
0	0	0
0	1	1
1	0	1
1	1	1

```
Operand 1                 ----   0000000000010100
Operand 2                 ----   0001000010001110
Bitwise Inclusive OR      ----   0001000010011110
```

This can be accomplished using the statement of type

```
c = a | b;
```

Using the value of *a* and *b*, as per type declaration statement, find out the value of *c*.

5.2.4.1.4 One's Complement Operator

One's complement operator is a unary operator. The One's complement is obtained by just inverting the individual bits.

That is, if the value of a is 0000000000010100, then its one's complement will be 1111111111101011

This can be achieved by using ~ operator as shown in the following statement

```
c = ~a;
```

5.2.4.2 Shift Operators

The shift operators move bits of the operand to the left or right. When applied to unsigned numbers, these operators are independent of the underlying implementation (hardware and software). When applied on signed numbers, however, the implementation is left to the software engineer who designs the system.

It is often known for which hardware and software platform the compiler is written. Therefore, the C standard leaves the implementation details up to the compiler writer (there is no standard), the code that shifts signed negative numbers may not be portable to other platforms.

5.2.4.2.1 Bitwise Shift-Right Operator

The shift-right operator is a binary operator where the value of the second operand specifies the by how many bit positions the value of the first operand is to be right shifted. The rightmost bit is discarded, whereas the 0 or 1 (depending on whether the operand is positive number or negative number) is inserted in the leftmost bit position.

Consider the following statement

```
c = a >> 1;
```

This statement causes the value of *a* (0000000000010100) shifted to right by one bit and 0 bit is added to the left, and finally the resultant value is assigned to variable *c*.

So the value of *c* will be 0000000000001010, *i.e.*, decimal 10. The right shift by 1 bit is equivalent to division by a factor of 2.

However, if the original left most bit is 1 (*i.e.*, the value is negative), then bit 1 is added to the left for every shift.

The signed numbers are stored in two's complement form. In two's complement form, the leftmost bit is used for sign. The sign bit is 0 for positive and 1 for negative numbers. For 2-byte integers, remaining 15 bits hold the magnitude of the number, whereas for 4-byte integers (*i.e.*, long integers) remaining 31-bits contain the magnitude of the number. But for unsigned integers, all the bits are used to hold the magnitude as all the numbers are positive and therefore no bit is used for sign to distinguish positive integers from negative integers.

5.2.4.2.2 Bitwise Shift-Left Operator

Like shift-right operator, shift-left operator is also a binary operator where the value of the second operand specifies the by how many bit positions the value of the first operand is to be left shifted. The leftmost bit is discarded, whereas 0 bit is inserted in the rightmost bit position.

Consider the following statement

```
c = a << 1;
```

This statement causes the value of *a* (0000000000010100) shifted to left by one bit and 0 bit is added to the right, and finally the resultant value is assigned to variable *c*.
So the value of *c* will be 0000000000101000, *i.e.*, decimal 40. The left shift by 1 bit is equivalent to multiplication by a factor of 2.

5.2.5 Special Operators

These operators were not supported by earlier high-level languages, and were incorporated for the first time in C. Most of the later high-level languages are influenced by C and almost all such languages, for example C, JAVA, etc., support these kind of operators. Let us discuss each of them briefly, and their utility you will be able to appreciate in the subsequent sections and chapters.

5.2.5.1 Increment & Decrement Operators

The C language supports two very useful operators – increment operator (++) and decrement operator (––). These operators can be used with all basic data types. The increment operator adds 1 to the operand, while decrement operator subtracts 1 from the operand.

Both operators are unary operators and can be used in the prefix as well as postfix notation as shown below:

```
++k;     or     k++;
--k;     or     k--;
```

Here, `++k` (as well as `k++`) is equivalent to `k = k + 1` or `k += 1`, and `--k` (as well as `k--`) is equivalent to `k = k - 1` or `k -= 1`.

These operators are most frequently used in *while* and *for* loops as control variables.

While `++k` or `k++` means the same thing when they form statements independently, they behave differently when they are used as part of other expressions.

Consider the following statements

```
int y, k = 5;
y = ++k;
```

In this case, first the value of *k* is incremented and then assigned to *y*, and hence y and k will have same value as 6. Thus, the above statements are equivalent to following statements

```
int y, k = 5;
++k;
y = k;
```

However, if we write the above statements as

```
int y, k = 5;
y = k++;
```

then first the current value of *k* (value 5) is assigned to y and then the value of k is incremented, and hence *y* will have value 5 while and *k* will have value as 6.

Thus, the above statements are equivalent to following statements

```
int y, k = 5;                  int y, k = 5;
y = k;              or         y = k;
++k;                           k++;
```

Similar is the case, when these operators are used in accessing elements of an array.

Therefore, the statement

```
a[i++] = 5;
```

is equivalent to

```
a[i] = 5;
i = i + 1;
```

These operators can also be used in complex expressions, such as

```
k = m++ - j / 10;
```

where exiting value of *m* is used in evaluating expression, and afterwards *m* is incremented.

Rules for Increment & Decrement Operators

1. These operators are unary operators and require variable as operand.
2. These operators mean the same thing, irrespective whether they are used in prefix or postfix notation, when they form statements independently.
3. When these operators are used in prefix notation in an expression, then first value of the variable is changed accordingly and the new value is used in evaluating the expression.
4. When these operators are used in postfix notation in an expression, then first existing value of variable is used in evaluating expression and then value is changed accordingly.

5.2.5.2 The *sizeof* Operator

The *sizeof* operator returns the size, in bytes, of the given operand. The syntax of *sizeof* operator is

```
sizeof( exp )
```

where *exp* represents a data type (built-in or user-defined), a literal/constant or a variable. For example

```
sizeof( float )
```

returns value 4.

Table 5.10: Use of *sizeof* Operator

sizeof Operator used as	Returns	Justification
`sizeof(12)`	2	Integer constant by default belong to data type *int*, and size for *int* is 2.
`sizeof('a')`	1	Size of Character constant is 1.
`sizeof(int)`	2	Size of data type *int* is 2.
`sizeof(125.25)`	8	Real constant by default belong to data type *double*.
`sizeof(125UL)`	4	Size of data type *unsigned long int* is 4.
`sizeof(125.25F)`	4	Real constant is of data type *float*, and size for *float* is 4.
`double x;` `sizeof(x);`	8	Variable *x* is of type *double*, and size for *double* is 8.

5.2.6.3 The *addressof* Operator

Character '&' when prefixed with a variable returns the address of the memory locations, hence its name *addressof* operator. It is used in the *scanf()* function and to initialize the pointers. The use of *addressof* (&) operator is demonstrated in Table 5.11.

5.2.6.4 Indirection Operator

Character '*' when prefixed with a pointer variable returns the value stored in a memory location whose address is held in pointer variable. That is, value is accessed through pointer variable indirectly, hence its name indirection operator. Consider following statements

```
int y = 10, x;
int *py;
py = &y;
x = *py;
```

The expression *"*py"* returns the values at address which held in pointer variable *py*, *i.e.*, value of variable y which is 10, and that get assigned to variable *x*.

Table 5.11: Use of *addressof* (&) Operators

addressof (&) Operator used as	Task Performed

`int y = 10;`
`int *py;`
`Py = &y;`
	First statement declares and defines variable *y* of type *int*. It also initializes it with value 10.
	Second statement declares and defines a pointer variable *py* that can hold an address of a memory location reserved for storing value of type *int*.
	Third statement assigns the address of variable y to pointer variable *py*.

`int x, y;` `scanf("%d %d", &x, &y);`	First statement declares and defines two variable *x* and *y* of type *int*. Second statement takes two integer values as input from the keyboard and stores those values at addresses of variable *x* and *y*, respectively.

5.2.6.5 Conditional/Ternary Operator

Consider the following segment, written in pseudocode

```
if ( a > b ) then
   set big = a
else
   set big = b
endif
```

This segments assigns the maximum of the values of *a* and *b* to *big*. As it is clear that the value assigned to variable *big* will depend on the outcome of the test condition "$a > b$".

Such expressions are known as *conditional expressions*, and can be written using conditional operator. The syntax of using ternary operator is

```
exp1 ? exp2 : exp3;
```

where *exp1*, *exp2*, *exp3* are expressions.

The expression *exp1* is evaluated first. If it is non-zero (true), then the expression *exp2* is evaluated, and that is the value of the conditional expression; otherwise expression *exp3* is evaluated, and that is the value of the conditional expression. Note that only one of the expression *exp2* and *exp3* is evaluated.

Thus, the above pseudocode segment using conditional operator can be written as

```
big = ( a > b ) ? a : b;
```

Parentheses are not necessary around the first expression of the conditional expression since the precedence of '?:' is very low, just above assignment operator. However, the use of parentheses is recommended as they make the condition part easier to see.

5.2.6.6 Comma Operator

A comma operator is used to combine several related expressions. The group of expressions separated by comma ',' are evaluated left-to-right and the result of the right-most expression becomes the value of the total comma separated expression.

For example, the statement

```
a = ( b = 10, b+1 );
```

first assigns the value 10 to *b*, and then adds 1 to *b* and finally the resultant value, *i.e.*, 11 is assigned to *a*. In the above expression, the use of parenthesis is mandatory because the comma operator has a lower precedence than the assignment operator. In fact comma operator has the lowest precedence among all the operators.

Comma operator can be used in *for loop* to separate expressions, as shown below:

```
for ( i = 0, j = n; i < j; i++, j-- )
{
    // some code
}
```

In the above *for loop*, first variable *i* is initialized to 0 and then variable *j* is initialized to *n*, and this happens only once. Then for every subsequent iteration, first variable *i* is incremented and then variable *j* is decremented before testing the terminating condition which states that the loop will execute till the value of *i* is less than *j*.

Comma operator can also be used in *while loop* to separate expressions such as

```
while ( ch = cin.get(), ch != '\r' )
{
    // some code
}
```

In the above *while loop*, the first expression accepts a character input from the user and assigns the same character to *char* variable *ch*, then variable *ch* is tested against carriage return. If the character input is not the carriage return, the loop continues to execute for next iteration.

Following is another example of use of the comma operator to swap the values of two integer variables *x* and *y*:

```
temp = x, x = y, y = temp;
```

Variable *temp* is used as a temporary variable just to facilitate swapping of values.

5.3 EXPRESSIONS

An *expression* is a formula consisting of one or more operands and zero or more operators linked together to compute a value. An operand may be a function reference, a variable, an array element or a constant. In the expression

```
a + b
```

plus character '+' is an operator and *a* and *b* are operands.

There are four types of expressions in C. These are

1. Arithmetic expressions
2. Relational expressions
3. Logical expressions
4. Conditional expressions

Each type of expression takes certain types of operands and uses a specific set of operators. Evaluation of every expression produces a value of specific type. Expressions are not statements, but may be components of statements.

For example, consider the line

```
x = 2.0/3.0 + a * b;
```

The entire line is a statement, but the portion after the equal sign is an expression.

5.3.1 Arithmetic Expressions

An arithmetic expression is made up of operands and arithmetic operators. It produces a value that is of type *int*, *float* or *double*. When an expression involve only integral operands, it is known as *pure integer expression*, when it involves only real operands it is known as a *pure real expression*, and when it involves both integral and real operands it is known as a *mixed mode expression*.

Following are some mathematical expressions and their corresponding C expressions. For simplicity, we have considered only those expressions that don't involve the use of any mathematical or trigonometric functions.

Table 5.12: Some Sample Arithmetic Expressions

Mathematical/Algebraic Expressions	The C Arithmetic Expressions
$\frac{a+b}{2}$	`(a+b)/2`
$a+\frac{b}{c}+d$	`a+b/c+d`
$\frac{ab}{c-d^2}+d$	`(a*b)/(c-d*d)+d`
$1+\frac{a}{b+\frac{1}{c}}$	`1+a/(b+1/c)`
$\frac{\frac{a}{c+b}}{d}-e$	`a/((c+b)/d)-e`
ax^2+bx+c	`a*x*x+b*x+c`
$ut+\frac{1}{2}at^2$	`u*t+0.5*a*t*t`
$m\left(a\times h+\frac{v^2}{2}\right)$	`m*(a*h+(v*v)/2)`

While converting mathematical/algebraic expressions into C arithmetic expressions, parentheses can be used to control associativity and the order in which operators are evaluated. If parentheses are present in an expression, then the expression within the parentheses is evaluated first and within the parentheses the implicit precedence is observed. If there is nesting of parentheses (parentheses inside parentheses), then the inner most parentheses are evaluated first.

5.3.1.1 Evaluation of Arithmetic Expressions

As you know expressions are evaluated by performing one operation at a time, and same is the case when expressions are evaluated by a computer. The order of evaluation of individual operations is governed by the precedence and associativity of operators. However, when individual operations are performed, the following cases can happen:

- When both of the operands are of type integer (*short*, *int*, *long*), the integer arithmetic will be performed and the results of the operation will an integer value. For example, 5/2 will yield 2 not 2.5 as the fractional part is ignored.
- When both of the operands are of type real (*float*, *double*), the real arithmetic will be performed and the results of the operation will a real value. For example, 5.0/2.0 will yield 2.5.
- However, when one operand is of type integer and the second one is of type real, the mixed mode arithmetic will be performed. In this case, first integer operand will be converted into real operand, and then real arithmetic will be performed and the results of the operation will a real value. For example, 5/2.0 or 5.0/2 will yield 2.5, because in the first case first value 5 will be converted to 5.0 and then division will be performed while in the second case value 2 will be converted to 2.0 and then the division will be performed.

To illustrate the way arithmetic expression are evaluated, consider the following expression

```
5 * 4 / ( 1 + 5 * 2 / 3 + 6 ) + 8 * ( 7 / 4 )
```

This expression is evaluated as depicted below:

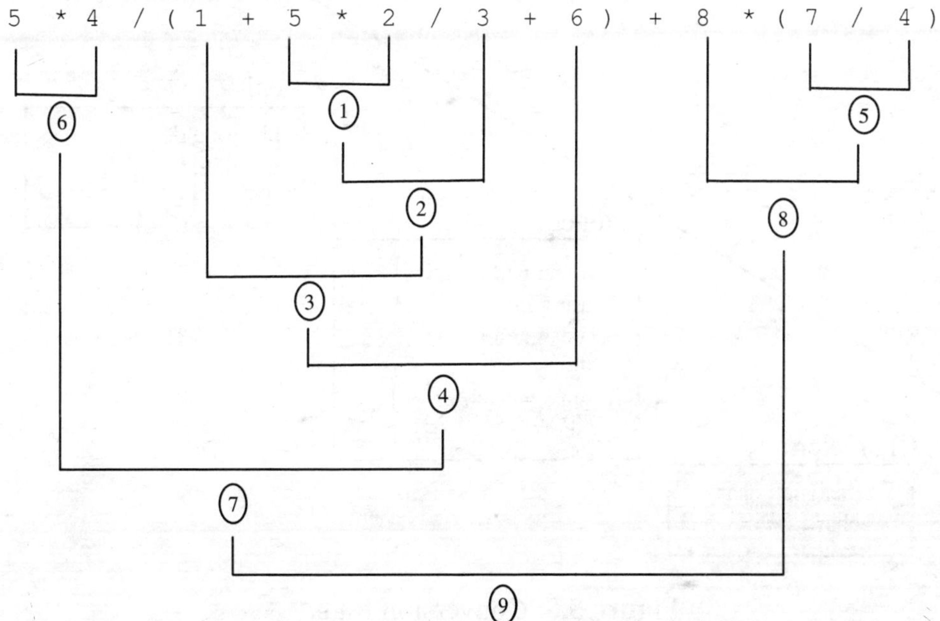

5.3.1.2 Type Conversion

When values of different types are mixed in an expression, then they are converted to the same type before performing the particular operation. This process of converting values for one type to another type is known as *type conversion*.
The C language facilitates type conversion in following two forms:

- Implicit type conversion
- Explicit type conversion

5.3.1.2.1 Implicit Type Conversion

An implicit type conversion is automatically performed by the compiler without programmer's intervention. This is done when expression is in mixed mode, so as not to lose the information. For implicit type conversion, the C language follows certain rules that are introduced next.

Conversion Rank

Before we discuss how conversions are handled, we will discuss the concept of conversion rank. We can assign a rank to integral and real types. The operand of larger size is given a higher rank than that of a smaller size.

Figure 5.5 shows the ranks that will be used for conversions. The *long double* has a higher rank than a *long int* while the *short* has a higher rank than a *char*.

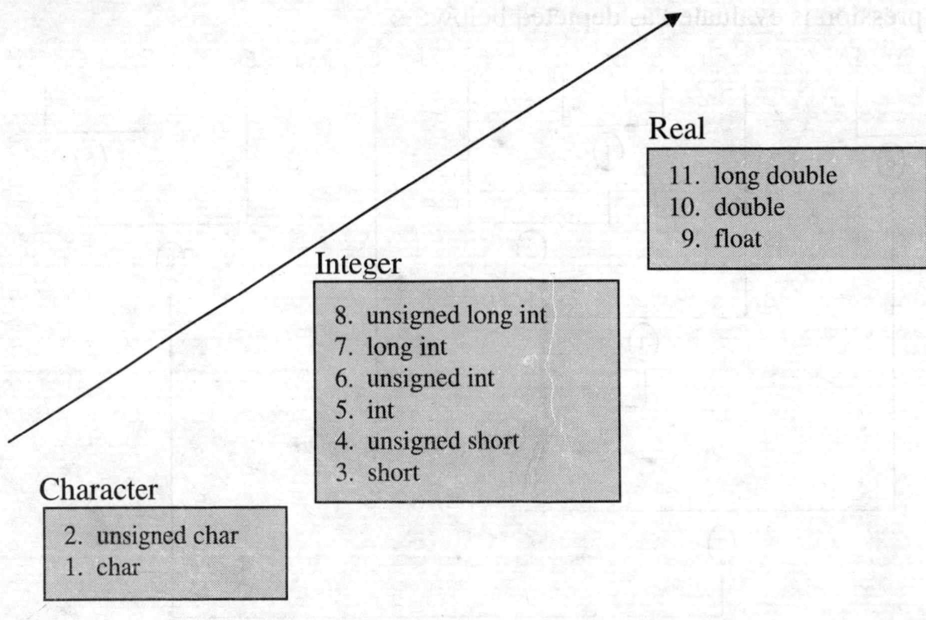

Figure 5.5: Conversion Rank

Conversion in Evaluation of Expressions

When an operation is to be performed on operands of different ranks, then operand of lower rank is converted to type so as to match the operand of higher rank. Once this conversion of the ranks is done, the operation is performed and the result of the operation will of type having higher rank.

For example, suppose variables *x* and *y* are of type *float* and variables `i` and `j` are of type *integer*, in expression:

```
x / i + y * j
```

While evaluating expression "*x/i*", integer operand *i* will be converted to *float* type and then operand *x* will be divided by *i*, and the result of the expression will be a value of type *float*.

Conversions in Assignment Statements

A simple assignment statement involves an assignment operator (=) and two operands. The operand on the left hand side of assignment operator is a variable while the operand on the right side can be a literal/constant, variable or a value after evaluation of an expression.

Depending on the difference of the rank of the operands, the C system tries to either *promote* or *demote* the right operand so as to make its rank match the left operand (variable). *Promotion* occurs if the right operand has lower rank; *demotion* occurs if the right operand has higher rank.

5.3.1.2.2 Explicit Type Conversion

An explicit type conversion is user-defined that forces an operand or expression to be of specific type. The process of explicit type conversion is also known as *type casting*. The syntax for explicit type conversion or type casting is

```
(type)operand    OR    (type)expression
```

Here *type* in the parentheses is referred to as *cast operator*, which has a precedence of 2. As shown in the syntax, to cast data from one type to another type, we specify the new type in parentheses before the value that we want to convert.

For example, to convert variable *x* of type *int* to *float* type, we write the expression as

```
(float)x
```

It is important to note here that in this operation like any other unary operation, the value stored in *x* is still of type *int*, but the value of the expression is promoted to type *float*.

One use of the cast operator is to ensure that the result of a divide operation on integer operands results in a real number.

For example, if we want to calculate the average, which may contain fraction as well, without a cast operator, the result will be an integer number because of integer division.

To force the real result, we can write the statement to compute average as

```
average = (float)total / n;
```

In this statement, there is explicit conversion of *total* to *float*, and then implicit conversion of *n* to *float*. Now real division will take place and the result will be a real number, as desired, which will be assigned to *average*.

5.3.2 Relational Expressions

Relational expressions compare the values of two operands. A relational expression is used to test a condition in order to decide whether an action should be taken or not. In relational expression, an arithmetic value should not be compared with a string value. Result of a relational expression is either zero or non-zero. Here, non-zero value is equivalent to a boolean *true* and zero to a boolean *false*.

Table 5.13: Sample Relational Expressions

Relation Expression	Description
`x % 2 == 0`	Used to test whether the operand *x* is an even number or not. The relational expression results in value 1 if *x* is even otherwise results in value 0.
`a != b`	Used to test whether the operands *a* is not the same as *b*. The relational expression results in value 1 if *a* is different from *b* otherwise results in value 0.
`a + b == x + y`	Used to test whether value of expression "*a+b*" is equal to the value of expression "*x+y*" or not.
`x >= 35`	Used to test whether the operand *x* is greater than or equal to 35. The relational expression results in value 1 if *x* is greater than or equal to 35 otherwise results in value 0.

Relational expressions with arithmetic operands may have one operand of *int* type and one of *float* type. In this case, the *int* operand is converted to *float* type before the relational expression is evaluated.

5.3.3 Logical Expressions

A logical expression also produces a zero or non-zero value. A logical expression is used to form a complex test condition in order to take a decision. For example, an assurance company has framed an assurance policy that states like

> *"Assure a person if he is married or his/her age is more than 30 years"*

Suppose marital status of a person is denoted by value '0' if unmarried and '1' if married. Let us further suppose that we use character variable *status* to store this value, and integer variable *age* is used to store age of the person.

Then, these two conditions separately can be written as

> **Condition 1** : status = '1'
>
> **Condition 2** : age ≥ 30

So according to the company's policy, the person is eligible for assurance, if at least one of the two conditions is satisfied.

In pseudo-language, the above decision criteria can written as

```
if ( ( status = '1' ) or ( age ≥ 30 ) ) then
   Insure the person
else
   Person not eligible for insurance
endif
```

The above test condition for Assurance Company can be written as

```
status == '1'  ||  age >= 30
```

Table 5.14: Sample Logical Expressions

Logical Expression	Description
`( x > 2 ) && ( x % 2 == 0 )`	Used to form a test condition to check whether the operand *x* is greater than 2 and x is even (divisible by 2). The result of this test condition will be 1 only if both the conditions hold true simultaneously.
`age >= 40 || sex == 'F'`	Used to form a test condition to check whether the operand *age* is greater than or equal to 40 or sex is 'F' (for female). The result of this test condition will be 1 only if either or both the conditions hold true.
`! ( x > 5 ) && ( y == 0 )`	Used to form a test condition to check whether the operand *x* is not greater than 5 and *y* is equal to 0. The result of this test condition will be 1 only if both the conditions hold true simultaneously.

5.3.4 Conditional Expressions

Note that the conditional expression is indeed an expression, and can be used wherever any other expression can be. A conditional expression looks like

```
exp1 ? exp2 : exp3
```

where *exp1* is a simple or compound condition; *exp2* and *exp3* are any other kind of expressions. The final value of this conditional expression depends on the outcome of condition represented by *exp1*. If condition *exp1* evaluates zero (true), then value of the condition expressional will be as represented by *exp2* otherwise as represented by *exp3*. Since, the final value of condition expression depends on the outcome of condition, hence its name.

For example, in the following condition expression

```
big  = a > b ? a : b;
```

it will assume value of *a* if *a* > *b* else as of *b*, and this value of conditional expression is then assigned to variable *big*.

Note that the use of parentheses are not necessary around the first expression of the conditional expression since the precedence of '?:' is very low, just above assignment operator. However, the use of parentheses is recommended as they make the condition part of the expression easier to see.

The above statement using parentheses can be written as

```
big  = ( a > b ) ? a : b;
```

Table 5.15: Sample Conditional Expressions

Logical Expression	**Description**
`(a%2==0)?1:0`	The conditional expression takes value 1 if *a* is divisible by 2 otherwise it takes value 0.
`(a>b)?((a>c)?a:c):((b>c)?b:c)`	The conditional expression takes largest of the three number (*a*, *b* and *c*) as its value.
`(x>y)? x--: y++`	Decrement *x* if *x* is greater than *y* else increment *y*.

5.4 ASSIGNMENT STATEMENTS

As we know that an expression is a formula consisting of one or more operands and zero or more operators linked together to compute a value. These computed values are to be stored in a variable for future use. This is done by using an *assignment statement*. The C language supports variety of assignment statements. These are

- Simple assignment statement
- Multiple assignment statement
- Shorthand assignment statement

5.4.1 Simple Assignment Statement

The syntax of simple assignment statement is

```
var = exp;
```

where *var* is a variable or array element reference; *exp* is any expression; and '=' is an *assignment operator.*
The unique feature about C is that it treats assignment operator in a way it treats other operators. For example, the following statement

```
printf( "\nValue of c = %d\n", c = a + b );
```

will be valid.

This statement first assigns the sum of variables *a* and *b* of type *int* to variable *c*, also of type *int*, and then the value of variable *c* is printed. The type of the variable or array element and the type of the expression must be compatible.

The following are valid assignment statements:

(Given *a, b, c, y, x, disc* are *float* variables)

(*a*) `a = 5.0;`
(*b*) `y = 4.0 * x + 7.5;`
(*c*) `disc = b * b – 4.0 * a * c;`
(*d*) `x = y;`

The following are invalid assignment statements with the reasons indicated:

(Given *a, b, c, y* are real variables)

(*a*) `b + c = a;` (*b* + *c* is not a valid variable)
(*b*) `abs(a) = x;` (abs(*a*) is not a valid variable)
(*c*) `y = 'R';` (*y* is not a character variable)

5.4.2 Multiple Assignment Statement

If the same value is to be assigned to different variable of same type, the usual way is to use different assignment statement for each variable. However, C language permits us to assign the same value to more than one variable using multiple assignment statement. The syntax of a multiple assignment statement is

```
v1 = v2 = v3 = exp;
```

where *v1*, *v2*, and *v3* are variables of same type; and *exp* is a literal, variable or expression.

The order of assigning the value is from right-to-left. For example, the following statement assigns value 0 to variables *a*, *b*, *c*, and *d*.

```
a = b = c = d = 0;
```

5.4.3 Shorthand Assignment Statement

Consider the following assignment statements

```
a = a + 5;
a = a / j;
```

Its purpose is to increase the value of *a* by factor of 5, *i.e.*, add value 5 to the current value of variable *a* and assign the new value to *a* itself.

Similarly, in the second assignment statement, the value of *a* is divided by *j*, and the result is again stored in *a*.

Such types of assignment statements are known as *self replacement* statements as the value of one of the variables of assignment statement is being replaced with a new value.

The same task can also be accomplished using following statements

```
a += 5;
a /= j;
```

The '+=' and '/=' (in general "op=" operator) are called *shorthand assignment operators*, and the entire statements is known as arithmetic assignment statements.

Table 5.16: Use of shorthand assignment operators

Assignment statement with simple assignment operator	Assignment statement with shorthand assignment operator
`k = k + m;`	`k += m;`
`k = k - m;`	`k -= m;`
`k = k * m;`	`k *= m;`
`k = k / ( n + 1 );`	`k /= ( n + 1 );`
`k = k % b;`	`k %= b;`

Only arithmetic and bitwise operators can be combined with the assignment operator.

5.5 PRECEDENCE AND ASSOCIATIVITY OF OPERATORS

Precedence is used to determine the order in which different operators are evaluated in an expression.

Associativity is used to determine the order in which different operators with same precedence are evaluated in an expression.

Precedence is applied before associativity to determine the order in which expressions are evaluated. Associativity is applied later, if necessary.

Precedence

The concept of precedence is well defined in the subject of mathematics. You must be aware of rule BODMAS – Brackets, Of, Division, Multiplication, Addition, and Subtraction. In algebra, division and multiplication is performed before addition and subtraction.

The following is a simple example of precedence:

```
5 + 3 * 4
```

This expression consists of one addition and one multiplication operator. As you know from the knowledge of algebra, multiplication has higher precedence than addition, and this multiplication is performed before addition.

Therefore, the expression will be evaluated as

```
( 5 + ( 3 * 4 ) ) → ( 5 + 12 ) → 17
```

giving value 17 as the value of the entire expression.

Associativity

Associativity is applied when more than one operator of same precedence is used in an expression. Associativity can be *left-to-right* or *right-to-left*. Left-to-right associativity evaluates the expression by starting on the left and moving to the right whereas the right-to-left associativity evaluates the expression by starting on the right and moving to the left.

The following is a simple example of associativity:

```
2 + 5 + 7
```

This expression consists of two addition operators. Here the associativity determines how the sub expressions are grouped together. Since the addition operator has left-to-right associativity, the expression will be grouped as

```
( ( 2 + 5 ) + 7 ) → ( 7 + 7 ) → 14
```

giving value 14 as the value of the entire expression.

We will learn more about precedence and associativity as we proceed with the discussion on various operators. *Table 5.17* summarizes the precedence and associativity of all the operators supported by C language.

Table 5.17: Precedence and Associativity of Operators

Operator	Description	Precedence Level (Rank)	Associativity
() [] –> .	Function call Array element reference Structure operator used with pointer to a structure Structure operator used with structure variable	1	Left-to-right
! ~ ++ — + – * & (type) sizeof	Logical NOT operator 1's complement Increment Decrement Unary plus Unary minus Pointer reference (indirection) Address of Type cast Size of an operand	2	Right-to-left
* / %	Multiplication Division Modulus (remainder)	3	Left-to-right
+ –	Addition Subtraction	4	Left-to-right
<< >>	Left shift Right shift	5	Left-to-right
< <= > >=	Less than Less than or equal to Greater than Greater than or equal to	6	Left-to-right
== !=	Equal to Not equal to	7	Left-to-right
&	Bitwise AND	8	Left-to-right
^	Bitwise exclusive OR	9	Left-to-right
|	Bitwise inclusive OR	10	Left-to-right
&&	Logical AND	11	Left-to-right
||	Logical OR	12	Left-to-right
?:	Condition (ternary)	13	Right-to-left
= += –= *= /= %= &= ^= != <<= >>=	Assignment operators	14	Right-to-left

5.6 LIBRARY FUNCTIONS

Quite often it is necessary to evaluate several mathematical functions that are used quite frequently, such as square root, logarithms, exponentials, etc. They are written and incorporated into the C language. Rather than each user writes these subprograms, C compiler provides these built-in library functions as a convenience.

The programmer has to include *math.h* header file in his program unit. In addition to these functions, there are many other library functions that perform specialized functions, and are defined in different header files. For example, library functions that manipulate strings are defined in *string.h.*

Arguments to certain library functions are limited by the definition of the function. For example, logarithm of a negative number is mathematically undefined. Similarly, square root of a negative number is mathematically undefined, and therefore, not permitted.

The following table describes some popular and frequently used mathematical functions. The user must include header file *math.h* in the program in order to use these functions.

Table 5.18: Some Commonly Used Mathematical Functions

Function	Description
`pow(x,y)`	Returns *x* raised to power *y, i.e.,* x^y .
`pow10(p)`	Returns the value 10^p .
`exp(x)`	Returns *e* to the *xth* power, *i.e.,* e^x .
`log(x)`	Returns the value of *ln*(*x*).
`log10(x)`	Returns the value of *log10*(*x*).
`sqrt(x)`	Returns the value of $\sqrt{x}$.
`sin(x)`	Returns the value sine of *x*, where the angle *x* is in radians.
`ceil(x)`	Returns the smallest integer greater than or equal to *x*.
`floor(x)`	Returns largest integer less than or equal to *x*.
`ploy(x,n, coeffs[])`	Returns the value of the polynomial in *x*, of degree *n*, with coefficients *coeffs*[0], *coeffs*[1], . . ., *coeffs*[*n*].

ILLUSTRATIVE EXAMPLES

In this section, let us consider few situations and/or problems that will give us an opportunity to use operators and expressions. Before that let us learn about the concept of *dry run*.

A *dry run* basically means we will imitate a computer and tracing the execution of the instructions using paper pencil. During the process, we record values of the variables as they change during the execution including the initial values. Note that if a variable is not initialized, its value will be shown with character '?'.

In the illustrative examples, you will see the dry run of the program and/or section of the program code, wherever it deemed necessary to clarify the key concept.

Example 5.1: Swap values of two variables 'a' & 'b' using third variable 't'.

Solution:

```
int a=10,b=20,t;    /* 1 */
t= a;               /* 2 */
a = b;              /* 3 */
b = t;              /* 4 */
```

Dry Run

Statement No.	*a*	*b*	*t*
1	10	20	?
2	10	20	10
3	20	20	10
4	10	20	10

The last three lines of the code can also be written in C as

```
t = a, a = b, b = t;
```

Example 5.2(*a*): Swap value of two variables 'a' & 'b' without using third variable and using arithmetic addition (+) and subtraction (–) operations only.

Solution:

```
int a=10,b=20;      /* 1 */
a = a + b;          /* 2 */
b = a – b;          /* 3 */
a = a – b;          /* 4 */
```

Dry Run

Statement No.	*a*	*b*
1	10	20
2	30	20
3	30	10
4	20	10

This task can also be implemented using following single statement:

```
a = ( a + b ) – (b = a );
```

Example 5.2(*b*): Swap value of two variables 'a' & 'b' without using third variable and using arithmetic multiplication (*) and division (/) operations only.

Solution:

```
int a=10,b=20;      /* 1 */
a = a * b;          /* 2 */
b = a / b;          /* 3 */
a = a / b;          /* 4 */
```

Dry Run

Statement No.	*a*	*b*
1	10	20
2	200	20
3	200	10
4	20	10

This task can also be implemented using following single statement:

```
a = ( a * b ) / (b = a );
```

Example 5.3: Swap value of two variables 'a' & 'b' without using third variable as well as without using any arithmetic operator.

Solution: This task can be accomplished using Exclusive OR (XOR) bit-wise (^) operator as illustrated as shown overleaf:

Dry Run

Statement No.	*a*	*b*
1	10 (00001010)	20 (00010100)
2	00011110 (30)	00010100 (20)
3	00011110 (30)	00001010 10
4	00010100 (20)	00001010 (10)

```
int a=10,b=20;      /* 1 */
a = a ^ b;          /* 2 */
b = a ^ b;          /* 3 */
a = a ^ b;          /* 4 */
```

Note that the XOR bit-wise operation is equivalent to addition operation.

Example 5.4: To find the largest of three numbers among *a*, *b*, and *c*.

Solution:

```
int a=10,b=30,c=25, big;                          /* 1 */

big = (a>b)?((a>c)?a:c)):((b>c)?b:c));            /* 2 */
```

Here, if *a* is greater than *b*, expression (*a*>*c*)?*a*:*c*) is evaluated; otherwise expression (*a*>*c*)?*a*:*c*).

Expression (*a*>*c*)?*a*:*c*) will return *a* if *a* is greater than *c* else returns *c*,

Expression (*b*>*c*)?*b*:*c*) will return *b* if *b* is greater than *c* else returns *c*,

Example 5.5: Write appropriate statements to find the sum of the right-most digit and left-most digit of a 4-digit number stored in *n*.

Solution:

```
left_most_digit = n / 1000;
right_most_digit = n % 1000;
sum = left_most_digit + right_most_digit;
```

Example 5.6: Write appropriate statements to convert temperature in Celsius scale into Fahrenheit scale.

Solution:

The relationship between temperatures in Celsius scale into Fahrenheit scale is

$$\frac{C}{100} = \frac{F-32}{180} \quad \Rightarrow \quad F-32 = \frac{180}{100}C \quad \Rightarrow \quad F = 1.8C + 32$$

```
float c = 65.25, f;
f = 1.8 * C + 32;
```

MULTIPLE CHOICE QUESTIONS

1. Which of the following is not an arithmetic operator?
   (*a*) + (*b*) &
   (*c*) % (*d*) *
2. Precedence of operators determines which operator
   (*a*) is used first (*b*) is important
   (*c*) operates on largest numbers (*d*) executes fast
3. The operator % can be applied to
   (*a*) Float values (*b*) Double values
   (*c*) Integral values (*d*) All
4. Expression *x* % *y* is equivalent to
   (*a*) (*x*–(*x*/*y*)) (*b*) (*x*–(*x*/*y*)***y*)
   (*c*) (*y*–(*y*/*x*)) (*d*) (*y*–(*x*/*y*)***x*)
5. Which of the following is not a bit-wise operator?
   (*a*) > (*b*) >>
   (*c*) ^ (*d*) &
6. What will be the value of expression 5 / 6 / 3 + 8 / 3?
   (*a*) 4 (*b*) 2
   (*c*) 2.333 (approx.) (*d*) None of above
7. In expression *x* / *y*, the value of *y* should be
   (*a*) integer (*b*) non-zero
   (*c*) non-negative (*d*) positive
8. If *x* is a variable of type *int*, which of the following is odd?
   (*a*) *x* << 1 (*b*) *x* = *x* * 2
   (*c*) *x* *= 2 (*d*) *x* <<= 1
9. Consider the following statement in C language

```
x = (a > b) ? ((a > c) ? a : c) : (b > c) ? b : c;
```

   What will be the value of *x* if *a* = 3, *b* = –5, *c* = 2?
   (*a*) 2 (*b*) 3
   (*c*) –5 (*d*) None of above
10. When a relational expression is false, it has the value _____.
   (*a*) 1 (*b*) 0
   (*c*) –ve (*d*) +ve

11. What will be the value of *y*?

```
int y;
y = 0x10 + 10 + 010;
```

(*a*) Code will not compile & compiler will complain about expression "0x10+10+010"
(*b*) Program will execute and output value 30
(*c*) Program will execute and output value 34
(*d*) Program will execute and output value 1010010

12. If *a* = 15 then the statement *b* = *a* << 2; will result in

(*a*) 30 (*b*) 60
(*c*) 7 (*d*) None of above

13. The result of expression ~~7 is

(*a*) 7 (*b*) 1
(*c*) 0 (*d*) None of above

14. Which of the following is used for exclusive OR operation?

(*a*) << (*b*) |
(*c*) ^ (*d*) &

15. What is the value of *x* after executing the statement?

```
x = 011 | 0x10;
```

(*a*) 25 (*b*) 50
(*c*) 10 (*d*) 15

16. Consider the following code segment

```
int i, j = 10;
i = ++j;
```

The value of *i* and *j* at the end of this segment will be

(*a*) 10, 10 (*b*) 10, 11
(*c*) 11, 10 (*d*) 11, 11

17. Consider the following code segment

```
int x = 6, y = 5, z = -3;
z = ( x == 6 || ++y );
```

The value of *x*, *y* and *z* at the end of this segment will be

(*a*) 6, 6, 1 (*b*) 6, 6, 0
(*c*) 6, 5, 1 (*d*) 6, 5, 0

18. Consider the following code segment

```
int x, y, z;
z = ( x = 5, y = 6, x * y );
```

The value of *z* at the end of this segment will be

(*a*) 5 (*b*) 6
(*c*) 30 (*d*) None of above

19. Consider the following code segment

```
int x = 5.234, y;
y = sizeof(x);
```

The value of *y* will be

(*a*) 4 (*b*) 2
(*c*) 8 (*d*) Syntax error

20. What will be the final value of *x*?

```
int x = 10;
x += (x++) + (++x) + x;
```

(*a*) 44 (*b*) 45
(*c*) 46 (*d*) 47

21. What will be the final value of *x*?

```
int x = 10;
x = (20 || 40 ) && (10);
```

(*a*) 60 (*b*) 70
(*c*) 0 (*d*) 1

22. What will be the final value of *x*?

```
int a=0;
a=5||2|1;
```

(*a*) 7 (*b*) 0
(*c*) 8 (*d*) 1

23. What will be the final value of *i, j, k, l, m*?

```
int i = -1, j = -1, k = 0, l = 2, m;
m = i++ && j++ && k++ || l++;
```

(*a*) 0, 0, 1, 2, 1 (*b*) 0, 0, 1, 3, 2
(*c*) 0, 0, 1, 3, 1 (*d*) 0, 1, 1, 3, 1

24. Which of the following operator can be used to perform division by 2?

(*a*) / (*b*) >>
(*c*) << (*d*) Both (*a*) & (*b*)

25. Predict the value of *ans*.

```
char val=250;
int ans;
ans = val + !val + ~val + ++val;
```

(*a*) –5 (*b*) –6
(*c*) 0 (*d*) 6

REVIEW EXERCISE

1. What would be the value of 'a' in following cases?
   (*a*) `int a = 45/10*23%45/(45%4*21);`
   (*b*) `float a = 45.0/10*23-45+(4*21.0);`
2. Which of following will be true?
   (*a*) `4>5 && 5>4`
   (*b*) `4>5 || 5>4`
   (*c*) `(232+23*1233) || 0`
   (*d*) `(232+23*1233) && 0`
3. If 'a' has value 15, then what would be the value of 'b' in following cases?
   (*a*) `b = ++a;`
   (*b*) `b = a++;`
   (*c*) `b = a—;`
   (*d*) `b = — a;`
4. What would be the value of 'a'?
   a = 1 == 5 == 5;
5. Will there be an error? If yes, identify the possible error.
   ```
   int a;
   3 = a;
   ```
6. Write appropriate statements to calculate the sum of the first and the second last digit of a 5-digit number.
7. Write appropriate statements to calculate the sum of the digits of a 4-digit number.
8. Write appropriate statements to calculate the total marks and %age of marks secured by Rohit. His score in subjects is *a*, *b*, and *c* respectively (each out of 100).
9. Write appropriate statements to convert length in meters to feet and inches.
10. Write appropriate statements to convert temperature in Fahrenheit scale into Celsius scale.
11. Suppose variable *a* is declared of type *int*. Does the following statements are permitted in C or not? If permitted, then what will be the value of 'a; in each case.
    (*a*) `a = 1, 2, 3;`
    (*b*) `b = ( 1, 2, 3 );`
12. Given the value of a variable *x* which is declared of type *int*. Write single line of statement to that prints EVEN if *x* is an even number otherwise prints ODD number?
    (*a*) `a = 1, 2, 3;`
    (*b*) `b = ( 1, 2, 3 );`

13. What do you have to say about the following statements?
```
int a;
a = 3 * 2--;
```
14. What do you have to say about the following statements?
```
int a;
a = 5.4 % 2;
```
15. What do you have to say about the following statements?
```
int i = 5, j, k;
l = i / -4;
k = i % -4;
```
16. What will be the value of a, b, c, and d once the following statements finish its execution?
```
int a = 1, b = 1, c, d;
c = a || --b;
d = a-- && --b;
```
17. Name the operator(s) that can be used to increment the value of a variable.
18. Name the operator(s) that can be used to divide the value of an integer variable by 2.
19. From the following statements that increment the value of *n* by 1, which one would you like to prefer, and why?
```
n = n + 1;       and        n++;
```
20. Are there any restrictions on modulus (%) operator?
21. Can we add two integer numbers without using arithmetic operators? If yes, how?
22. Find the value of each of the following expressions.
    (*a*) 3 – 5 * 2.5
    (*b*) 10 / (2 * 3)
    (*c*) 22 % 3 * 12 / 3
    (*d*) (3 / 7) / (6 / 5)
23. What is the role played by parentheses in expressions?
24. Write C expressions equivalent to the following algebraic expressions:

(*a*) $a - \frac{a}{a^{-x}} + b$ (*b*) $1 + \frac{1}{a + \frac{1}{a}} + b$ (*c*) $x + \frac{y^{-z}}{k} - \sqrt{x}$

(*d*) $x^3 - \frac{y^7}{-z} + e^x$ (*e*) $\left(\frac{1-x}{1+x}\right) / \left(\frac{1+y}{1-y}\right)$ (*f*) $\frac{1}{r} = \frac{1}{r_1} + \frac{1}{r_2}$

(*g*) $\sqrt{\left(x^2 + y^2 + z^2\right)^3}$ (*h*) $\frac{-b - \sqrt{b^2 - 4ac}}{2a}$

❑ ❑ ❑

Chapter 6

Console Input/Output

Learning Outcomes

After reading this chapter, students will be able to

- explain the Console I/O facility in C
- explain the working of character I/O functions
- explain the working of string I/O functions
- explain the working of formatted I/O functions
- describe various features for formatting the output

6.1 INTRODUCTION

In almost every computer program, the users have to input some data that will be stored in computer's memory, and then subsequently processed. This input data may be from a keyboard, a disk file, or from another computer via a communication channel. The intermediate and final results of the computations are also stored in computer's memory. And finally, there is need to output the results of the computations so that user can use them in their day-to-day work. The output can be displayed on the computer screen, stored on a disk, or transmitted via a communication channel to other computer.

In microcomputers (PCs), the common input device is the keyboard and the output device is a computer screen, sometimes called the *monitor*. This subsystem comprising keyboard and monitor is referred to as a *console*.

The various forms of input and output, abbreviated as I/O, obviously are very different. It would have been easier if all these forms of I/O can be handled in a unified manner as they are handled in other high level languages, such as in Pascal and FORTRAN. But one of the objectives of Dennis Ritchie, developer of C, was to have C language as compact as possible. Therefore, he deliberately omitted I/O statements that generate extra overhead. Instead, every C compiler provides several I/O functions that can be linked with any C program. Though, these I/O functions are not part of C's formal definition, they have become a standard feature of C language.

This chapter describes the various I/O functions that perform the input from a standard input device and output to a standard output device. To use entire range of I/O functions, we need to include header file named *stdio.h* in all our programs.

6.2 CONSOLE INPUT/OUTPUT

When a C program begins execution, it has access to the following pre-defined files:

- The first one, known as the *standard input* (*stdin*), is the file from which input is received. This file is usually a keyboard, but it can also be redirected to obtain input from a disk file or other device.
- The second one, known as *standard output* (*stdout*), is the file to which output is directed. The file is usually a computer screen, but it can be redirected to a disk file or to another device.

- The third one, known as *standard error* (*stderr*), is the file used to keep error messages separate from program's output. For example, if output is going to a disk file, the standard error file can be a computer screen, so that the user sees error messages immediately.

6.3 CONSOLE I/O FUNCTIONS

There are many library functions available for console I/O. These functions are divided into two categories:

- unformatted
- formatted

The only difference between these functions is that the formatted functions permit the input from a standard input device or output sent to a standard output device to be formatted as per requirements. For example, if different values are to be displayed, how much field width i.e. how many columns on screen, is to be used, and how much space between two values is to be given. If a value to be displayed is of real type, then how many decimal places to output.

6.3.1 Unformatted I/O Functions

In the unformatted I/O category, we have functions for performing input/output of one character at a time, known as *character I/O functions* and functions that perform input/output of one string at a time, known as *string I/O functions*. These functions are described in the next sections.

6.3.1.1 Character I/O Functions

For character input, the functions is *getchar()* that is available in all versions of C language. In addition, Turbo C provides two more versions of character input functions in the form of *getche()*, and *getch()*. For character output, the function available is *putchar()*.

***getchar() function*:** This function returns a character that has been recently typed. The typed character is echoed to the computer screen. After typing the appropriate character, the user is required to press *Enter* key.

***putchar() function*:** This function output a character constant or a character variable to the standard output device.

Following character input functions are only supported by Turbo C/C++ Compilers

getche() function: Like *getchar()* function, this function also returns a character that has been recently typed. The typed character is also echoed to the computer screen. But the user is not required to press *Enter* key after typing a character. The advantage of this function over *getchar()* function is that as soon as the user types a character, the character is immediately accepted. In reduces the user's action to single key.

getch() function: This function too returns a character that has been recently typed. But, neither the user is required to press *Enter* key after entering the character nor the typed character is echoed to the computer screen. This function has advantages over the other functions in the applications where the user wants to hide the input.

This input is usually the password used for system security. A *password* is a secret word or a code set by the programmer in case of software or assigned by a system administrator of a multi-user system or a computer network to the individual users. Whenever they want to use that particular software or want to work on that particular system, they have to enter their password. If their password is valid, they will be permitted to use the software or system.

The next program illustrates the working of all these functions. In this program, to display user prompt, the *printf()* function, which will be described in the later sections, is used.

Listing 6.1

```
/* Program to illustrate working of character I/O functions  */

#include <stdio.h>

int main(void)
{
   char ch;

   printf( "\nEnter any character: " );
   ch = getchar();
   printf( "\nCharacter you typed is " );
   putchar(ch);

   printf( "\nEnter any other character: " );
   ch = getche();
   printf( "\nCharacter you typed is " );
   putchar(ch);
```

```
    printf( "\nEnter any other character: " );
    ch = getch();
    printf( "\nCharacter you typed is " );
    putchar(ch);
    return 0;
}
```

Test Run

```
Enter any character: A
Character you typed is A
Enter any other character: x
Character you typed is x
Enter any other character:
Character you typed is H
```

6.3.1.2 String I/O Functions

For string input, the function is *gets()* is used, and for string output, the function is *puts()* is used.

gets() function: This function accepts a string from the standard input device. The length of the string is limited by the declaration of the string variable. The input string may consist of multiple words. Typing Enter key completes the input.

puts() function: This function outputs a string constant or a string variable to the standard output device. The following program illustrates the working of these string I/O functions.

Listing 6.2

```
/* Program to illustrate working of string I/O functions    */
#include <stdio.h>
void main()
{
    char str[31];
    printf( "\nEnter string of length <= 30: " );
    gets( str );
    printf( "\nString you typed is " );
    puts( str );
}
```

Test Run

```
Enter string of length <= 30: string input
String you typed is string input
```

6.3.2 Formatted I/O Functions

In the formatted I/O category, we have *scanf()* function for input and *printf()* function for output. The working of these functions is described in this section.

6.3.2.1 The *scanf()* Function

This function allows us to input data in a fixed format. Its syntax is

```
scanf( "format string", list of addresses of variables );
```

where the *format string* contains the format specifiers that begin with character '%' and are, usually, separated by space or comma.

It returns the number of items read from the keyboard.

Table 6.1 lists various format specifiers used with *scanf()* function. The list of addresses of variables are used so that *scanf()* function can place the data received from a standard input device. The address of a variable is obtained by using address operator, character '&' and pronounced as *addressof* operator.

Table 6.1: List of Format Specifiers

Format Specifier	Used For
%d	signed decimal integer
%ld	signed long decimal integer
%u	unsigned decimal integer
%f	single precision floating real number
%lf	double precision floating real number
%e	floating point (exponential notation)
%x	unsigned hexadecimal integer
%o	unsigned octal integer
%c	single character
%s	single-word string
%[^\n]s	multi-word string

On execution, the input data must be supplied strictly according to the specified format string otherwise results can be very strange. This strange behavior is illustrated next.

Consider the following segment:

```
int a;
float b;
char c;
  :
scanf( "%d,%f,%c", &a, &b, &c );
  :
```

Suppose that the input data is 20 for *a*, 50.25 for *b*, and character 'A' for *c*.

The data must be entered or typed in the following format

```
20,50.25,A
```

In case, if your enter the data as

```
20 50.25 A
```

results can be very surprising.

6.3.2.1.1 Limitation of *scanf()* Function

Consider the following statement

```
scanf( "%d", &intVariable );
```

If we input the value as

```
75.50
```

the *scanf()* function will accept 75 only.

Similarly, in the following statement

```
scanf( "%f", &floatVariable );
```

If we input the value as

```
25
```

the *scanf()* function will store this value as 25.0, *i.e.*, integer value is converted to a real value.

However, in the following statement

```
scanf( "%d,%f", &intVariable, &floatVariable );
```

If we input the above values as

```
75.50,25
```

the *scanf()* function will store a value 75 in *intVariable* but the value of the *floatVariable* will be unpredictable.

6.3.2.2 The *printf()* Function

The *printf()* function allows to output the data in a fixed format. Its general form looks like

```
printf( "format string", list of variables );
```

where the *format string* contains the format specifiers, the sequences that begin with character '\' (back slash), known as *escape sequences*, and the text to be output along with the output data.

It returns the number of characters displayed in the output.

All the format specifiers used with *scanf()* function are valid for *printf()* function. Another format specifier "%p" is used to print the value of a pointer variable in hexadecimal notation. Table 6.2 lists common *escape sequences*.

Table 6.2: Common Escape Sequences

Escape Sequence	Represents	Effect
\n	newline	has the effect of both a carriage return and linefeed, *i.e.*, the subsequent output starts from new line.
\t	tab	moves over to the next eight-space-wide field.
\b	backspace	moves the cursor one space left.
\r	carriage return	carriage return.
\f	formfeed	advances to the top of the next page on printer.
\a	beep	produces one audio tone of the speaker
\'	single quote	insert character ' in the output
\"	double quote	insert character " in the output
\\	Backslash	insert character \ in the output

Consider the following segment:

```
int a = 15;
float b = 23.75;
char c = 'V';
     :
printf( "%d,%f,%c", a, b, c );
     :
```

The output of the above segment will look like

```
15,23.75,V
```

But if the segment is written like

```
int a = 15;
float b = 23.75;
char c = 'V';
     :
printf( "%d%f%c", a, b, c );
     :
```

The output of the above segment will look like

```
1523.75V
```

Needless to say, the output in this form is very difficult to understand.

The above segment can also be written as

```
int a = 15;
float b = 23.75;
char c = 'V';
    :
printf("\nValue of a=%d\nValue of b=%f\nValue of c=%c",
       a, b, c );
    :
```

The output of the above segment will look like

```
Value of a=15
Value of a=23.75
Value of a=V
```

You must have noticed that, in the later case the output is self-explanatory and is the preferred way for the output.

Let us see how the *printf()* function interprets the format string.

We will consider the following example of *printf()* function

```
printf( "\nValue of a = %d\nValue of b = %f\n", a, b );
```

Let us further assume that $a = 15$, and $b = 23.46$.

The *printf()* function examines the format string from left to right. It keeps on sending the characters to the standard output device until it encounters either character '%' or character '\'. The moment it encounters character %, it picks up the corresponding variable and outputs its value in the specified format. Similarly, the moment the character '\' is encountered, it indicates the presence of escape sequence, and the action taken will depend on the next character.

In our example, the first character is '\', and the next character is *n*, so the action of '\n' is to insert *newline* character so that subsequent output can continue from the next line. So the string

```
Value of a =
```

is send to the standard output device, and then character % is encountered following by character *d*, so it picks up the variable *a* and print its value. The output up to this stage appears as

```
Value of a = 15
```

As the scanning of the format string continues, it finds another *newline* character, and the subsequent output is sent to the next line up to % character. Up to this, output appears as

```
Value of a = 15
Value of b =
```

Now it picks up the second variable and outputs its value, and after outputting the value of variable *b*, it finds another *newline* character that causes the subsequent output to be sent to the next line. The final output appears as

```
Value of a = 15
Value of b = 23.46
```

6.3.2.2.1 More About *printf()* Function

As you must have observed in the output of the above program that the *printf()* function outputs six decimal digits for a real value, because in the format specifier we have not specified any additional information. In general, to have a control over the output, we can pass additional information to the *printf()* function.

With this additional information, the format specifier for a real value takes the form

```
%w.df
```

where *w* is the total width or space used for outputting a real value including decimal point and the number of decimal digits; and *d* is the number of decimal digits to be outputted.

This field width is the minimum width to be used. If the real value is small, the output value is right justified in the field and if the value can not be accommodated in the specified field width, then the field width is assumed to be large enough to accommodate the output value.

Similarly for a integer value, it takes the form

```
%wd
```

For a string value, it takes the form

```
%ws
```

By default, the output value is always right justified in the field.

The output value can be left justified in the field by writing hyphen '–' after character % as shown below

```
%-5d
```

In this case, the corresponding value will appear left justified i.e. if the value is small, it will be padded with blanks on the right. To fix these ideas, let us consider an example.

Consider the following segment

```
      :
      :
int a = 123;
float b = 123.564;
char name[20] = "Language";
      :
      :
printf( ":%d:  :%f:  :%s:\n", a, b, name);
printf( ":%6d:  :%8.2f:  :%20s:\n", a, b, name);
printf( ":%-6d:  :%-8.2f:  :%-20s:\n", a, b, name);
      :
      :
```

The output of this segment will appear as

```
:123:  :123.564:  :Language:
:   123:  :  123.56:  :            Language:
:123   :  :123.56  :  :Language            :
```

ILLUSTRATIVE EXAMPLES

Let us write a few example programs to use features that we have learned so far.

Example 6.1: Program to find the sum of digits of a given 4-digit positive integer number.

Listing 6.3

```
/* Program to find sum of digits of a 4-digit positive
 *  integer number */
#include<stdio.h>
int main(void)
{
  int number, sum, digit1, digit2, digit3;
  int digit4, temporary;
  printf( "\nEnter 4-digit positive number: ");
  scanf( "%d", &number );
  temporary = number;
  digit1 = temporary % 10;
  temporary /= 10;
  digit2 = temporary % 10;
  temporary /= 10;
  digit3 = temporary % 10;
  temporary /= 10;
  digit4 = temporary % 10;
  sum = digit1 + digit2 + digit3 + digit4;
  printf( "\n\nSum of digits of %d = %d\n", number, sum );
  return 0;
}
```

Test Run

```
Enter 4-digit positive number: 2375

Sum of digits of 2375 = 17
```

Example 6.2: Program to reverses the digits of a given 4-digit positive integer number.

Listing 6.4

```
/* Program to print 4-digit positive integer number in
 *  reverse order  */

#include<stdio.h>
int main(void)
{
  int number, sum, digit1, digit2, digit3;
  int digit4, temporary;
  printf( "\nEnter 4-digit positive number: ");
```

```
  scanf( "%d", &number );
  temporary = number;
  digit1 = temporary % 10;
  temporary /= 10;
  digit2 = temporary % 10;
  temporary /= 10;
  digit3 = temporary % 10;
  temporary /= 10;
  digit4 = temporary % 10;
  sum = digit1*1000+digit2*100+digit3*10+digit4;
  printf( "\nNumber after reversing digits of %d is %d\n",
          number, sum );
  return 0;
}
```

Test Run

```
Enter 4-digit positive number : 2375
Number after reversing the digits of 2375 is 5732
```

Example 6.3: Program to convert time in seconds to time in hours, minutes and seconds.

Listing 6.5

```
/*  Program to convert time in seconds to time in hours,
*   Minutes and seconds  */

#include<stdio.h>

int main()
{
  int seconds,minutes, hours, total_seconds, temp;
  printf( "\nEnter time in seconds: " );
  scanf( "%d", &total_seconds );

  temp = total_seconds;
  hours = temp / 3600;
  temp %= 3600;
  minutes = temp / 60;
  seconds = temp % 60;

  printf( "\nTime in %d seconds = ", total_seconds);
  printf( "%d hour(s),%d minute(s) & %d secs.\n", hours,
          minutes,seconds );

  return 0;
}
```

Test Run

```
Enter time in seconds: 3678
Time in 3678 seconds=1 hour(s),1 minute(s) & 18 secs.
```

Example 6.4: Program to compute the area of a triangle whose sides are given.

Suppose that length of the sides is a, b and c, then area of triangle is given by formula

$$\text{Area of triangle} = \sqrt{s(s\text{-}a)(s\text{-}b)(s-c)}\text{, where } s = \frac{a+b+c}{2}$$

Listing 6.6

```
/*  Program to find area of triangle whose sides are given  */
#include<stdio.h>
#include<math.h>
void main()
{
  float sideA, sideB, sideC, s, area;
  printf( "Enter side A: " );
  scanf( "%f", &sideA );
  printf( "Enter side B: " );
  scanf( "%f", &sideB );
  printf( "Enter side C: " );
  scanf( "%f", &sideC );
  s = ( sideA + sideB + sideC ) / 2;
  area = sqrt((double)(s*(s-sideA)*(s-sideB)*(s-sideC)));
  printf( "\nArea of triangle: %.3f\n", area );
}
```

Test Run

```
Enter side A: 3
Enter side B: 3
Enter side C: 3

Area of triangle: 3.897
```

The program is run for an equilateral triangle with each side equal to 3 units.

Example 6.5: An amount of rupees, say P, program to compute simple interest and compound interest if the interest rate is $R\%$ annually and the time is N years.

We know that formulae to compute simple interest and compound interest are

$$\text{Simple interest } (si) = \frac{P \times R \times N}{100}$$

$$\text{Compound interest } (ci) = P\left(1+\frac{R}{100}\right)^{N} - P$$

Listing 6.7

```
/*  Program to compute simple interest & compound interest  */
#include<stdio.h>
#include<math.h>
void main()
{
  float p, r, n, si, ci;
  printf( "Enter amount: " );
  scanf( "%f", &p );
  printf( "Enter rate of interest: " );
  scanf( "%f", &r );
  printf( "Enter time: " );
  scanf( "%f", &n );
  si = ( p * r * n ) / 100;
  ci = p * pow((double)1.0+r/100, (double)n ) - p;
  printf( "\nSimple Interest   : %.2f\n", si );
  printf( "\nCompound Interest: %.2f\n", ci );
}
```

Test Run

```
Enter amount: 100
Enter rate of interest: 5
Enter time: 5

Simple Interest   : 25.00
Compound Interest: 27.63
```

Example 6.6: Program that reads a lowercase alphabet and prints it in uppercase.

Note that characters are internally represented using ASCII code. The ASCII for lowercase alphabet letters (*a-z*) is 97 decimal to122 and for uppercase alphabet letters (*A-Z*) it is 65 to 90. It means that the ASCII code of lowercase alphabet letters is more by a factor of 32 than corresponding uppercase alphabet letters. Therefore, given a lowercase alphabet letter, by subtracting a value of 32, we get corresponding uppercase alphabet letter.

The following program does this.

Listing 6.8

```
/*  Program to reads a lowercase character and outputs it as
*   uppercase character  */
#include<stdio.h>
void main()
{
   char ch;
   printf( "Enter lowercase alphabet letter: " );
   ch = getchar();
   ch = ch - 32;
   printf( "\nUppercase alphabet letter:  %c\n", ch );
}
```

Test Run

```
Enter lowercase alphabet letter: a
Uppercase alphabet letter: A
```

Example 6.7: Program that reads a decimal integer number and prints it equivalent octal number without doing any explicit conversion.

Note that integer numbers (in general all kind of data) are internally represented as binary numbers whether you input as decimal number, octal number or hexadecimal number. The stored binary integer number can be printed either as decimal number, octal number or hexadecimal number using *d*, *o* and *x* specifier in the *printf()* function. The following program does this.

Listing 6.9

```
/* Program that reads decimal integer number and
*  prints it as octal integer number  */
#include<stdio.h>
void main()
{
   int number;
   printf( "Enter decimal integer number: " );
   scanf("%d", &number );
   printf( "\nEquivalent octal number:  %o\n", number );
}
```

Test Run

```
Enter decimal integer number: 250
Equivalent octal number: 372
```

MULTIPLE CHOICE QUESTIONS

1. What will be the output of the following code?

```
printf( "%d", 'B' );
```

(*a*) 66 (*b*) B
(*c*) b (*d*) Error

2. What will be the output of the following code?

```
float x = 1 / 2.0 - 1 / 2;
printf( "%.2f", x );
```

(*a*) 0.00 (*b*) 0.50
(*c*) 0.25 (*d*) 1.00

3. What is the function of character hyphen '–' when used after the character % in the field specifier?
(*a*) To print the output left justified
(*b*) To print '–' character before output
(*c*) To print the output right justified
(*d*) None of above

4. Which of the following function will prefer to use in case you want to provide the password facility in your program for security purposes?
(*a*) getchar() (*b*) getch()
(*c*) scanf() (*d*) gets()

5. Every escape sequence begins with character
(*a*) % (*b*) /
(*c*) \ (*d*) –

6. If x = 12.4568, what value will be printed if the *printf()* function is used as?

```
printf( "%.3f", x );
```

(*a*) 12.457 (*b*) 12.000
(*c*) 12.456 (*d*) 12.4568

7. What is the value returned by *printf()* function?
(*a*) It returns the number of items printed.
(*b*) It returns the number of character positions printed.
(*c*) It returns the number of bytes printed.
(*d*) None of above.

8. What will be the output of the following segment of code?

```
int i = -3, j = 2, k = 0, m;
m = ++i && j++ && ++k;
printf("\n%d, %d, %d, %d\n", i, j, k, m);
```

(*a*) –2, 3, 1, 1 (*b*) –3, 2, 0, 1
(*c*) –2, 3, 0, 1 (*d*) –2, 3, 1, 0

9. What will be the output of the following segment of code?

```
int i = -3, j = 2, k = 0, m;
m = ++i || j++ || ++k;
printf("\n%d, %d, %d, %d\n", i, j, k, m);
```

(*a*) -2, 3, 0, 0 (*b*) -2, 2, 0, 1
(*c*) -2, 2, 1, 1 (*d*) -2, 3, 1, 0

10. What will be the output of the following segment of code?

```
printf("%c", 7);
```

(*a*) Syntax error (*b*) Run time error
(*c*) Prints value 7 (*d*) None

11. Which would be the output of the following segment of code?

```
int a = 4;
printf("\n%d, %d\n", a++, ++a );
```

(*a*) 4, 5 (*b*) 5, 4
(*c*) 4, 4 (*d*) 5, 5

12. What is the value returned by *scanf()* function?
(*a*) It returns the number of items read.
(*b*) It returns the number of character in the items read.
(*c*) It returns the number of bytes read.
(*d*) None of above.

13. What will be the output of the program?

```
void main() {
   int a=1,b=2,c=3;
   printf("%d %d %d");
}
```

(*a*) Garbage value (*b*) Syntax error
(*c*) 3 2 1 (*d*) None of above

14. What will be the output of the program?

```
void main() {
   printf("%d",printf("Hello"));
}
```

(*a*) 5Hello (*b*) Hello5
(*c*) Error (*d*) None of above

15. Find the output of the program?

```
void main() {
   float a = 3.25, b= 4.34;
   printf("%.0f, %.1f", b, a);
}
```

(*a*) 0, 3.2 (*b*) 4.4, 3.0
(*c*) 4, 3,3 (*d*) None of above

REVIEW EXERCISE

1. How many decimal places are displayed when a number has a corresponding format specifier "%f"?
2. How does the %u format specifier compare with %d format specifier?
3. What is the difference between 'x' and "x"?
4. How does the %15s format specifier compare with %–15s format specifier?
5. How the % sign can be printed with a *printf()* function as a part of the output?
6. How the output will appear as a result of the following statement?

```
printf("\n:%20s:\n:% -20s:\n", "Test String", "Test String" );
```

7. Find the errors, if any, in the following segment of code.

```
float a, b;
printf("\nEnter value of a: ");
scanf( "%f", a );
b = a * 3;
printf("\nValue of b = %f\n", b);
```

8. Find errors, if any, in the following program. What will be the output of the C program if it is error free?

```
int i = 3, j = 2, k;
k = 2 * i + j;
i += 3;
j = i + 2 * j - k;
k += 2 * i;
printf("\ni=%d\tj=%d\tk=%d\n",i,j,k);
```

9. What will the output of the following code?

```
printf(" %d\n", printf("\nGood Bye\n"));
```

10. What will be the output of the following segment of code?

```
int m, j = 1, k = 3;
m = 2 * j + k;
j = 2 * m + j / 3;
k /= 2;
m = j + k+ m;
printf ( "\nj = %d\tk = %d\tm = %d\n", j, k, m );
```

11. What will the output of the following code?

```
int intVar = 20;
printf("\n%d,%d,%d\n", intVar == 20,intVar = 35,intVar < 50);
```

PROGRAMMING EXERCISE

1. Write a program that accepts a decimal integer and prints it in the octal form.
2. Write a program that accepts a character in uppercase and prints it as lowercase.
3. Write a program that accepts a temperature in Fahrenheit scale and prints its equivalent temperature in Centigrade scale.

4. An amount of rupees, say *P*, is deposited in a commercial bank for *N* years, which pays simple interest at the rate of *R*% annually. Write a C program that prints the amount after *N* years.
5. An amount of rupees, say *P*, is deposited in a commercial bank for *N* years, which pays interest at the rate of *R*% annually. Write a C program that prints the difference between the compound interest and simple interest after *N* years.
6. Write C program to find real roots of a quadratic equation of type $ax^2+bx+c = 0$.
7. Basic salary (BS) of an employee is input through the keyboard. The dearness allowance (DA) is 25% of the basic salary while the house rent allowance (HRA) is 15% of the basic salary. Provident fund (PF) is deducted at the rate of 10% of the gross salary (BS+DA+HRA). Write a program to calculate the net salary (NS).
8. A department store places an order with a company for *n* pieces of miners, *m* pieces of toasters, and *p* number of fans. The cost of items are as follows:

Item Description	Price per Unit (in Rupees)
Miners	1,500
Toaster	200
Fan	450

The discount allowed for various items are 5% for miners, 15% for fan, and 10% for toaster. The company charge 10% as sales tax on all items on net value after deducting the discount. Formulate an algorithm that reads *m*, *n*, and *p* and computes the amount to be paid by the store.

9. Write a set of programs to demonstrate the swapping of two values using different methods discussed.

❑ ❑ ❑

Chapter 7

Decision Making and Branching

Learning Outcomes

After reading this chapter, students will be able to

- explain the need for control transfer
- explain the kind of control statement available in C language
- explain the working of *if* and *if-else* statements for decision making
- explain the working of nested *if* statements
- explain the usefulness of *if - else if* ladder in multiple alternatives in decision making
- explain the working of *switch* statement
- use these decision making statements to write programs to solve problems requiring decision making

7.1 INTRODUCTION

A computer program is a set of instructions for a computer. These instructions, also known as *statements*, are executed sequentially, *i.e.*, one after the other as they appear in a program. This order of executing the statement works well for simple problems where neither any decision making process is involved nor any repetition of statements is involved. However, in practice, it is often required to change the order of execution of statements or repeat a group of statements for known number of times or until certain specified expressions are met. In such cases, the order in which these statements will be executed is to be controlled.

Changing the order of execution of the statements *i.e.* changing the flow of control, might involve one of the following situations.

- *Jumping* — transferring control from one point to another point in the same program unit.
- *Branching* — executing one group of instructions depending on the outcome of a decision.
- *Looping* — executing a group of instructions repetitively either for a given number of times or until the required expressions is met.

The C Language provides facilities for controlling the order of execution of the statements, which are referred to as *flow control statements* or simply as *control statements*. The various flow control statements are clubbed in the following categories:

1. *Decision Making Statements* — In this category, C language provides the following statements
   - *if* statement
   - *if – else* statement
   - *else if* ladder
   - *switch* statement
2. *Looping Statements* — In this category, C language provides the following statements
   - *for* statement
   - *while* statement
   - *do – while* statement
3. *Jumping Statements* — In this category, C language provides the following statements
   - *break* statement
   - *continue* statement

This chapter describes various decision making statements, and the *break* statement. The remaining statements will be discussed in *Chapter 8*.

7.2 DECISION MAKING STATEMENTS

These statements allow the execution of selective statements based on certain decision criteria. We will see their working supported by well-designed illustrative examples.

7.2.1 The *if* Statement

The syntax of *if* statement is

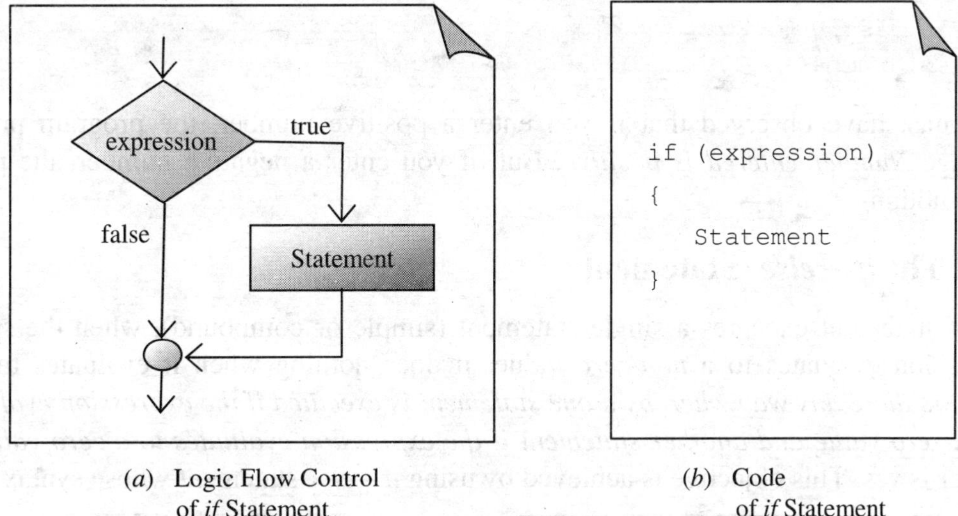

(*a*) Logic Flow Control of *if* Statement

(*b*) Code of *if* Statement

Figure 7.1: Logic flow control and code of *if* statement

The *expression* may represent a relation expression, a logical expression, a numeric variable or a numeric constant. The specified expression may be a simple expression or compound expression.

The expression in C++ evaluates to a *zero* or *non-zero* value. If expression evaluates to a *non-zero* value, then the statements in the statement-block are executed; otherwise they are bypassed.

Listing 7.1

```
/* Program to find largest of three numbers using 'if' statement  */
#include <stdio.h>
int main()
{
   int a, b, c, max;
   printf( "\nEnter value for a : " );
   scanf( "%d", &a );
   printf( "\nEnter value for b : " );
   scanf( "%d", &b );
   printf( "\nEnter value for c : " );
   scanf( "%d", &c );
   max = a;
```

```
    if ( b > max )
        max = b;
    if ( c > max )
        max = c;
    printf("\nLargest number = %d\n", max);
    return 0;
}
```

Test Run

```
Enter three numbers : 10 25 22
Largest number = 25
```

You must have observed that if you enter a positive number, the program prints the message *Number entered is positive*. But, if you enter a negative number, the program prints nothing.

7.2.2 The *if – else* Statement

The *if* statement executes a single statement (simple or compound), when the specified expression evaluates to a *non-zero* value. It does nothing when it evaluates to a *zero* value. *Is there any way whereby a one statement is executed if the expression evaluates to a non-zero value and another statement if the expression evaluates to a zero value*? The answer is yes. This objective is achieved by using *if – else* statement whose syntax is

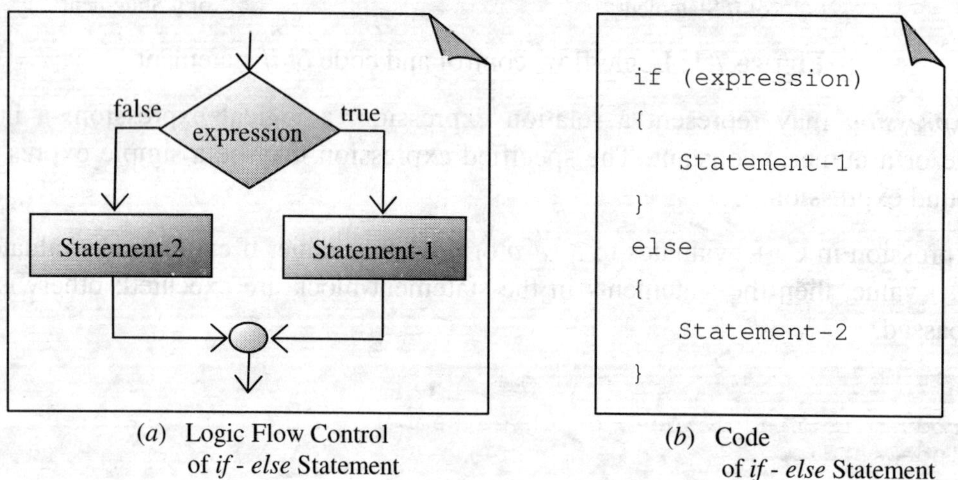

(*a*) Logic Flow Control of *if - else* Statement

(*b*) Code of *if - else* Statement

Figure 7.2: Logic flow control and code of *if-else* statement

If expression evaluates to a non-zero value, the *statement-1* is executed and *statement-2* is bypassed. However, if the expression evaluates to a zero value, the *statement-1* is bypassed and *statement-2* is executed.

Listing 7.2

```
/*
*  Program to demonstrate the use of if-else statement
*/
#include<stdio.h>
int main()
{
   int n;
   printf( "\nEnter any whole number : " );
   scanf( "%d", &n );
   if ( n % 2 == 0 )
      printf( "\nNumber entered is EVEN\n" );
   else
      printf( "\nNumber entered is ODD\n" );
   return 0;
}
```

Test Runs

First Run

```
Enter any whole number: 22
Number entered is EVEN
```

Second Run

```
Enter any number: 15
Number entered is ODD
```

You must have observed that if you enter a even number, then the program prints the message *Number entered is EVEN*. If you enter odd number, then the program prints the message *Number entered is ODD*.

Listing 7.3

```
/*
*  Program to find the largest of two numbers 'a' and 'b'
*/
#include<stdio.h>
int main()
{
    int a, b;
    printf( "\nEnter value for a : " );
   scanf( "%d", &a );
    printf( "\nEnter value for b : " );
    scanf( "%d", &b );
    if ( a > b )
       printf( "\n%d is the largest\n", a );
    else
       printf( "\n%d is the largest\n", b );
    return 0;
}
```

Test Runs

First Run

```
Enter value for a : 20
Enter value for b : 10
20 is the largest
```

Second Run

```
Enter value for a : 15
Enter value for b : 25
25 is the largest
```

7.2.3 Nested *if* and *if – else* Statements

If statements can be nested, *i.e.*, an *if* statement can be contained within another *if* statement. The inner *if* statement will be executed if expression of outer *if* statement evaluates to *non-zero* value.

Likewise, *if-else* statements can also be nested. Here, *if* statement can be nested in *if* part *or else* part. Likewise, *if - else* statement can be nested in *if* part *or else* part.

Listing 7.4

```
/*
*  Program to create equivalent of four function ( +, -, *, / )
*  calculator using nested if statements
*/
#include<stdio.h>
int main()
{
   int a, b;
   int result;
   float result2;
   char op;
   printf("\nEnter expression such as 5+3 : ");
   scanf("%d%c%d", &a, &op, &b );
   if ( op == '+' ) {
      result = a + b;
      printf("\n%d%c%d=%d\n",a,op,b,result);
   } else {
      if ( op == '-' ) {
         result = a - b;
         printf("\n%d%c%d=%d\n",a,op,b,result);
      } else {
         if ( op == '*' ) {
            result = a * b;
            printf("\n%d%c%d=%d\n",a,op,b,result);
         } else {
            if ( op == '/' ) {
               result2 = (float)a / b;
               printf("\n%d%c%d=%.2f\n",a,op,b,result2);
```

```
            } else {
                printf("\nWrong Operator...\n");
            }
        }
    }
  }
  printf("\n");
  return 0;
}
```

Test Run

```
Enter expression such as 5+3 : 10-3
10 - 3 = 7
```

Listing 7.5

```
/*
*  Program to find the largest of three numbers a, b, and c
*  using nested if-else statements
*/
#include<iostream.h>
int main()
{
    int a, b, c;
    printf( "\nEnter value for a : " );
    scanf( "%d", &a );
    printf( "\nEnter value for b : " );
    scanf( "%d", &b );
    printf( "\nEnter value for c : " );
    scanf( "%d", &c );
    if ( a > b ) {
        if ( a > c )
           printf("\nLargest number = %d\n", a );
        else
           printf("\nLargest number = %d\n", c );
    } else {
        if ( b > c )
           printf("\nLargest number = %d\n", b );
        else
           printf("\nLargest number = %d\n", a );
    }
    return 0;
}
```

Test Run

```
Enter three numbers : 12 25 13
Largest number = 25
```

Listing 7.6

```
/* Program to print the name of the day of week when we are given the
 * day of week as number (0 for Sunday, 1 for Monday, ..., 6 for
 * Saturday) using nested if-else statements (Example deep nesting)
*/
#include <stdio.h>
int main()
{
    int day;
    printf("\nEnter day of week as number ( 0 - 6 ) : ");
    scanf("%d", &day);
    if ( day == 0 )
       printf("\nDay of week is Sunday");
    else
       if ( day == 1 )
          printf("\nDay of week is Monday");
       else
          if ( day == 2 )
            printf("\nDay of week is Tuesday");
          else
             if ( day == 3 )
                printf("\nDay of week is Wednesday");
             else
                if ( day == 4 )
                   printf("\nDay of week is Thursday");
                else
                   if ( day == 5 )
                      printf("\nDay of week is Friday");
                   else
                      if ( day == 6 )
                         printf("\nDay of week is Saturday");
                      else
                         printf("\n Wrong input");
    printf("\n");
    return 0;
}
```

Test Run

First Run

```
Enter day week as number ( 0 - 6 ) : 2
Day of week is Tuesday
```

7.2.4 The *if - else if* Ladder

Consider the listing 7.6, which demonstrates how *if-else* statements are nested. As the resulting indentation gets deep and deep, the program becomes more and more difficult to read. There is another way to write and think about this situation.

This is using *if - else-if* construct/ladder whose syntax is

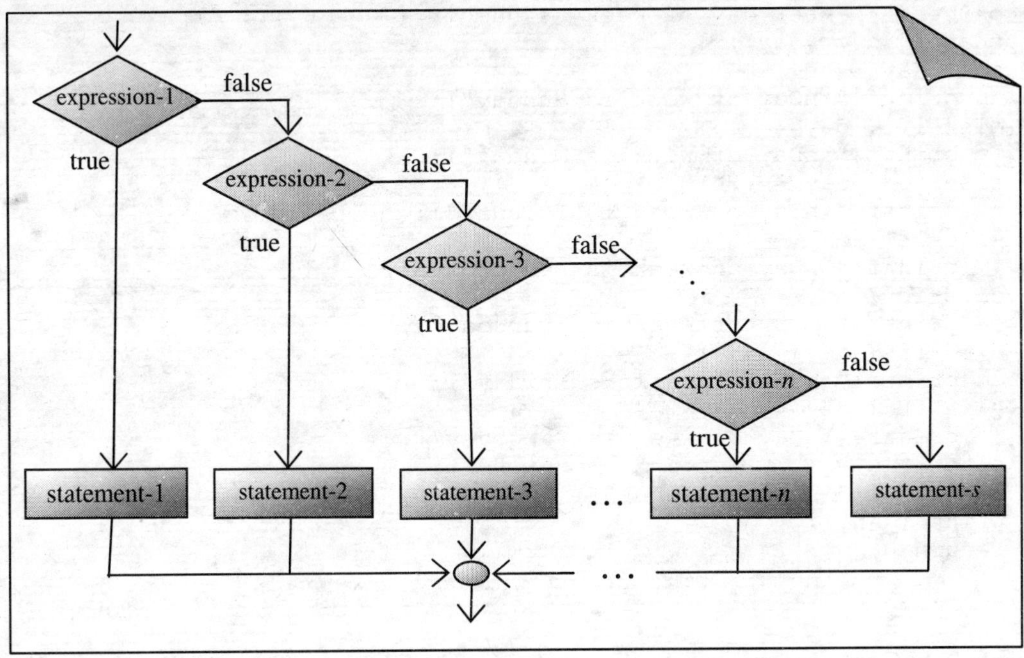

Figure 7.3: Logic Flow Control of *if - else if* Ladder

The expressions are evaluated in order, and if any expression is true then the statement block associated with it is executed, and this terminates the whole chain. The last *else* part handles *none of the above* or default case where none of the specified expressions are satisfied.

This sequence of *if* statements is the most general way of writing a multi-way decision. The program listing 7.6 can be rewritten using *if - else if* construct/ladder as shown in program listing 7.7.

```
if ( expression-1 )
   statement-1
else if ( expression-2 )
   statement-2
else if ( expression-3 )
   statement-3
      :
      :
else if ( expression-n )
   statement-n
else
   statement-s
```

Figure 7.4: Code of *if - else if* Ladder

Listing 7.7

```
/* Program to print the name of the day of week when we are given the
 * day of week as number (0 for Sunday, 1 for Monday, ..., 6 for
 * Saturday) using if - else if ladder.
 */
#include <stdio.h>
int main()
```

```
{
    int day;
    printf("\nEnter day of week as number ( 0 - 6 ) : ");
    scanf("%d", &day);
    if ( day == 0 )
       printf("\nDay of week is Sunday");
    else if ( day == 1 )
       printf("\nDay of week is Monday");
    else if ( day == 2 )
       printf("\nDay of week is Tuesday");
    else if ( day == 3 )
       printf("\nDay of week is Wednesday");
    else if ( day == 4 )
       printf("\nDay of week is Thursday");
    else if ( day == 5 )
       printf("\nDay of week is Friday");
    else if ( day == 6 )
       printf("\nDay of week is Saturday");
    else
       printf("\n Wrong input");
    printf("\n");
    return 0;
}
```

Example 7.1: Commission on sales by a salesman is calculated as per following policy:

Amount of Sale (in Rs.)	Commission Rate
0 – 5000	Nil
5001 – 10000	5 %
10001 – 15000	7.5 %
15001 – 20000	9 %
> 20000	10 %

Write a program that accepts sales made by a salesman and displays the commission.

Listing 7.8

```
/* Program to compute the commission as per the stated policy */
#include <stdio.h>
int main()
{
    float sales, comm;
    printf("\nEnter sales made by the salesman (in Rs.) : ");
    scanf("%f", &sales);
    if ( sales <= 5000 )
       comm = 0;
    else if ( sales <= 10000 )
       comm = ( sales - 5000 ) * 0.05;
    else if ( sales <= 15000 )
       comm = 250 + ( sales - 10000 ) * 0.075;
    else if ( sales <= 20000 )
```

```
        comm = 675 + ( sales - 15000 ) * 0.09;
    else
        comm = 1125 + ( sales - 20000 ) * 0.1;
    printf("\nThe computed commission = Rs. %.f", comm);
    return 0;
}
```

Test Runs

First Run

```
Enter sales made by the salesman (in Rs.) : 4500
The computed commission = 0
```

Second Run

```
Enter sales made by the salesman (in Rs.): 17500
The computed commission = Rs. 900
```

7.2.5 The *switch* Statement

The *switch* statement provides an alternative to *else if* construct. The *switch* statement has more flexibility and a clearer format than *else if* construct. The syntax of *switch* statement is shown in Figure 7.5.

If *expression* takes any value from *val-1*, *val-2*, *val-3*, ..., *val-n*, the control is transferred to that appropriate case. In each case, the statements are executed and then the *break* statement transfers the control out of *switch* statement.

```
switch ( expression )
{
    case val-1 :
                    statement-1
                    break;
    case val-2 :
                    statement-2
                    break;
    case val-3 :
                    statement-3
                    break;
         :
    case val-n :
                statement-n
                break;
    default    :
                statement-d
}
```

Figure 7.5: Code of *switch* Statement

If no *break* statement is used following a case, except the last one in the absence of *default* keyword, the control will fall through to the next case. If the value of the *expression* does not match any of the case values, control goes to the *default* keyword, which is usually at the end of the *switch* statement. The use of the *default* keyword can be of a great convenience. If there is no *default* keyword, the whole *switch* statement simply terminates when there is no match.

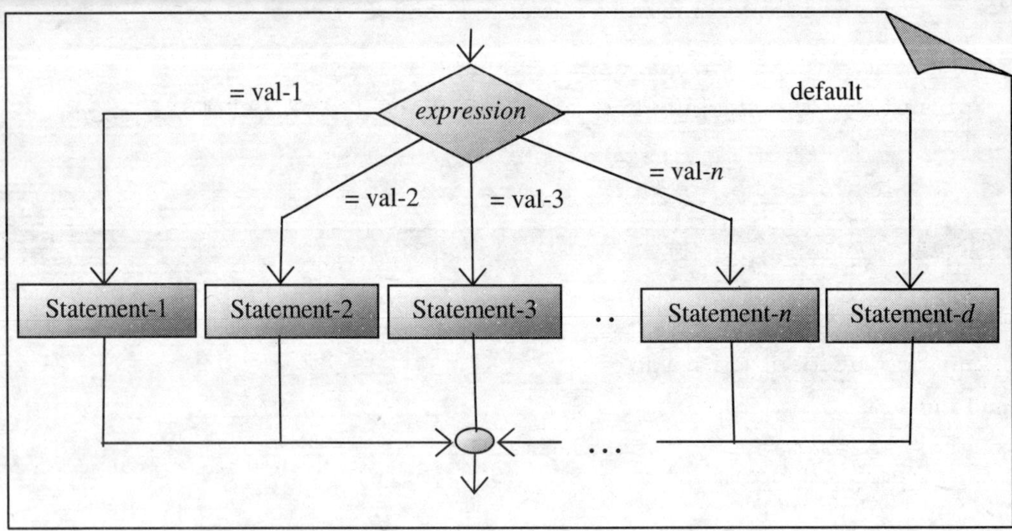

Figure 7.6: Logic Flow Control of *switch* Statement

Listing 7.9

```
/*  Program to demonstrate the use of switch statement  */
#include<stdio.h>
int main()
{
   int day;
   printf( "\nEnter day of week as number: " );
   scanf( "%d", &day );
   switch ( day )
   {
     case 0  : printf("\nDay of week is Sunday" );
               break;
     case 1  : printf("\nDay of week is Monday" );
               break;
     case 2  : printf("\nDay of week is Tuesday" );
               break;
     case 3  : printf("\nDay of week is Wednesday" );
               break;
     case 4  : printf("\nDay of week is Thursday" );
               break;
     case 5  : printf("\nDay of week is Friday" );
               break;
     case 6  : printf("\nDay of week is Saturday" );
               break;
     default : printf("\nWrong input" );
   }
   printf( "\n" );
   return 0;
}
```

You will find that program listing 7.9 is more readable that program listings 7.6 and 7.7.

Listing 7.10

```
/* Program to create equivalent of four function ( +, -, *, / )
 * calculator using switch statement
 */
#include<stdio.h>
int main()
{
    float a, b, result;
    char op;
    printf("\nEnter expression such as 5+3 : ");
    scanf("%d%c%d", &a, &op, &b );
    switch ( op )
    {
       case '+' : result = a + b;
                  break;
       case '-' : result = a - b;
                  break;
       case '*' : result = a * b;
                  break;
       case '/' : result = a / b;
                  break;
       default :  printf("\nWrong input\n");
                  return 0;
    }
    printf("\n%.2f%c%.2f = %.2f\n", a, op, b, result);
    return 0;
}
```

However, if you want that a same statement is to executed for more than one value of the *expression*, then we have to code these *cases* one after the other, and then the specified statement as shown below:

```
switch ( expression )
{
     :
   case val-4 :
   case val-5 :
   case val-6 :
                 statement-block
                 break;
     :
}
```

The *statement-block* will be executed whenever the value of the *expression* is *val-4*, *val-5*, or *val-6*.

7.2.5.1 More About *switch* Statement

Following are some important points about switch statement.

1. The *expression* of switch statement must be of type integer or character type.
2. The *default* case need not be used as the last case. It can be placed at any place.
3. The case values must be integer constants or character constants. These values can also be constant expressions as these are evaluated at compilation time. For example, if expression 5 + 7 appears as case value, it will be evaluated to value 12 at compilation. Similarly, if expression 5 < 7 appears as the case value, it will evaluate to 1 by compiler at compilation time.
4. The case statements can only be used within the scope of switch statement otherwise the compiler will display en error message similar to *"Case outside of switch statement in function ..."*.
5. The case values need not be in any specific order.

ILLUSTRATIVE EXAMPLES

Let us write few more C programs to use the features that we have learned in this chapter.

Example 7.2: Program to find roots of a quadratic equation of type
$ax^2 + bx + c = 0$ provided $a \neq 0$.

Solution: The roots of the quadratic equation are given by the formula:

$x_{1,2} = \dfrac{-b \pm \sqrt{b^2 - 4ac}}{2a}$, where expression $b^2 - 4ac$ is known as the discriminant.

Depending on the sign of the discriminant, there are three possibilities for the roots:

1. If $b^2 - 4ac < 0$, then the root are imaginary, and we can compute real part and imaginary part separately as

$$\text{real part} = -\frac{b}{2a} \qquad \text{imaginary part} = \frac{\sqrt{-(b^2 - 4ac)}}{2a}$$

2. If $b^2 - 4ac = 0$, then the roots are real and equal, and root is simply computed as

$$\text{root} = -\frac{b}{2a}$$

3. If $b^2 - 4ac > 0$, then the roots are real and distinct and are computed as

$$\text{root one} = \frac{-b - \sqrt{b^2 - 4ac}}{2a} \qquad \text{root two} = \frac{-b + \sqrt{b^2 - 4ac}}{2a}$$

The complete program will look like as shown below:

Listing 7.11

```
/*  Program to find the roots of an quadratic equation of type
*   of  type a*x^2+b*x+c = 0, where a is not equal to zero
*/
#include<stdio.h>
#include<math.h>
int main()
{
   float a, b, c, disc, root1, root2, twoA;
   float realPart, imagPart;
   printf( "\nEnter values of a,b,c: " );
   scanf( "%f,%f,%f", &a, &b, &c);
   if ( a == 0 ) {
      printf( "\nIt is not an quadratic equation\n" );
      return 0;
   }
   disc = b * b - 4.0 * a * c;
   twoA = a + a;
   if ( disc < 0 ) {
      realPart = - b / twoA;
      imagPart = sqrt((double)-disc) / twoA;
      printf( "\nRoots are imaginary" );
      printf( "\nReal part = %.3f", realPart );
      printf( "\nImaginary part = %.3f\n", imagPart );
   } else if ( disc == 0 ) {
      root1 = - b / twoA;
      printf( "\nRoots are real and equal" );
      printf( "\nEach root = %.3f\n", root1 );
   } else {
      root1 = (-b-sqrt((double)disc ))/twoA;
      root2 = (-b+sqrt((double)disc ))/twoA;
      printf( "\nRoots are real and distinct" );
      printf( "\nRoot 1 = %.3f", root1 );
      printf( "\nRoot 2 = %.3f\n", root2 );
   }
   return 0;
}
```

Test Run

```
Enter values of a,b,c: 2,3,5

Roots are imaginary

Real part = -0.750
Imaginary part = 1.392
```

Example 7.3: Write a program to determine whether the given year is a leap year.

Under the current (Gregorian) calendar, a year is said to be a leap year if it is divisible by 400 or by 4 but not by 100

Listing 7.12

```
/* Program to determine whether the given year is a leap year  */
#include<stdio.h>
int main()
{
   int year;
   printf("\nEnter any year : ");
   scanf("%d",&year);
   if ((year % 400==0)||((year % 4==0) )&&(year % 100!=0))
      printf("\nYear %d is a leap year.\n", year);
   else
      printf("\nYear %d is not a leap year.\n", year);
   return 0;
}
```

Test Runs

First Run

```
Enter any year : 2000
Year 2000 is a leap year.
```

Second Run

```
Enter any year : 2008
Year 2000 is a leap year.
```

Third Run

```
Enter any year : 2006
Year 2000 is not a leap year.
```

Example 7.4: Income tax for individuals is computed on slab rates as follows:

Income	Tax Payable
Upto Rs. 1,00,000/-	Nil
From Rs. 1,00,001/- to Rs. 2,00,000/-	10% of the excess over Rs. 1,00,000/-
From Rs. 2,00,001/- to Rs. 3,00,000/-	20% of the excess over Rs. 2,00,000/-
Above Rs. 3,00,000/-	30% of the excess over Rs. 3,00,000/-

Write a program that reads the income and prints the income tax due.

Listing 7.13

```
/*  Program to compute income tax for given gross income   */
#include<stdio.h>
void main()
{
    float income, tax;
    printf( "\nEnter gross income (in Rs.): " );
    scanf( "%f", &income );
    if ( income <= 100000.0 )
        tax = 0;
    else if ( income <= 200000.0 )
        tax = ( income - 100000.0 ) * 0.1;
    else if ( income <= 300000.0 )
        tax = 10000 + ( income - 200000.0 ) * 0.2;
    else
        tax = 30000 + ( income - 300000.0 ) * 0.3;
    printf( "\nTax due = Rs. %.2f\n", tax );
}
```

Test Run

```
Enter gross income : 250000
Tax due = Rs. 20000.00
```

Example 7.5: Given roll number, name and marks in four subjects. Write a program to print the roll number, name and grade, where grade is computed as follows:

Percentage of Marks	Grade
percentage $\geq$ 90	A
90 > percentage $\geq$ 75	B
75 > percentage $\geq$ 60	C
60 > percentage $\geq$ 50	D
percentage < 50	F

Listing 7.14

```
/*  Program to compute grade of a student   */
#include<stdio.h>
int main()
{
    int   rollno, i;
    char  name[21], grade;
    float marks[4], total, percent;
    printf("\nEnter name (max 20 characters) : ");
    fgets(name, 21, stdin);
    printf("Enter rollno number : ");
    scanf("%d", &rollno);
    printf("\nEnter marks in FOUR subjects\n");
```

```
    for (i = 0; i < 4; i++ )
        scanf("%f", &marks[i]);
    total = 0;
    for ( i = 0; i < 4; i++ )
        total += marks[i];
    percent = total/4;
    if ( percent >= 90 )
        grade = 'A';
    else if ( percent >= 75 )
        grade = 'B';
    else if ( percent >= 60 )
        grade = 'C';
    else if ( percent >= 50 )
        grade = 'D';
    else
        grade = 'F';
     printf("\nParticulars of Student\n\n");
     printf("\nRoll number : %d", rollno);
     printf("\nName        : %s", name);
     printf("\nGrade       : %c\n", grade);
     return 0;
}
```

Test Run

```
Enter name (max 20 characters) : Ram Murti
Enter rollno number : 1200
Enter marks in FOUR subjects
75 80 90 85

Particulars of Student

Roll number : 1200
Name        : Ram Murti
Grade       : B
```

MULTIPLE CHOICE QUESTIONS

1. Consider the following code:

```
switch(ch)
{
    case 'a': printf("a");
    case 'b': printf("b");
    default:  printf("error");
}
```

If value 'a' is given to character variable *ch*, then the output will be

(*a*) a
(*b*) ab
(*c*) aberror
(*d*) error

2. Consider the following code:

```
int a = 6, b = 6;
if ( b == 6 || --a )
{
    statement1;
    statement2;
}
```

Suppose *statement1* and *statement2* does not modify the value of *a* and *b*, then what is the value of *a* after executing the above code?

(*a*) 6 (*b*) 5
(*c*) 7 (*d*) error

3. Which of the following statement is true about *switch* statement?
   (*a*) It may contain zero or more cases.
   (*b*) Constant expressions are valid case values.
   (*c*) Statement block of every case must have *break* statement as its last statement to avoid the control to fall through the later cases.
   (*d*) All of above.

4. If default statement is omitted in *switch* statement and there is no match with case values, then
   (*a*) No statement in the switch block is executed.
   (*b*) Execute all statements in the switch block.
   (*c*) Execute the statements in the last case block only.
   (*d*) A run time error occurs.

5. What will be the result of attempting to compile and run the following program?

```
#include <stdio.h>
void main()
{
    int c;
    printf( "\nEnter value of c: " );
    scanf( "%d", &c);
    switch(c);
    {
          case 1 : printf( "\nHello 1\n" );
                   break;
          case 2 : printf( "\nHello 2\n" );
                   break;
        default : printf( "\nInvaild value\n" );
                   break;
          case 3 : printf( "\nHello 3\n" );
                   break;
          case 4 : printf( "\nHello 4\n" );
    }
}
```

(*a*) Program will fail to compile because *default* can only appear at the end after all valid cases.

(*b*) Program will compile and execute.

(*c*) Program will fail to compile and compiler will report the syntax error as *"Case outside of switch statement in function main"* followed *"Misplaced break in function main"* for all cases and the *break* statements.

(*d*) None of above.

6. Consider the following program

```
#include <stdio.h>
void main()
{
    int a, b = 10;
    scanf( "%d", &a );
    if ( a = 0 )
        b *= 2;
    else
        b /= 2;
    printf( "%d  ", b );
}
```

What will be result of executing the above program with user input of number 6?

(*a*) 20 (*b*) 10
(*c*) 5 (*d*) None

7. Which of the following is not a valid type of expression in the *switch* statement?

(*a*) character (*b*) integer
(*c*) float (*d*) enum

8. What will be the output of following program?

```
void main() {
   char val=1;
   if(val--==0)
     printf("TRUE");
   else
     printf("FALSE");
}
```

(*a*) TRUE (*b*) FALSE
(*c*) Error (*d*) None of above

9. What will be the output of following program?

```
#define TRUE 1
void main()
{
    if(TRUE)
        printf("1");
        printf("2");
    else
        printf("3");
        printf("4");
}
```

(*a*) Error (*b*) 1
(*c*) 12 (*d*) 34

10. What will be the output of following program?

```
void main()
{
    int a=10;
    switch(a) {
        case 5+5:
            printf("Hello\n");
        default:
            printf("OK\n");
    }
}
```

(*a*) Hello (*b*) OK
(*c*) Hello
OK (*d*) Error

11. What will be the output of following program?

```
void main()
{
    int a=2;
    switch(a)
    {
        printf("Message\n");
        default:
            printf("Default\n");
        case 2:
            printf("Case-2\n");
        case 3:
            printf("Case-3\n");
    }
    printf("Exit from switch\n");
}
```

(*a*) Case–2 (*b*) Message
(*c*) Message
Case–2 (*d*) Case–2
Case–3
Exit from switch

12. What will be the output of following program?

```
void main()  {
     short day=2;
     switch(day) {
          case 2: || case 22:
               printf("%dnd",day);
          break;
          default:
               printf("%dth",day);
          break;
     }
}
```

(*a*) 2nd (*b*) 22nd
(*c*) Error (*d*) 2nd
22nd

13. What will be the output of following program?

```
void main()
{
        int a=2;
        switch(a) {
                case 1L:
                        printf("One\n"); break;
                case 2L:
                        printf("Two\n"); break;
                default:
                        printf("Else\n"); break;
        }
}
```

(*a*) One (*b*) Two
(*c*) Error (*d*) Else

14. What will be the output of following program?

```
#define TRUE 1
void main()    {
        switch(TRUE)   {
                printf("Hello");
        }
}
```

(*a*) Hello (*b*) No output
(*c*) Garbage value (*d*) Error

15. Can we use string value/variable in switch test condition?

```
void main()
{
      int x;
      float y = 5.5;
      switch(x=y+1)
      {
            case 6: printf("It's Eight."); break;
            default: printf("Oops No choice here.");
      }
}
```

(*a*) Oops No choice here. (*b*) It's Eight.Oops No choice here!!!
(*c*) It's Eight. (*d*) Error

REVIEW EXERCISE

1. If the *if* statement do not have associated *else*, then what happens if the condition evaluates to zero value?
2. What is the advantage of *switch* statement over *else-if* construct?
3. What does nesting mean?
4. Is any thing wrong with the following program?

```
void main()
{
        char ch;
        if ( ( ch = cin.get() ) == 'a' )
                printf("\nYou typed character a\n");
}
```

5. Where the *break* statement is used, and what is its function?

6. Rewrite the following code fragment using the *switch* statement:

```
char code;
code = getchar();
if ( code == 'A' )
    printf("\nAccountant\n");
else if ( code == 'C' || code == 'G' )
    printf("\nGrade IV\n");
else if ( code == 'F' )
    printf("\nFinancial Advisor\n");
else
    printf("\nIncorrect code\n");
```

7. Rewrite the following code fragment using the *if-else* statements:

```
int month;
scanf("%d", &month);
switch ( month )
{
   case 1 :
   case 3 :
   case 5 :
   case 7 :
   case 8 :
   case 10:
   case 12: printf("\nIt is a month having 31 days\n");
            break;
   case 4 :
   case 6 :
   case 9 :
   case 11: printf("\nIt is a month having 30 days\n");
            break;
   case 2:  printf("\nIt is a month having 28/29 days\n");
   default: printf("\nIncorrect month\n");
}
```

8. Rewrite the following code segments using the *switch* statement:

```
if ( ch == 'N' )
    north++;
if ( ch == 'S' )
    south++;
if ( ch == 'E' )
    east++;
if ( ch == 'W' )
    west++;
else
    unknown++;
```

```
if (ch == 'O')
    Outstanding++;
else if (ch == 'E')
    Excellent++;
else if (ch == 'G')
    Good++ ;
else if (ch == 'P')
    Poor++;
else
    Unknown++ ;
```

9. Write any two point of difference in operation in *switch* and *if- else*.
10. What is the effect of the absence of *break* statement in *switch* statement.

PROGRAMMING EXERCISE

1. Write a program to check whether the given natural number is EVEN or ODD without using any arithmetic operator.
2. Given measure of three line segments as *a*, *b*, and *c*. Write a program to check whether these line segments can be used to form a triangle or not.
3. Given measure of three angles of a triangle ABC as *a*, *b*, and *c* degrees, respectively, write a program to check whether a triangle can be formed with these angles or not.

4. Write a program, using switch statement, to check whether the given alphabet is a vowel or a consonant.
5. Given a measure of an angle in degrees and in anti-clockwise direction, write a program to find the type of the angle.
6. Given three points $A(x_1, y_1)$, $B(x_2, y_2)$ and $C(x_3, y_3)$, determine whether they are collinear, i.e., lie on the same line.
7. Given points (x_1, y_1) & (x_2, y_2) on line *AB*, and points (x_3, y_3) & (x_4, y_4) on line *CD*, write a program to determine whether lines AB & CD intersect each other.
8. The monthly telephone bill is to be computed as follows:

   Minimum Rs. 200 for upto 100 calls
   plus Rs. 0.60 per call for next 50 calls
   plus Rs. 0.50 per call for next 50 calls
   plus Rs. 0.40 per call for any call beyond 200 calls.

   Write a program to compute monthly bill for given number of calls.
9. Suppose the postal rates for mailing letters are as follows:

   Rs. 0.50 per 10 grams for the first 50 grams,
   Rs. 0.40 per 10 grams for the next 100 grams,
   Rs. 0.25 per 10 grams for the next 250 grams, and
   Rs. 25 per kilogram for letters weighing more than 400 grams.

   Write a program that prompts for weight of a letter and prints the postage to be paid.
10. Write a program to check whether the given date in format *dd/mm/yyyy* is valid or not.
11. Write a program to find the day of the week on a given date.
12. Write a program to find your age when your date-of-birth and today's date is given, both in the format *dd/mm/yyyy*.
13. Write a program to find convert a time in 12 hours system to 24 hours system.
14. Write a program to find convert a time in 24 hours system to 12 hours system.
15. Write a program to find difference in time when the start time and ending time is given, both in the format *hh:mm:ss*.
16. You know that an even number is a number which is divisible by 2. However there is another approach that can be used to check whether the given number is even or not. The approach is by inspecting the unit digit - *if the unit digit is 0, 2, 4, 6, or 8, the number is even*. So you are required to write a program to check whether the given number is even or not without performing the division by 2.

❑ ❑ ❑

Chapter 8

Decision Making and Looping

Learning Outcomes

After reading this chapter, students will be able to

- explain the working of *while* statement
- explain the working of *for* statement
- explain the working of *do-while* statement
- explain the nesting of looping statements
- explain the working of *break*, *continue* statements
- explain the usefulness of *else if* ladder in multiple alternatives in decision making
- use these statements to write programs to solve problems requiring iteration

8.1 INTRODUCTION

As discussed in the last chapter, a computer program is a set of instructions for a computer that are executed sequentially, *i.e.*, one after the other as they appear in a program.

However, following situations may arise:

- There may be situations when we need to alter this sequence by choosing a particular set of statements (*selection*) among two or more alternatives, based on certain criteria (*i.e.*, when certain conditions are satisfied/not satisfied) is met.
- There may also be situations when we need to repeat (*iteration*) certain group of statements over either fixed number of times or till certain criteria (*i.e.*, when certain conditions are satisfied/not satisfied) is met.
- There may also be situations when we need to transfer the control from one part of the program to another part (*jumping*) in order to skip certain instructions or go back to repeat certain instructions either depending on the satisfaction of certain condition (*conditional transfer*) or without the involvement of any condition (*unconditional transfer*).

This means, we may need to control the execution of the instructions.

In the previous chapter, we had discussed the control statements that allow us to perform the action of selection. This chapter focuses on the control statements that allow us to perform the action of iteration. In addition, the statements that allow us to perform the action of transfer of control within the program conditionally or unconditionally are also discussed.

8.2 THE *while* STATEMENT

The *while* statement is suited for problems where it is not known in advance that how many times a statement or a statement-block will be executed.

Te logic flow and the syntax of the while statement is shown in Figure 8.1. Here *expression* can be a constant, a variable or an expression. The *statement* is executed repeatedly till the *exp* evaluates to a *non-zero* value. Whenever *exp* evaluates to a *zero* value, the execution of *while* statement will terminate and control will pass to a statement immediately following it.

Though in C language there are certain exceptions, but in general, the following points must be kept in mind while using the *while* statement:

- There must be a statement prior to *while* statement that initializes the *exp*.
- In the statement-block, there must be a statement that modifies the expression.

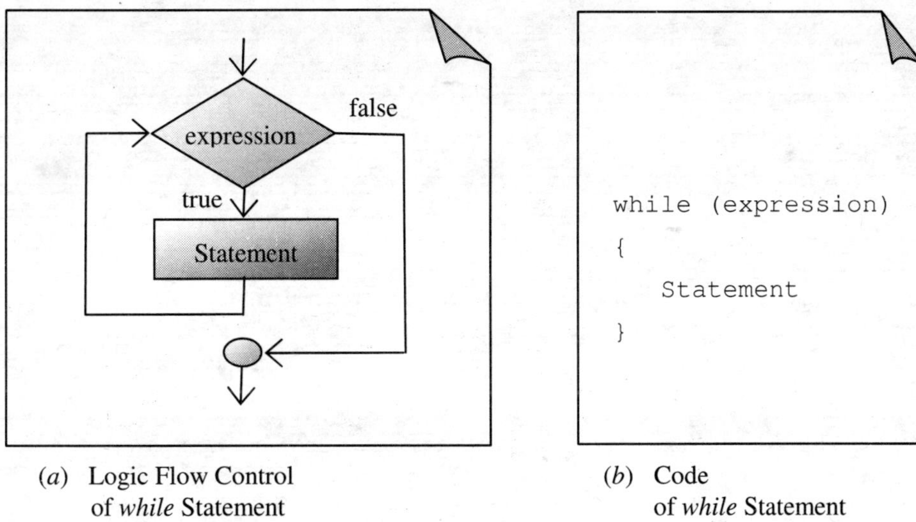

(*a*) Logic Flow Control of *while* Statement

(*b*) Code of *while* Statement

Figure 8.1: Logic flow control and code of *while* statement

To illustrate the use of *while* statement, let us develop some programs.

Listing 8.1

```
/* Program to find sum of first 100 positive integer numbers */
#include <stdio.h>
int main()
{
   int sum = 0, number;
   number = 1;
   while ( number <= 100 )
   {
      sum += number;
      number++;
   }
   printf( "Sum of first 100 positive " );
   printf( "integer numbers = %d\n", sum );
   return 0;
}
```

Test Run

```
Sum of first 100 positive integer numbers = 5050
```

Listing 8.2

```
/* Program to find sum of even and odd numbers from
*  first 100 positive integer numbers   */
#include<stdio.h>
int main()
{
   int sumOdd = 0, sumEven = 0, number;
```

```
    number = 1;
    while ( number <= 100 )
    {
       if ( ( number % 2 ) == 0 )
          sumEven += number;
       else
          sumOdd += number;
       number++;
    }
    printf( "Sum of even numbers = %d\n", sumEven );
    printf( "Sum of odd numbers  = %d\n", sumOdd );
    return 0;
}
```

Test Run

```
Sum of even numbers = 2550
Sum of odd numbers  = 2500
```

Listing 8.3

```
/* Program to find sum of digits of a positive
   integer number     */
#include<stdio.h>
int main()
{
   int sum = 0, digit;
   long number, temp;
   printf( "Enter any positive integer number: " );
   scanf( "%ld", &number );
   temp = number;
   while ( temp > 0 )
   {
      digit = temp % 10;
      temp /= 10;
      sum += digit;
   }
   printf( "\nSum of digits of %ld = %d\n", number, sum );
   return 0;
}
```

Test Run

```
Enter any positive integer number : 3265
Sum of digits of 3265 = 16
```

Listing 8.4

```
/*
* Program to count number of characters, vowels & words
* in a paragraph
*/

int main()
{
   int vowelCount=0,characterCount=0,wordCount=0;
   char ch;

   printf( "Type in the paragraph and terminate");
   printf( "\nby ENTER key\n\n" );

   while ( ( ch = getche() ) != '\r' )
   {
      characterCount++;
      if ( ( ch == ' ' ) || ( ch == '\t' ) )
         wordCount++;
      if ( (ch=='a')||(ch=='A')||(ch=='e')||
           (ch=='E')||(ch=='i')||(ch=='I')||
           (ch=='o')||(ch=='O')||(ch=='u')||
           (ch == 'U') )
        vowelCount++;
   }

   wordCount++;

   printf( "\nCharacter count = %d", characterCount);
   printf( "\nVowel count = %d", vowelCount);
   printf( "\nWord count = %d\n", wordCount);
   return 0;
}
```

Test Run

```
Type the paragraph and terminate by ENTER key

i enjoy programming in c++ like anything ↵

Character count = 40
Vowel count = 11
Word count = 7
```

8.3 THE *for* STATEMENT

The *for* statement is suited for problems where the number of times a statement or statement-block will be executed is known in advance.

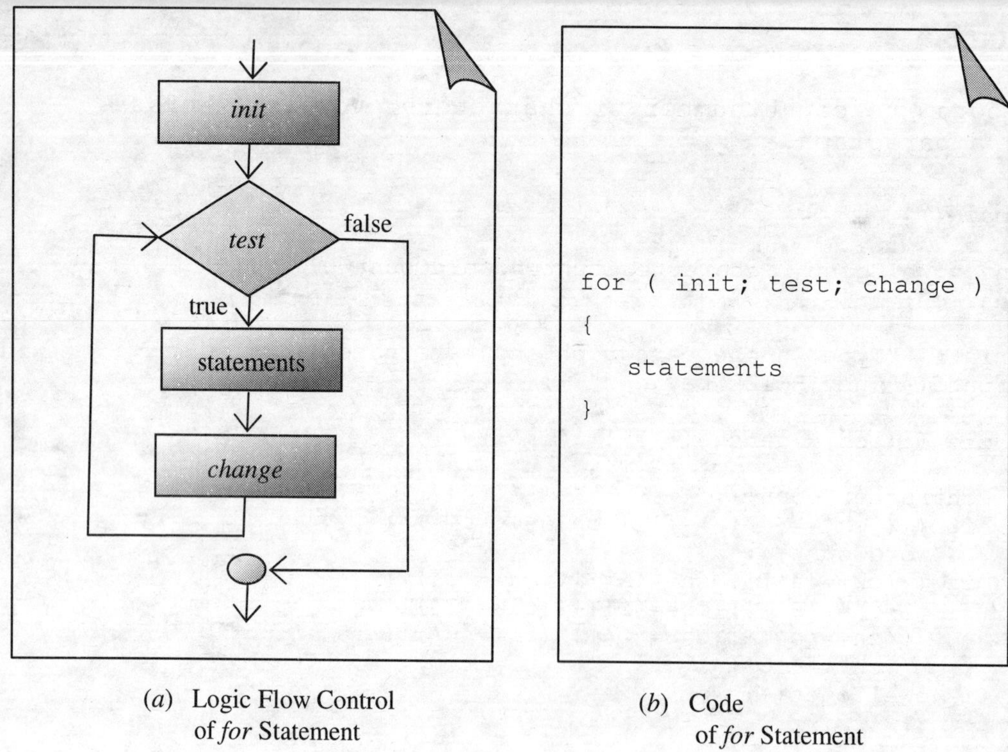

(*a*) Logic Flow Control of *for* Statement

(*b*) Code of *for* Statement

Figure 8.2: Logic flow control and code of *for* statement

where *init* is an expression to initialize the counter, the *test* is an expression to see when to stop iterating, and *change* is an expression to change the counter for each pass of the loop. The *init* and *change* parts can have more than one statement separated by a comma.

To demonstrate the use of *for* statement, let us develop some program making using of *for* statement.

Listing 8.5

```
/*
 *  Program to find sum of first 100 positive numbers using for loop
 */
#include<stdio.h>
int main(void)
{
   int i, sum = 0;
   for ( i = 1; i <= 100; i++ ) {
     sum = sum + i;
   }
   printf( "\nSum of first 100 positive\n" );
   printf( "integer numbers = %d\n", sum );
   return 0;
}
```

Listing 8.6

```
/* Program to find sum of even and odd numbers from first 100
  Positive numbers using 'for' loop    */
#include<stdio.h>
int main()
{
   int sumOdd = 0, sumEven = 0, number;
   for ( number = 0; number <= 100; number++ ) {
      if ( ( number % 2 ) == 0 )
         sumEven += number;
      else
         sumOdd += number;
   }
   printf( "\nSum of even numbers = %d\n", sumEven );
   printf( "Sum of odd numbers  = %d\n", sumOdd );
   return 0;
}
```

Listing 8.7

```
/*  Program to compute average of 'n' real numbers    */
#include<stdio.h>
int main()
{
   int i, n;
   float number, sum = 0, average;
   printf( "Enter value of n: " );
   scanf( "%d", &n );
   for ( i = 1; i <= n; i++ ) {
      printf( "Enter real number %d: ", i );
      scanf ( "%f", &number );
      sum += number;
   }
   average = sum / n;
   printf( "\nAverage of numbers = %.2f\n", average );
   return 0;
}
```

Test Run

```
Enter value of n: 8

Enter real number 1: 2.5
Enter real number 2: 12.0
Enter real number 3: 10.25
Enter real number 4: 8.75
Enter real number 5: 11.0
Enter real number 6: 20.25
Enter real number 7: -7.75
Enter real number 8: 9.0

Average of numbers = 8.25
```

8.4 THE *do* – *while* STATEMENT

The *do-while* statement, like *while* statement, is also suited for problems where it is not known in advance that how many times a statement will be executed.

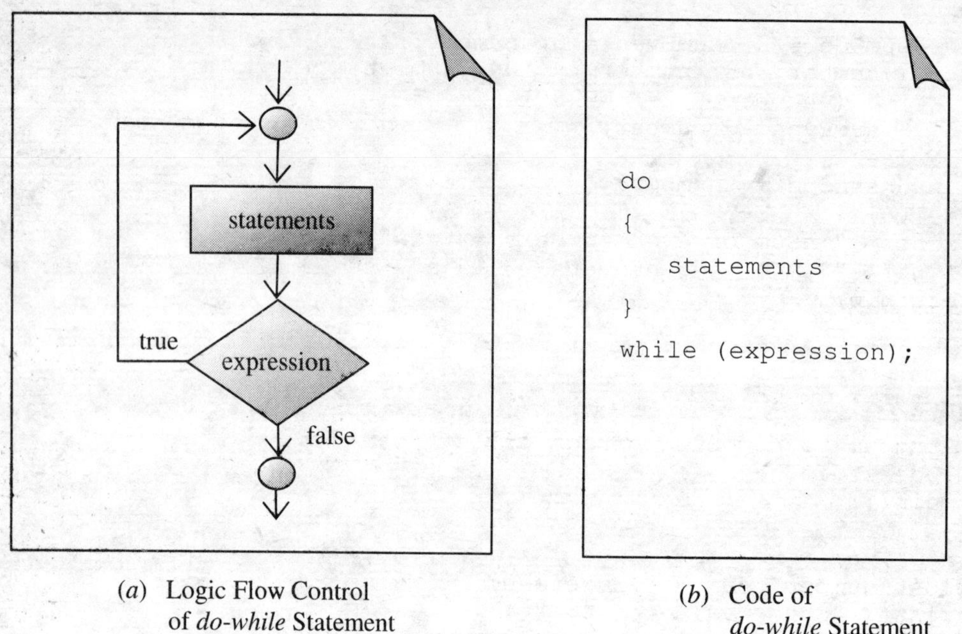

(*a*) Logic Flow Control of *do-while* Statement

(*b*) Code of *do-while* Statement

Figure 8.3: Logic flow control and code of *do-while* statement

where *expression* is a constant, a variable or an expression. The *statement-block* is executed repeatedly till the *exp* evaluates to a *non-zero* value.

To demonstrate the use of *do* – *while* statement, let us develop a program to find the average of unknown real numbers.

[illegible]

```
/*
*  Program to compute average of unknown real numbers
*/
#include<stdio.h>
int main()
{
   int n = 0;
   char yesno;
   float number, sum = 0, average;
   do {
      printf( "\nEnter number %d: ", n+1 );
      scanf( "%f", &number );
      n++;
```

```
        sum += number;
        printf( "\nAny more number [yn]?: " );
        yesno = getchar();
    }
    while ( ( yesno == 'y' ) || ( yesno == 'Y' ) );
    average = sum / n;
    printf( "\n\nAverage of given numbers = %.2f\n", average );
    return 0;
}
```

Test Run

```
Enter number 1: 2.5
Any more number [yn]?: y
Enter number 2: 12.0
Any more number [yn]?: y
Enter number 3: 10.25
Any more number [yn]?: y
Enter number 4: 8.75
Any more number [yn]?: y
Enter number 5: 11.0
Any more number [yn]?: n
Average of given numbers = 8.90
```

8.5 NESTED *while*, *for* and *do —while* STATEMENTS

Just as *if* statements can be nested, these statements can also be nested. The inner loop is executed from the beginning, for each iteration of the outer loop. The following sections of code show the nesting of looping statements within their own types.

```
for(i=0;i<m;i++)
{
    :

    for(j=0;j<n;j++)
    {
        :

    }
}
```

```
i=0;
while(i<m)
{
    :
    j=0;
    while(j<n)
    {
        :
        j++;
    }
    i++;
}
```

```
i=0;
do
{
    :
    j=0;
    do
    {
        :
        j++;
    } while(j<n);
    i++;
} while(i<m);
```

In general, it is possible to nest while and do-while statements inside *for* statement, *for* statement inside the *while* and *do-while* statements, *i.e.*, all sort of nesting combinations are permitted.

Listing 8.9

```
/* Program to print following pattern using nested for loops
         1
         2   2
         3   3   3
         4   4   4   4
         5   5   5   5   5
*/
#include <stdio.h>
int main()
{
    int i, j;
    printf("\n");
    for ( i = i; i < 5; i++ )
    {
        for ( j = 1; j <= i; j++ )
        {
            printf("%4d", i);
        }
        printf("\n");
    }
    return 0;
}
```

Listing 8.10

```
/* Program to print following pattern using nested for loops
         5   5   5   5   5
         4   4   4   4
         3   3   3
         2   2
         1
*/
#include <stdio.h>
int main()
{
    int i, j;
    printf("\n");
    for ( i = 5; i >= 1; i-- )
    {
        for ( j = 1; j <= i; j++ )
        {
            printf("%4d", i);
        }
        printf("\n");
    }
    return 0;
}
```

Listing 8.11

```
/* Program to print following pattern using nested for loops
        1
        1   2
        1   2   3
        1   2   3   4
        1   2   3   4   5
*/

#include <stdio.h>

int main()
{
    int i, j;
    printf("\n");
    for ( i = i; i < 5; i++ )
    {
        for ( j = 1; j <= i; j++ )
        {
            printf("%4d", j);
        }
        printf("\n");
    }
    return 0;
}
```

Listing 8.12

```
/* Program to print following pattern using nested for loops
        *
        *   *
        *   *   *
        *   *   *   *
        *   *   *   *   *
*/

#include <stdio.h>

int main()
{
    int i, j;
    printf("\n");
    for ( i = i; i < 5; i++ )
    {
        for ( j = 1; j <= i; j++ )
        {
            printf("   *");
        }
        printf("\n");
    }
    return 0;
}
```

Listing 8.13

```
/*
*  Program to demonstrate the nesting of 'for' statements
*/
#include <stdio.h>
#include <ctype.h>
int main()
{
   int j, i, jj;
   for ( i = 2; i <= 10; i++ ) {
      printf ( "\n\t\t   Table of %d", i );
      printf ( "\n\t\t   ==========\n" );
      for ( j = 1; j <= 10; j++ ) {
         jj = j + 10;
         printf("\n\t%3d x %2d = %3d\t%3d x %3d = %3d",
                i, j, i*j, i, jj, i*jj );
      }
      printf( "\n\n\tPress any key to continue...");
      getch();
   }
   return 0;
}
```

When you run this program, it will print the tables from 2 to 10, one table per screen. After printing one table, it will ask the user to press any key to continue.

Test Run

```
               Table of 2
               ==========

     2 x  1 =   2   2 x  11 =  22
     2 x  2 =   4   2 x  12 =  24
     2 x  3 =   6   2 x  13 =  26
     2 x  4 =   8   2 x  14 =  28
     2 x  5 =  10   2 x  15 =  30
     2 x  6 =  12   2 x  16 =  32
     2 x  7 =  14   2 x  17 =  34
     2 x  8 =  16   2 x  18 =  36
     2 x  9 =  18   2 x  19 =  38
     2 x 10 =  20   2 x  20 =  40

   Press any key to continue...
```

8.6 JUMPING STATEMENTS

These statements transfer the control from one part of the program to another part. In this section, we will see their working.

8.6.1 The *break* Statement

The *break* statement is always used inside the body of the *switch* statement, and looping statements.

```
switch ( expression )
{
    case val-1 : statement-1
                 break;
    case val-2 : statement-2
                 break;
    case val-3 : statement-3
                 break;
       :
       :
    case val-n : statement-n
                 break;
    default    :
                statement-d
}
```

Figure 8.4: Action of *break* statement in *switch* statement

In *switch* statement, it is used as the last statement of every case except the last one. When executed, it transfers the control out of *switch* statement and the execution of the program continues from the statement following *switch* statement.

In *for, while* and *do-while* statements, it is always used in conjunction with *if* statement. Note that it is never used with *if* statement if it is not part of the body of the looping statement. When executed, it transfers the control out of looping statement and the execution of the program continues from the statement following looping statement.

```
for(i=0;i<m;i++)
{
   :
   if (expression)
      break;
   :
}
```

```
i=0;
while(i<m)
{
   :
   if (expression)
      break;
   :
}
```

```
i=0;
do
{
   :
   if (expression)
      break;
   :
} while(i<m);
```

Figure 8.5: Action of *break* statement in *for, while* and *do-while* statements

In simple terms, we can say that the *break* when executed terminates the execution of the loop.

Consider the following code segment

```
   :
 i = 0;
 while ( i <= 100 )
 {
     . . . . . .
     scanf("%d", &k);
     if ( k == 0 )
         break;
     m = i / k;
      . . . . . .
 }
   :
```

The execution of the *while* loop will be terminated as soon as the user enters value 0 for variable *k* as input or the value of variable *i* exceeds 100, which ever happens earlier.

8.6.2 The *continue* Statement

The *continue* statement is always used inside the body of the looping statements. In looping statements, sometimes a situation may arise where we want that from a given statement onward, the rest of the statement up to the last statement of the loop are to be bypassed. This task is accomplished by using *continue* statement.

The *continue* statement transfers the control to the beginning of the next iteration of the loop thus bypassing the statements which are not yet executed. Note that the *continue* statement is always used in conjunction with the *if* statement.

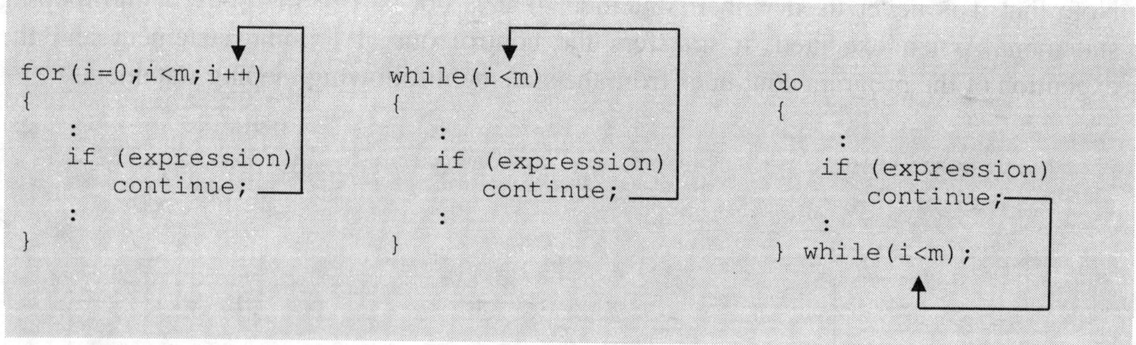

Figure 8.6: Action of *continue* statement in *for, while* and *do-while* statements

In simple terms, we can say that the continue statement when executed terminates the current iteration of the loop.

Consider the following segment

```
    :
i = 0;
while ( i <= 100 )
{
    . . . . . . .
    if ( i % 10 == 0 )
        continue;
    . . . . . .
}
    :
```

The use of *continue* statement in conjunction with *if* statement skips the rest of the statements of the *while* loop for values of *i* divisible by 10.

ILLUSTRATIVE EXAMPLES

Let us write a few more C programs to use the features that we have learned in this chapter.

Example 8.1: Program to find whether the given positive number *n* is a prime number or composite.

I hope you know that a positive integer number is said to be prime if it divisible by 1 and itself only, *i.e.*, it cannot be factorized. In addition, to this definition, an even number except 2, is not a prime number.

Therefore, our test criteria becomes

1. If *n* is greater than 2 and is even then *n* is not a prime number.
2. If test at step 1 fails, then we try to divide number *n* by factors $k = 3, 5, 7, \ldots \sqrt{n}$. Therefore, if *n* is divisible by any value of *k*, number *n* is not a prime number.
3. If test at step 2 also fails, then *n* is a prime number.

The following program implements this criterion.

Listing 8.14

```
/* Program to find whether given positive number 'n' is prime  */
#include<stdio.h>
#include<math.h>
int main()
{
    int n, k, m;
    printf( "\nEnter a positive integer number: " );
    scanf("%d", &n);
    if ( ( n > 2 ) && ( ( n % 2 ) == 0 ) ) {
        printf( "\n%d is not a prime number.\n", n );
    } else {
        m = sqrt( n );
```

```
        for ( k = 3; k <= m; k += 2 ) {
            if ( ( n % k ) == 0 )
                break;
        }
        if ( k > m )
            printf( "\n%d is a prime number.\n", n );
        else
            printf( "\n%d is not a prime number.\n", n );
    }
    return 0;
}
```

Test Runs

First Run

```
Enter a positive integer number: 43
43 is a prime number.
```

Second Run

```
Enter a positive integer number: 92
92 is not a prime number.
```

Example 8.2: Program to find the *n*th prime number.

We know that first prime number is 2, second prime number is 3, fourth prime number is 5, and so on.

Therefore, we start with number 2 and generate the subsequent numbers. On each number we apply primality test, if is passes the test, we count it. After every count, we see do we have reached the desired prime number, if yes we print that number and stop.

The following program implements this criterion.

Listing 8.15

```
/*  Program to find the nth prime number      */
#include<stdio.h>
#include<math.h>
int main()
{
    int n, k, m, count=0, num = 2;
    printf("Program to find nth prime number\n\n");
    printf("Enter positive value for n: ");
    scanf("%d", &n);
    while ( 1 )
    {
        if ( ( num > 2 ) && ( ( num %2 ) == 0 ) ) {
            num++;        /* move to next number, it is even    */
            continue;
        } else {
```

```
            m = (int) sqrt ( (double) n );
            for ( k = 3; k <= m; k+= 2 ) {
               if ( ( num % k ) == 0 ) {
                 num++;              /* move to next number,     */
                 break;              /* as it is divisible by k */
               }
            }
            if ( k > m ) {
               count++;                  /*  increase the count      */
               if ( count == n ) {      /* is it desired number     */
               {
                   printf("\n%dth prime number = %d\n", n, num);
                   return;
               }
               num++;                    /*  move to next number     */
            }
        }
    }
    return 0;
}
```

Test Run

```
Program to find nth prime number
Enter positive value for n: 10
10th prime number = 29
```

Example 8.3: Program to check whether the given positive integer number is an Armstrong number.

Note that a positive integer number is an Armstrong number if the sum of cubes of its digits equals the number itself. For example, number 153 is an Armstrong number as

$$1^3 + 5^3 + 3^3 = 1 + 125 + 27 = 153$$

Listing 8.16

```
/* Program to check whether the given positive integer
   number is an Armstrong number
*/
#include<stdio.h>
int main()
{
  int intNumber, intTempNumber, intSum = 0, digit;
  printf( "Enter positive integer number : " );
  scanf("%d", &intNumber);
  intTempNumber = intNumber;   /* create copy */
  while ( intTempNumber > 0 ) {
     digit = intTempNumber % 10;
     intTempNumber /= 10;
     intSum += digit * digit * digit;
  }
```

```
    if ( intSum == intNumber )
       printf("\n%d is an Armstrong Number\n", intNumber);
    else
       printf( "\n%d is not an Armstrong Number\n", intNumber);
    return 0;
}
```

Test Runs

First Run

```
Enter positive integer number: 153
153 is an Armstrong number
```

Second Run

```
Enter positive integer number: 120
120 is not an Armstrong number
```

Example 8.4: A Fibonacci sequence is defined as follows: the first and second terms in the sequence are 0 and 1. Subsequent terms are found by adding the preceding two terms in the sequence. Write a program to generate the first *n* terms of the sequence.

Listing 8.17

```
/*  Program to generate the first 'n' terms of Fibonacci sequence   */
#include<stdio.h>
int main()
{
   int previous = 0, current = 1, next, n, count;
   printf( "\nEnter value for n (>2) : " );
   scanf( "%d", &n );
   printf( "%d  %d", previous, current );
   count = 2;
   while ( count < n )
   {
       next = previous + current;
       printf( "%d  ", next );
       count++;
       previous = current;
       current = next;
   }
   printf( "\n" );
   return 0;
}
```

Test Run

```
Enter value for n : 10
0  1  1  2  3  5  8  13  21  34
```

Example 8.5: Write a program to generate all the prime numbers between *m* and *n*, where value of *m* and *n* is supplied by the user.

Listing 8.18

```
/*
* Program to generate all prime numbers between 1 and n, where
* value for n is supplied by the user
*/

#include<stdio.h>

int main()
{
   int i, m, n, k, flag;
   printf( "\nEnter value for m,n : " );
   scanf( "%d,%d", &m, &n );
   for ( i = m+1; i < n; i++ )
   {
      if ( ( i > 2 ) && ( ( i % 2 ) == 0 ) )
         continue;
      else
      {
         flag = 1;
         m = (int) sqrt ( (double) i );
         for ( k = 3; k <= m; k += 2 )
         {
            if ( ( n % k ) == 0 )
            {
               flag = 0;
               break;
            }
         }
         if ( flag == 1 )
            printf( "%d   ", i );
      }
   }
   printf( "\n" );
   return 0;
}
```

Test Runs

First Run

```
Enter value for n : 0

Incorrect value for n, re-run the program with correct value for n
```

Second Run

```
Enter value for n : 10

2  3  5  7  9
```

Example 8.6: Write a program to print a pyramid of numbers with *n* rows. An example pyramid with 5 rows is shown below:

```
                1
            2   3   2
        3   4   5   4   3
    4   5   6   7   6   5   4
5   6   7   8   9   8   7   6   5
```

Listing 8.19

```
/*  Program to print pyramid of numbers with 'n' rows    */
#include<stdio.h>
int main()
{
   int i, j, k, n;
   printf("\nEnter number of rows for the pyramid (n) : ");
   scanf("%d",&n);
   printf("\n");

   for ( i = 1; i <= n; i++ )
   {
      for ( j = 1; j <= n-i; j++ )
         printf("    ");

      k = i;

      for ( j = 1; j <= i; j++ )
         printf("%4d", k++);

      k = k-2;

      for ( j = 1; j < i; j++ )
         printf("%4d", k--);

      printf("\n\n");
   }
  return 0;
}
```

Test Run

```
Enter number of rows for the pyramid : 5

                1
            2   3   2
        3   4   5   4   3
    4   5   6   7   6   5   4
5   6   7   8   9   8   7   6   5
```

Example 8.7: Write a program to find greatest common divisor (GCD), of two positive integer *m* and *n*.

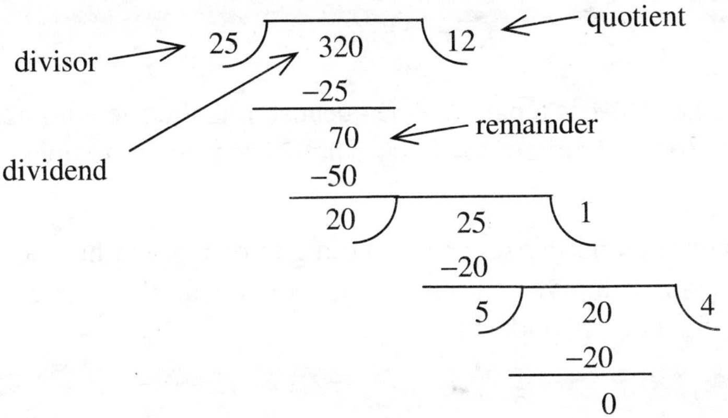

Figure 8.5: Illustration of computational procedure for GCD

From above figure, you must have observed that in successive divisions, the divisor of the previous division becomes dividend, remainder of becomes divisor, and the division is again carried out. This process is carried out till the reminder becomes zero, and the current divisor is taken as GCD of the given integer numbers.

This process can be implemented by using the following steps

1. Perform division.
2. If remainder is zero, then stop and take the divisor as HCF.
3. Replace dividend by divisor.
4. Replace divisor by remainder.
5. Repeat from step 1.

Listing 8.20

```
/*  Program to compute GCD of two positive Integer numbers  */
#include<stdio.h>
void main()
{
    int m, n, r;
    printf( "\nEnter value for m,n : " );
    scanf( "%d,%d",&m, &n );
    while ( 1 ) {
        r = n % m;
        if ( r == 0 ) {
            printf( "\nGCD = %d\n", n );
            return;
        }
        n = m;
        m = r;
    }
}
```

Test Run

```
Enter value for m,n : 35,155
GCD = 5
```

Example 8.8: The number 1991 is a palindrome because it is the same number when read forward or backward. Write a program to check whether the given number is palindrome or not.

Here, we first find a new number by reversing the digits of a given number and then we compare the given number with reversed number. If they match, then the given number is palindrome otherwise it is not a palindrome.

Listing 8.21

```
/* Program to check whether given number is palindrome or not   */
#include<stdio.h>
int main()
{
   int sum = 0, digit;
   int number, temp;
   printf( "\nEnter any positive integer number: " );
   scanf( "%d", &number );
   temp = number;
   while ( temp > 0 )
   {
      digit = temp % 10;
      temp /= 10;
      sum = sum * 10 + digit;
   }
   if ( number == sum )
      printf( "\n%d is a palindrome number.\n", number );
   else
      printf( "\n%d is not a palindrome number.\n", number );
   return 0;
}
```

Test Runs

First Run

```
Enter any positive integer number: 1991
1991 is a palindrome number.
```

Second Run

```
Enter any positive integer number: 1234
1234 is not a palindrome number.
```

MULTIPLE CHOICE QUESTIONS

1. Consider the following code:

```
while ( ++i <= n );
```

What will be the value of *i* when the loop completes, initial value of *i* being 1?

(*a*) n (*b*) n-1
(*c*) n+1 (*d*) n+2

2. Consider the following code. What is its output?

```
for ( i = 1;i <= 5; i++ ) {
   printf( "%d", i += 2 );
}
```

(*a*) 1 2 3 4 5 (*b*) 3 4 5 6 7
(*c*) 3 5 7 9 11 (*d*) None

3. Which of the following statement is true about *continue* statement?
   (*a*) It can be used in conjunction with *switch* statement.
   (*b*) It terminates the loop.
   (*c*) It terminates the current iteration.
   (*d*) None of above.

4. Which of the following statement is true about *switch* statement?
   (*a*) It may contain zero or more cases.
   (*b*) Constant expressions are valid case values.
   (*c*) Statement block of every case must have *break* statement as its last statement to avoid the control to fall through the later cases.
   (*d*) All of above.

5. Which of the following statement is true about *break* statement?
   (*a*) The *break* statement can be used with *if* statement to terminate the execution of its block.
   (*b*) The *break* statement can only be used with loop statements and *switch* statement.
   (*c*) The *break* statement terminates current iteration of the loop.
   (*d*) None of above

6. The *continue* statement is used to
   (*a*) continue the next iteration of the loop statement.
   (*b*) exit the block of loop statement .
   (*c*) exit from the outermost block even it is used in the innermost block.
   (*d*) continue execution of the program even error occurs.

7. Consider the following program

```
#include <stdio.h>
void main()
{
    int i = 3;
    while ( i > 0 ) {
        int i = 75;
        printf( "%d\n", i );
        i--;
    }
}
```

What will be the result of attempting to compile and execute above program?

(*a*) Program will fail to compile because variable *i* is duplicated.
(*b*) Program will fail to compile because variables can only be declared in the beginning of the program.
(*c*) Program will compile, execute successfully and will print 3, 2, 1 on different lines.
(*d*) Program will enter into infinite loop.

8. Consider the following program

```
void main()
{
    int k = 0;
    for ( ;; )
    {
        printf( "%d  ", k++ );
        if ( k > 5 )
            break;
    }
}
```

What will be the result of attempting to compile and execute above program?

(*a*) Program will fail to compile because of incorrect use of *for* statement.
(*b*) Program will fail to compile because expression cannot be included in list of variables for the *printf()* function.
(*c*) Program will compile and execute successfully and will print 0 1 2 3 4 5.
(*d*) None of above.

9. Minimum number of times the *do-while* loop will be executed?

(*a*) 0 (*b*) 1
(*c*) Cannot be predicted (*d*) None

10. Consider the following program

```
void main()
{
        int i, j;
        for ( i = 0; j = 10; i < j;  i++, j-- );
            printf( "x" );
}
```

How many times letter 'x' will be printed?

(*a*) 5 (*b*) 1
(*c*) 10 (*d*) 4

11. Consider the following program

```
void main()
{
        unsigned char ch;
        for ( ch = 0; ch <= 256; ch++ )
            printf( "%d = %c", ch, ch );
}
```

How many times *for* loop will be executed?

(*a*) 256 (*b*) 255
(*c*) 257 (*d*) Infinitely

12. Consider the following program

```
void main()
{
       int i;
       for ( i = 1; i <= 10; i++ )
       {
          if ( i % 5 )
             continue;
           else
              break;
          printf("\nKeep it up!");
       }
       printf( "%d  ", i );
}
```

How many times the message "Keep it up!" will be printed?

(*a*) 5 times (*b*) 10 times
(*c*) Will not be printed at all (*d*) Once

13. How many times the while loop will be executed in the following program?

```
void main()
{
    int j = 1;
    while ( j <= 100 );
    {
        printf("\nj = %d, j*j = %d", j, j*j );
        j++;
    }
}
```

(*a*) Infinite times (*b*) 100 times
(*c*) 99 times (*d*) 101 times

14. What will be the output of following program?

```
void main()
{
    char cnt=0;
    for(;cnt++;printf("%d",cnt)) ;
    printf("%d",cnt);
}
```

(*a*) 0 1 2 . . . infinite times (*b*) 0 1 2 . . . 127
(*c*) 0 (*d*) 1

15. What will be the output of following program?

```
void main()
{
    int a[5]={0x00,0x01,0x02,0x03,0x04},i;
    i=4;
    while(a[i])
    {
        printf("%02d  ",*a+i);
        --i;
    }
}
```

(*a*) 00 01 02 03 04 (*b*) 004 03 02 01 00
(*c*) 04 03 02 01 (*d*) 01 02 03 04

REVIEW EXERCISE

1. What happens if the expression of a *while* statement is initially false?
2. What is the difference between *while* and *do - while* statements?
3. When is the *do - while* statement preferred over *while* statement?
4. Can there be multiple increment expressions in increment part of the *for* statement?
5. When is the *while* statement preferred over a *for* statement?
6. Where the *break* statement is used, and what is its function?
7. What is the action of *continue* statement?
8. Differentiate between the working of *break* and *continue* statement.

PROGRAMMING EXERCISE

1. Two prime are said to be twin prime numbers if they differ by 2, such as 11 and 13, 17 and 19, etc. Write a program to print first *n* pairs of twin prime numbers.
2. Write programs to generate following patterns:

```
*
*   *
*   *   *
*   *   *   *
*   *   *   *   *
```

```
*   *   *   *   *
*   *   *   *
*   *   *
*   *
*
```

```
*
*   *
*   *   *
*   *   *   *
*   *   *   *   *
*   *   *   *
*   *   *
*   *
*
```

```
*   *   *   *   *
*   *   *   *
*   *   *
*   *
*
*   *
*   *   *
*   *   *   *
*   *   *   *   *
```

```
*                               *
*   *                       *   *
*   *   *               *   *   *
*   *   *   *       *   *   *   *
*   *   *   *   *   *   *   *   *
*   *   *   *       *   *   *   *
*   *   *               *   *   *
*   *                       *   *
*                               *
```

```
*   *   *   *   *   *   *   *   *
*   *   *   *       *   *   *   *
*   *   *               *   *   *
*   *                       *   *
*                               *
*   *                       *   *
*   *   *               *   *   *
*   *   *   *       *   *   *   *
*   *   *   *   *   *   *   *   *
```

3. Write programs to generate following patterns:

```
1
2   2
3   3   3
4   4   4   4
5   5   5   5   5
```

```
5   5   5   5   5
4   4   4   4
3   3   3
2   2
1
```

```
1
2 2
3 3 3
4 4 4 4
5 5 5 5 5
4 4 4 4
3 3 3
2 2
1
```

```
5 5 5 5 5
4 4 4 4
3 3 3
2 2
1
2 2
3 3 3
4 4 4 4
5 5 5 5 5
```

4. A factorial chart from 1 to 5 looks like the following:

```
1! = 1
2! = 1×2
3! = 1×2×3
4! = 1×2×3×4
5! = 1×2×3×4×5
```

   Write a program that will generate a factorial chart from 1 to *n* for any ≤ 9.
5. Write a program to print all prime numbers between *m* and *n*. The value for *m* and *n* will be supplied by the user.
6. Suppose an amount (say *amount*) is deposited in a commercial bank, which pays compound interest at the rate of *r*% annually, for *n* years. Write a program that prints the amount in account after each year.
7. A sample inverted pyramid of numbers with 5 rows is shown:

```
5 6 7 8 9 8 7 6 5
  4 5 6 7 6 5 4
    3 4 5 4 3
      2 3 2
        1
```

   Write a program to print an inverted pyramid of numbers with n rows, where the value of *n* will be supplied by the user at execution time.
8. A positive integer number IJK is said to be *well-ordered* if I < J < K. For example, number 138 is called *well-ordered* because the digits in the number increase from left to right, *i.e.*, 1 < 3 < 8. Number 365 is not *well-ordered* because 6 is larger than 5. Write a program that displays all possible three digit *well-ordered* numbers.
9. The numbers in the triangles below reveal a pattern for a *number triangle*.

```
1
2  6
3  7  10
4  8  11  13
5  9  12  14  15
```

Discover this pattern and write a program to generate *number triangles* for any size (number of rows) *n* between 1 and 9.

10. A perfect number is a number in which the sum of its proper divisors is equal to the number itself. Proper divisors of a number are all divisors of a number excluding itself.

    For example, 6 is a strong number, since 1, 2, 3 are its proper divisors, and 1+2+3=6. Write a program to find whether a given number 'n' is a perfect number or not.

11. A strong number is a number whose sum of the factorial of its each digit equals to the number itself.

    For example, 145 is a strong number since

    $$1! + 4! + 5! = 1 + 24 + 120 = 145.$$

    Write a program to find whether a given number 'n' is a strong number or not.

12. A 2-digit number is said to be a special number if the sum of the sum of its digits and the products of its digits is equal to the number itself.

    For example, 19 is a special number. The digits in 19 are 1 and 9. The sum of the digits is 10 and the product of the digits is 9, sum of sum of digits and product of digits is 19.

    Write a program to find all special numbers.

13. Consider the following series:

    1, 1, 2, 5, 4, 9, 8, 13, 16, 17, . . .

    If you note carefully, you will find that odd term represent GP with common ratio of 2, and even term represent AP with a common difference of 4.

    Write a program to generate first 'n' terms of this series when 'r' as common ratio of GP, and 'd' is the common difference of AP.

14. Consider the following series:

    1, 1, 1, 5, 2, 9, 3, 13, 5, 17, 8, 21, 13,. . .

    If you note carefully, you will find that odd term represent terms of Fibonacci sequence, and even term represent AP with a common difference of 4.

    Write a program to generate first 'n' terms of this series with 'd' as the common difference of AP.

15. Given a positive integer 'n'. Write a program to check whether this integer has an alternate pattern of 0 and 1 in its binary representation or not.

16. Write a program that counts the bits that are 1 in the binary representation of natural number 'n'.

17. Write a program that prints all the Armstrong numbers.
18. Write a program to find the second largest digits in a number. For example, if the given number is 12057, the output should indicate the 5 is the second largest digit in 12057.
19. Write a program to find the *n*th term of the Fibonacci sequence.
20. Write a program to generate the Fibonacci sequence upto the term not exceeding the given *limit*, where the value for the limit will be supplied by the user at run time.
21. Write a program to find the sum of first n natural numbers excluding those numbers that are multiples of 5.
22. Write a program to print the calendar of given month of a particular year. For example, if the month is October of year 2020, the output should be

```
        --- Calendar for the Month of October ---

        Sun  Mon  Tue  Wed  Thu  Fri  Sat
                              1    2    3
          4    5    6    7    8    9   10
         11   12   13   14   15   16   17
         18   19   20   21   22   23   24
         25   26   27   28   29   30   31
```

23. Write a program to print the calendar of entire given year (1700..2023).

❑ ❑ ❑

Chapter 9

User-Defined Functions

Learning Outcomes

After reading this chapter, students will be able to

- explain the concept of top-down design and its advantages
- use structure charts to depict the top-down design of their software solution
- explain the concept of local data and global data
- explain the way functions are declared, defined, and accessed
- explain the different ways in which arguments can be passed to a functions
- explain the use of *return* statement
- explain the concept of recursion, and will be able to write recursive functions
- explain and use command-line arguments
- explain the concept of storage classes and their usage

9.1 INTRODUCTION

The programs we have discussed so far were very simple and straightforward. They could solve problems that can be understood without much effort. As we move to larger and larger and more complex problems and hence programs, you will discover that it is not possible to understand all aspects of such problems and hence programs without some how reducing them to more elementary parts.

The best way to handle large and complex problems is to split the problem into small problems, called *subproblems* that can be handled easily. This we are used to do in our day-to-day working. Once all these sub problems are solved separately, solutions of these subproblems can be synthesized to generate the solution of the given complex problem.

Like wise, if program to be developed is very large and complex, it is always recommended to break into two or more functions (modules) that can be developed separately, tested separately and then integrated.

The planning for large and complex program starts with understanding the problem as a whole; then breaking it into smaller, simpler and understandable parts. We will call each of these parts of a program a *module* and the process of subdividing a problem into manageable parts *top-down design*.

The principles of top-down design and structured programming dictate that a program should be divided into main module and its related modules. Each module further be divided into submodules and this process should be continued until each submodule performs some specialized task and cannot be further subdivided. This process is known as *factoring*.

In top-down design, a program is divided into a main module and its related modules. Each module, in turn, is divided into submodules until the resulting modules are small enough, specialized and cannot be further subdivided.

Top-down design is usually done using a visual representation of the modules, known as *structure chart*. The structure chart shows the relation between each module and its submodules. The main module represents the entire set of code to solve the problem.

Moving down and left, we then read Module 1. One the same level with Module 1 are Module 2 and Module 3. The main module consists of three submodules. At this point, however, we are dealing only with Module 1. We now note that Module 1 is further subdivided into three modules, Module 1*a*, Module 1*b* and Module 1*c*.

The Main Module is known as the *calling* module because it has submodules. Each of its submodules is known as *called* module. Since Module 1, Module 2 and Module 3 also have submodules, they are also called calling modules; they are both *called* and *calling* modules.

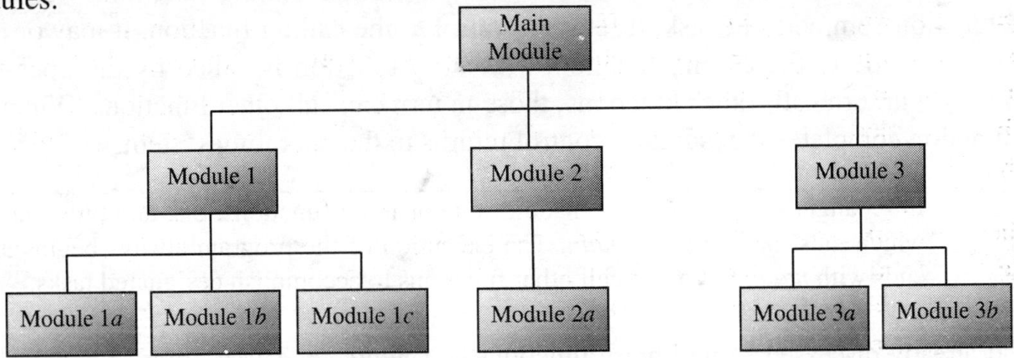

Figure 9.1: Structure Chart

Communication between modules in a structure chart is allowed only through a calling module. If Module 1 needs to send data to Module 2, the data must be passed through the calling module, Main Module. No communication can take place directly between modules that do not have a calling relationship.

The technique used to pass data to a module is known as *argument passing*. The arguments are contained in a list that is a definition of the data passed to the module by the caller. The list serves as the formal declaration of the data types and names.

9.2 FUNCTIONS IN C LANGUAGE

In C language, the idea of top-down design is done using functions. A C program is made of one or more functions, one and only one of which must be named as *main*. The execution of the program always begins and ends with *main*, but it can call other functions to accomplish designated tasks.

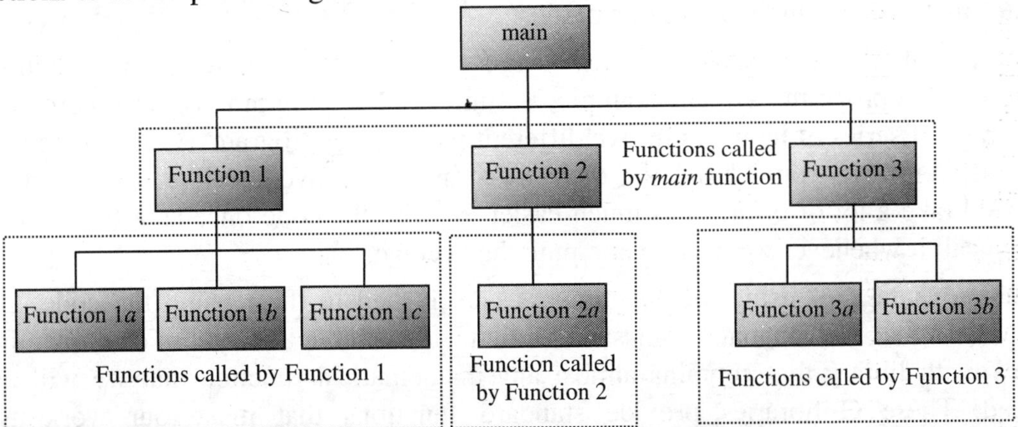

Figure 9.2: Structure Chart for a C Program

A function in C (including *main*) is an independent module that will be called to do a designated task. A *called function* receives control from a *calling function*. When the called function completes its task, it returns control to the calling function. It may or may not return a value to the calling function. The *main* function is called by the operating system; *main* in turn calls other functions, those in turn can call other functions. When the *main* function completes it execution, control returns to the operating system.

In C language, a program is made of one or more functions, one and only one of which must be named as *main*. The execution of the program always begins and ends with *main*, but it can call other functions to accomplish designated tasks.

We had already discussed some library functions in *Chapter 4*. These library functions are actually subprograms, which are written and incorporated into the C compiler because they are used in many situations. Rather than each user writes these subprograms, the C compilers provide these library functions as a convenience. Obviously, the C library cannot contain all the functions used in all application areas. However, it is possible for users to write their own functions whenever such a need arises.

In this chapter, we will study the way functions in C language are declared, defined and accessed, arguments are passed, data types of arguments is specified, the concept of recursion, and building your own library of user-defined functions.

9.3 ADVANTAGES OF USING FUNCTIONS

Several advantages are associated with the use of functions.

Some of the major advantages are

1. The use of functions allows managing large and complex problems by factoring them into small, simple and manageable parts.
2. The use of functions provides a way to reuse code that is required n more than one place in a program. As an example, assume that a program requires computing average of series of numbers in five different parts of the program. Each time the data is different. We could write the code to compute the average five times, but this would take a lot of effort. It is much easier to write the code once as a function and then call it whenever we need to compute the average.
3. This advantage of using the functions is closely tied to the reusing the code. Like many language, C language comes with a rich and valuable library. For example, there is a math library that contains almost any mathematical function that we will ever need. These C libraries provide standard functions that make our work as a programmer much easier. We can also create our personal libraries that make developing new systems much easier.

4. The use of functions can protect data. This is done using the concept of *local data.* Local data consist of data described in a function. This data is available only to the function and only while the function is executing. When the function is not executing, the data is not accessible. Data in one function, then, cannot be seen or changed by another function.

9.4 CONCEPT OF LOCAL DATA & GLOBAL DATA

The formal arguments and variables declared in the body of a function are unknown outside the function. The factor that determines which function recognizes a variable and which don't is called the *visibility* of variable. A local variable will be visible to the function in which it is declared, but not to others.

A local variable used in this way in a function is known as *automatic* variable, because it is automatically created when a function is called and destroyed when the function returns except variable having *static* storage class. The length or duration of time a variable lasts is called its *lifetime*.

In contrast to a local variable, a global variable is visible to more functions, and is declared outside all functions. The usual place of declaration of a global variable is in the beginning of the program file, *i.e.*, before all function definitions. In this case, it will be visible to all functions and exists till the program terminates. If the variable is not declared before all the functions, then it will be visible to only those functions following its declaration i.e. global in the restricted sense.

9.5 DECLARING AND DEFINING FUNCTIONS

Like every other data item in C, function also need to declared and defined. The syntax for function declaration and definition is illustrated in Figure 9.3 and Figure 9.4, respectively.

```
returnType functionName(formal argument list);
```

Figure 9.3: Syntax for function declaration

The function declaration, which needs to be done before the function call, gives the whole picture of the function that needs to be defined later. The declaration mentions the name of the function, the return type, and the type and order of formal arguments. In other words, the declaration uses only the header of the function definition ended in a semicolon.

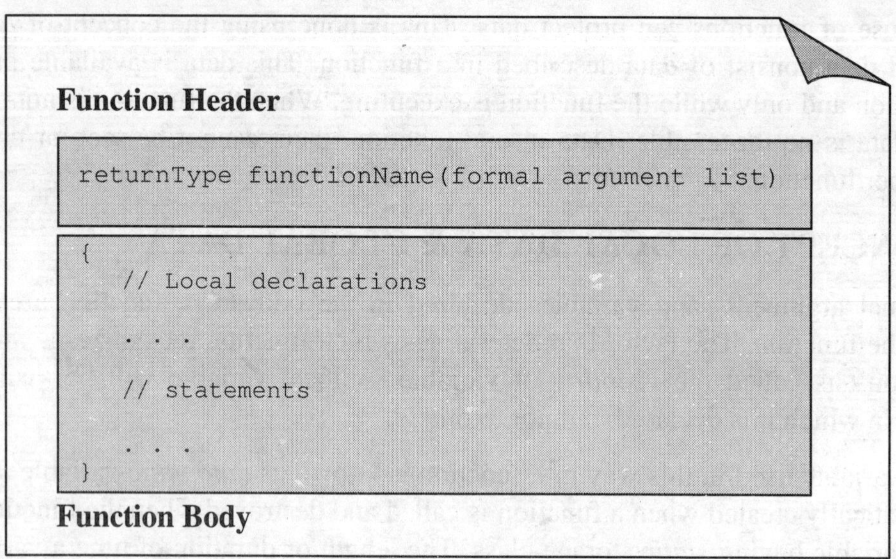

Figure 9.4: Syntax for function definition

The function definition, which is traditionally coded after the function that makes the call, contains the code needed to complete the task.

9.5.1 Declaring a Function

Function declaration consists only of a function header; it contains no code. Function header consists of three parts: *the return type*, *the function name*, and *the formal argument list*. A semicolon follows the function header.

If the function has no arguments, then write *void* in the parentheses. If the function has two or more arguments, each is separated by comma.

It is not mandatory (essential) to include the names of the arguments in the argument list. The purpose of the argument list is to tell the compiler *the number*, *data type* and *order of arguments*. However, readability and understandability are improved if the names of arguments are used. One point to note, however, is that the names of arguments do not need to be the same in the function declaration and the function definition. On the other hand, if the data types do not match, the compiler will flag an error. The compiler checks the data types of the arguments in the function call to ensure that they are same or at least compatible.

A function declaration statement is also known as *prototype* of the function or as *function prototype*.

The following function declaration

```
int larget(int a, int b, int c);
```

tells the compiler that the return type of the function is *int*, name of the function is *largest*, and it has three argument, all of type *int*. The names of argument are not significant.

The following function declaration

```
int larget(int, int, int);
```

is equally good and serves the purpose as earlier declaration

9.5.2 Defining a Function

Function definition consists two parts; *function header* and *function body.*

Function Header

Function header consists of three parts: *the return type*, *the function name*, and *the formal argument list*. A semicolon is not used at the end of function header. With the implementation of new standard of C (new standard is C99), the return type is mandatory. In the earlier implementations, if the return type was omitted in the function header, it was taken as *int*, by default. When nothing is to be returned, the return type *void* is used in the function header.

The formal argument list declares the variable that will receive the data from the calling function and may also contain variable that can be used to send the data back to the calling function.

Function Body

The function body contains the local declarations and the statements that define the task to be accomplished by the function. The body starts with local declarations that specify the variables needed by the function. After the local declarations, the function statements, followed by the *return* statement, are written.

If the return type of function is *void*, it can be written without *return* statement.

It is recommended that the default statements should be explicitly written for clarity. Therefore, even for a function with *void* as its return type, the return statement must be written.

SOME RULES FOR DEFINING FUNCTIONS

Following are some points to remember while defining functions in C:

- The first statement of the function must be the function defining statement specifying the return type, name and formal arguments.
- If the type of the function is omitted, then it is assumed of type integer (*int*). But it is recommended that the return type must be explicitly specified.
- Rules for naming a function are same as for variable names.
- The formal arguments should neither be constants or expressions.
- If a formal argument is an array name, it must be declared as an array in a type declaration statement before the body of the function starts and its maximum size must be equal to the actual size in the calling function.

As an example, let consider a task where we are concerned to find the largest of three integer numbers.

Listing 9.1

```
/*
*    Function that returns largest of three integer numbers (Ver 1)
*/
int largest(int x, int y, int z)
{
    if ( ( x > y ) && ( x > z ) )
        return x;
    else if ( y > z )
        return y;
    else
        return z;
}
```

The logic of the above function can also be written as shown below.

Listing 9.2

```
/*
*    Function that returns largest of three integer numbers (Ver 2)
*/
int largest(int x, int y, int z)
{
    int big;
    big = x;
    if ( y > big ) big  = y;
    if ( z > big ) big  = z;
    return big;
}
```

9.6 INVOKING/CALLING/ACCESSING A FUNCTION

Like library functions, the user-defined functions are accessed from a function simply by its name, including the actual arguments, if any, enclosed within parentheses. However, parentheses must follow the function name even if there is no actual argument to be passed to the function. The actual arguments, if any, must correspond in *number*, *type*, and *order* with formal arguments. The actual arguments can be constants, variable names, subscripted variables, or expressions.

When the name of the function is encountered, the control is transferred to the called function. The formal arguments are replaced by the actual arguments and the execution of the function is carried out. When the *return* statement is executed or last statement has finished its execution, the control is transferred back to the calling function.

The *largest()* function will be accessed/invoked/called using a statement of type

```
big = largest(a, b, c);
```

9.7 THE *return* STATEMENT

The syntax of *return* statement is

```
return;        or        return (exp);
```

where *exp* can be a constant, variable or expression. Use of parentheses around *exp* is optional. The first form is used with functions defined with return type as *void*.

The *return* statement serves two purposes:

- Execution of *return* statement immediately transfers control from the function back to the calling function.
- Whatever is following the *return* statement is returned as a value to the calling function.

In the functions described in Listing 9.2, the statement

```
return big;
```

return the value of variable *big*.

The *return* statement need not be at the end of the function. It can be used any where in the function. As soon as it is executed, the control will return to the calling function. A function can contain any number of *return* statements.

Consider the following function:

Listing 9.3

```
/* Function that returns value 1 if number is even else return 0  */
int oddOrEven( int number )
{
   if ( number % 2 == 0)
      return 1;
   else
      return 0;
}
```

This function returns value 1 if the *number* is not divisible by 2, *i.e.*, *number* is odd; otherwise it returns value 0. So here, different *return* statement will be executed depending on whether the *number* is odd or even.

There is key limitation of *return* statement — it can return only one value. If you want your function to return two or more values to the calling function, you need another mechanism.

9.8 PASSING ARGUMENTS TO A FUNCTION

The mechanism used to pass data to a function is via argument list, where individual arguments are called *actual arguments*. These arguments are enclosed in parentheses after the function name. The actual arguments must correspond in *number*, *type*, and *order* with formal arguments specified in the function definition. The actual arguments can be *constants*, *variables*, *array names*, or *expressions*.

There are two approaches to passing arguments to a function:

- Call by value
- Call by reference

Let us describe these one by one.

9.8.1 Call by Value

In this approach, the actual arguments are used in the function call. The actual argument can be a variable, a constant, or an expression. The values of the actual arguments are passed to the function. When control is transferred to the called function, the values of the actual arguments are substituted to the corresponding formal arguments and the body of the function is executed. If the called function is supposed to return a value, it is returned via *return* statement.

The following points must be noted about passing arguments using call by value mechanism:

1. The actual arguments can be constants, variables, or expressions.
2. When the control is transferred from the calling function to the called function, the memory for formal arguments and local variables is allocated, values of the actual arguments are substituted in the corresponding formal arguments, and the statements in the function body are executed.
3. As soon as the called function finishes its execution, the memory allocated for it is de-allocated, *i.e.*, the values of formal arguments and local variable are destroyed, and finally the control is transferred back to the calling function.
4. Any change made to the formal arguments will have no effect on actual arguments, since the function will only be using the local copy of the arguments.

Following function definition illustrates the mechanism of passing arguments by value.

```
void swap( int a, int b )
{
    int temp;
    temp = a;
    a = b;
    b = temp;
}
```

The above function will be accessed as

```
swap(x,y);
```

where *x* and *y* are actual arguments in the calling function.

Listing 9.4

```
/* Program to illustrate the passing of arguments by value. It calls a
 * function swap() that interchanges values of arguments.
 */
#include <stdio.h>
void swap(int a, int b);      /* declaration of function */
int main()
{
    int x, y;
    printf("\nEnter value for x : ");
    scanf("%d", &x);
    printf("\nEnter value for y : ");
    scanf("%d", &y);
    printf("\nBefore calling swap function\n");
    printf("\nValue of x = %d, Value of y = %d\n", x, y);
    swap(x,y);  /* function call   */
    printf("\nAfter returning from swap function\n");
    printf("\nValue of x = %d, Value of y = %d\n", x, y);
    return 0;
```

```
}
void swap( int a, int b )
{
    int temp;
    temp  = a;
    a = b;
    b = temp;
}
```

Test Run

```
Enter value for x : 10
Enter value for y : 12
Before calling swap function
Value of x = 10, Value of y = 12
After returning from swap function
Value of x = 10, Value of y = 12
```

You must have observed that any change made to the formal arguments have not effect on the actual arguments.

9.8.2 Call by Reference

In this approach, the addresses of the actual arguments are used in the function call. In this way the addresses of the actual arguments are passed to the function. When control is transferred to the called function, the addresses of the actual arguments are substituted to corresponding formal arguments and the body of the function is executed. The formal arguments are declared as pointers to *types* that match the data types of the actual arguments. This approach is of practical importance while passing arrays and structures among function, and also for passing back more than one value to the calling functions.

If the called function is supposed to return a value, it is returned via *return* statement. Additional values can be returned to the calling function via formal arguments which will be of pointer type.

Following points mu· noted about passing arguments using call by reference mechanism:

1. The actual arguments can only be variables.
2. When the control is transferred from the calling function to the called function, the memory for formal arguments and local variables is allocated, addresses of the actual arguments are substituted in the corresponding formal arguments, and the statements in the function body are executed.

3. As soon as the called function finishes its execution, the memory allocated for it is de-allocated, *i.e.*, the values of formal arguments and local variable are destroyed, and finally the control is transferred back to the calling function.
4. Any change made to the formal arguments will have immediate effect on actual arguments, since function will be working on actual arguments through pointers.

Following function illustrates mechanism of passing arguments by reference (address).

```
void swap( int *a, int *b )
{
   int temp;
   temp   = *a;
   *a = *b;
   *b = temp;
}
```

The above function will be accessed as

```
swap(&x,&y);
```

where *x* and *y* are actual arguments in the calling function.

Listing 9.5

```
/* Program to illustrate the passing of arguments by reference. It
*  calls a function swap() that interchanges values of arguments.
*/
#include <stdio.h>
void swap(int *a, int *b);    /* declaration of function */
int main()
{
  int x, y;
  printf("\nEnter value for x : ");
  scanf("%d", &x);
  printf("\nEnter value for y : ");
  scanf("%d", &y);
  printf("\nBefore calling swap function\n");
  printf("\nValue of x = %d, Value of y = %d\n", x, y);
  swap(&x,&y);  /* function call    */
  printf("\nAfter returning from swap function\n");
  printf("\nValue of x = %d, Value of y = %d\n", x, y);
  return 0;
}
void swap( int *a, int *b )
{
  int temp;
  temp   = *a;
  *a = *b;
  *b = temp;
}
```

Test Run

```
Enter value for x : 10
Enter value for y : 12
Before calling swap function
Value of x = 10, Value of y = 12
After returning from swap function
Value of x = 12, Value of y = 10
```

You must have observed that any change made to the formal arguments is reflected back in the actual arguments.

Next program is another example to demonstrate the passing of arguments using call by reference.

Listing 9.6

```
/* Program that calls function example() to find the average of
*  two numbers, largest & smallest of these numbers. The
*  average is retuned via return statement while largest and
*  smallest numbers are returned via formal arguments
*/
#include <stdio.h>
/* function declaration  */
float example(float x, float y, float *big, float *small);
int main()
{
   float x, y, avg, larger, smaller;
   printf("\nEnter value for x : ");
   scanf("%f", &x);
   printf("\nEnter value for y : ");
   scanf("%f", &y);
   avg = example(x,y,&larger,&smaller); /*  function call  */
   printf("\nAverage of %.2f and %.2f is %.2f\n", x, y, avg);
   printf("\nLarger number is %.2f\n", larger);
   printf("\nSmaller number is %.2f\n", smaller);
   return 0;
}
/*
   Function to compute the average of two numbers and find the
   largest and smallest of these numbers
*/
float example(float x, float y, float *big, float *small)
{
    float average;
    average = ( x + y ) / 2;
    if ( x > y )
    {
        *big = x;
        *small = y;
```

```
    }
    else
    {
        *big = y;
        *small = x;
    }
    return average;
}
```

Test Run

```
Enter value for x : 10.5
Enter value for y : 12.5
Average of 10.50 and 12.50 is 11.50
Larger number is 12.50
Smaller number is 10.50
```

9.8.3 Comparison between Call by Value and Call by Reference

Table 9.1 highlights the key points of differences regarding the prototype, definition and function call.

Table 9.2 highlights the key points of differences regarding the nature of actual arguments, nature of formal arguments and the impact of any change/modification of formal arguments on actual arguments.

Table 9.1

Call by Value	Call by Reference
Function Prototype: `void swap(int a, int b);`	**Function Prototype:** `void swap(int *a, int *b);`

Function Definition:
`void swap(int a, int b)`
`{`
`    int temp;`
`    temp = a;`
`    a = b;`
`    b = temp;`
`}`
	Function Definition:
	`void swap(int *a, int *b)`
	`{`
	`    int temp;`
	`    temp = *a;`
	`    *a = *b;`
	`    *b = temp;`
	`}`

Function Call: `swap(x, y);`	**Function Call:** `swap(&x, &y);`

Table 9.2

Call by Value	Call by Reference
Actual arguments can be constants, variables, or expressions.	Actual arguments can only be variables.
Formal arguments are ordinary variables.	Formal arguments are pointer variables.
Values of actual arguments are substituted in formal arguments.	Addresses of the actual arguments are substituted in formal arguments.
Any change made to formal arguments will have no effect on actual arguments, since the function will only be using the local copy of the arguments.	Any change made to pointer variables will have immediate effect on actual arguments, since the function will be working on actual arguments through address.

ILLUSTRATIVE EXAMPLES

To have hands on defining functions, let use consider some examples.

Example 9.1: An integer function to compute the factorial of a positive integer number *k*.

Listing 9.7

```
/*
 *    Function to compute factorial of a +ve integer number
 */

int factorial ( int k )
{
   int prod = 1, i;
   for ( i = 2; i <= k; i++ )
      prod *= i;
   return prod;
}
```

Example 9.2: Using the definition of factorial function of Example 9.1, let us write a complete program to compute the Binomial coefficient ${}^{n}c_{r}$ defined as

$$ {}^{n}c_{r} = \frac{n!}{r!(n-r)!} $$

Listing 9.8

```
/*  Program to compute Binomial coefficient (Version 1)  */
#include <stdio.h>
int factorial ( int k );  /*  function prototype  */
int main()
{
    int ncr, n, r;
    printf( "\nEnter value of n: " );
```

```
    scanf( "%d", &n );
    printf( "\nEnter value of r: " );
    scanf( "%d", &r );
    ncr = factorial(n)/(factorial(r)*factorial(n-r));
    printf( "\nValue of ncr = %d\n" , ncr);
    return 0;
}
```

This program calls the factorial function thrice with argument *n*, *r*, (*n*–*r*) respectively, and then using these values computes the binomial coefficient.

Example 9.3: A function to find highest common factor (HCF), also known as greatest common divisor (GCD), of two positive integers *m* and *n*.

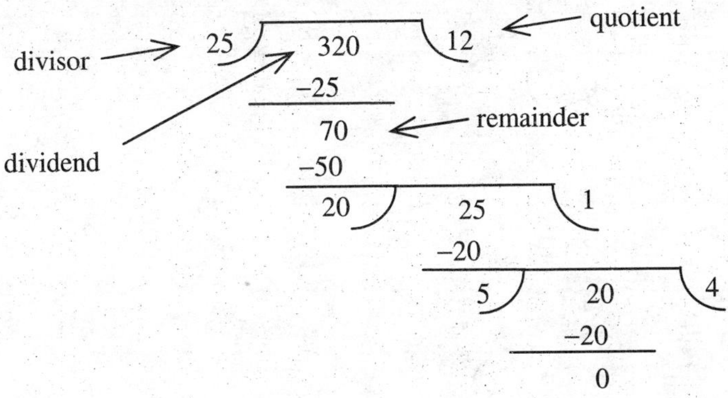

Figure 9.5: Illustration of computational procedure for HCF

From Figure 9.5, you must have observed that in successive divisions, the divisor of the previous division becomes dividend, remainder of becomes divisor, and division is again carried out. This process is carried out till the reminder becomes zero, and the current divisor is taken as HCF of the given integer numbers.

This process can be implemented by using the following steps

1. Perform division.
2. If remainder is zero, then stop and take the divisor as HCF.
3. Replace dividend by divisor.
4. Replace divisor by remainder.
5. Repeat from step 1.

Listing 9.9

```
/*  Function to compute HCF of two +ve integer numbers    */
int computeHCF ( int m, int n )
{
    int r;
    while ( 1 )
```

```
    {
        r = m % n;
        if ( r == 0 )
            return n;
        n = m;
        m = k;
    }
}
```

Example 9.4: Using the definition of function to compute HCF of Example 9.3, let us write a complete program to compute the HCF of two given integers.

Listing 9.10

```
/* Program to compute HCF of two given integers (m and n)  */
#include <stdio.h>
int computeHCF( int, int );   /*  function prototype */
int main()
{
    int m, n, hcf;
    printf( "\nEnter value of m: " );
    scanf( "%d", &m );
    printf( "\nEnter value of n: " );
    scanf( "%d", &n );
    hcf = computeHCF(m,n);
    printf( "\nValue of HCF = %d\n" , hcf);
    return 0;
}
/*  Function to compute HCF of two +ve integer numbers   */
int computeHCF ( int m, int n )
{
   int r;
   while ( 1 )
   {
      r = m % n;
      if ( r == 0 )
         return n;
      n = m;
      m = k;
   }
}
```

Test Run

```
Enter value of m: 125
Enter value of n: 35
Value of HCF = 5
```

9.9 RECURSIVE FUNCTIONS

Recursion is an important concept in computer science and is convenient for a variety of problems that would be difficult to solve using *iterative* constructs such as *for, while* and *do* loops. Recursive functions can be directly implemented in C language.

9.9.1 Factorial Function

The factorial of a positive number n, written as $n!$, is the product of the positive integers from 1 to n:

$$n! = 1\times2\times3\times\ldots\times(n-2)\times(n-1).n \qquad \text{where as } 0! = 1$$

Thus, definition of the iterative version of the factorial function can be written as

$$n! = \begin{cases} 1 & \text{if } n = 0 \\ \prod_{i=1}^{n} i & \text{if } n > 0 \end{cases}$$

From above definition, we have

$$\begin{aligned} 0! &= 1 \\ 1! &= 1 \\ 2! &= 1\times2 = 2 \\ 3! &= 1\times2\times3 = 6 \\ 4! &= 1\times2\times3\times4 = 24 \\ 5! &= 1\times2\times3\times4\times5 = 120 \end{aligned} \qquad \text{and so on.}$$

Observe that

$$4! = 4\times3! = 4\times6 = 24 \quad \text{and} \quad 5! = 5\times4! = 5\times24 = 120$$

And this is true for every positive integer n; that is

$$n! = n\times(n-1)!$$

Thus the factorial function can be defined as

$$n! = \begin{cases} 1 & \text{if } n = 0 \\ n\times n-1)! & \text{if } n > 0 \end{cases}$$

This definition of $n!$ is recursive, since it refers to itself when it uses $(n-1)!$.

Listing 9.11

```
/*
*   Recursive function to compute factorial of a number
*/
int factorial( int n )
{
  if ( n == 0 )
     return 1;
  else
     return ( n * factorial( n - 1 ) );
}
```

Listing 9.12

```
/*
*  Program to print the factorials of first 'n' natural numbers
*/
#include <stdio.h>
int factorial( int n );      /*  function prototype   */
int main()
{
   int i, n;
   printf( "\nEnter value of n: " );
   scanf( "%d", &n );
   for ( i = 1; i <= n; i++ )
   {
      printf( "\n%d! = %d", i, factorial(i) );
   }
   return 0;
}

int factorial( int n )
{
  if ( n == 0 )
     return 1;
  else
     return ( n * factorial( n - 1 ) );
}
```

Test Run

```
Enter value of n: 6
1! = 1
2! = 2
3! = 6
4! = 24
5! = 120
6! = 720
```

9.9.2 Fibonacci Numbers

A very important sequence, *Fibonacci sequence,* usually denoted by $F_0, F_1, \ldots, F_n$, is as follows:

0, 1, 1, 2, 3, 5, 8, 13, . . .

That is, $F_0 = 0$ and $F_1 = 1$ and subsequent terms are sum of the two preceding terms.

For example, the next term in the above sequence is

$8 + 13 = 21$

A formal definition for Fibonacci number F_n is:

$$fib(n) = \begin{cases} n & \text{if } n \leq 1 \\ fib(n-1) + fib(n-2) & \text{if } n > 1 \end{cases}$$

Notice that the recursive definition of the Fibonacci numbers differs from the recursive definitions of the factorial function and positive exponential power function in the sense that it refers to itself twice.

Listing 9.13

```
/*  Recursive function to compute Fibonacci number 'n'   */
int fibNumber( int n )
{
   if ( n <= 1 )
      return n;
   else
      return ( fibNumber(n-1) + fibNumber(n-2) );
}
```

Listing 9.14

```
/* Program to print first 'm' terms of Fibonacci sequence  */
#include <stdio.h>
int fibNumber( int n );          /*  function prototype    */
int main()
{
    int i, m;
    printf( "\nEnter value of m: " );
    scanf( "%d", &m );
    for ( i = 0; i < m; i++ )
    {
        printf( "%d ", fibNumber(i) );
    }
    return 0;
}
```

```
int fibNumber( int n )
{
  if ( n <= 1 )
     return n;
  else
     return ( fibNumber(n-1) + fibNumber(n-2) );
}
```

Test Run

```
Enter value of m: 8
0   1   1   2   3   5   8   13
```

9.9.3 Greatest Common Divisor

Computation of greatest common divisor (*GCD*) or highest common factor (*HCF*) of two natural numbers *m* and *n* is demonstrated in Figure 9.5, which is reproduced here.

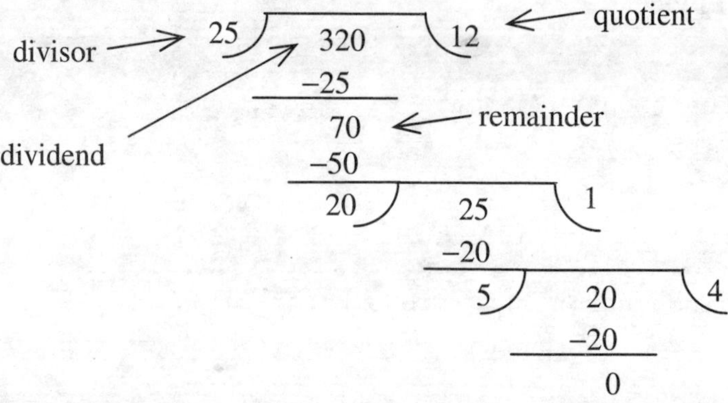

Figure 9.6: Illustration of computational procedure for *GCD*

You must have observed that in successive divisions, the divisor of the previous division becomes dividend, remainder of becomes divisor, and division is again carried out. This process is carried out till the reminder becomes zero, and the current divisor is taken as *GCD* of the given integer numbers.

The G*CD* function can be formally defined in the recursive manner as

$$\text{gcd}(m,n) = \begin{cases} n & \text{if } m \,\%\, n = 0 \\ \text{gcd}(n, m\%n) & \text{if } m \,\%\, n \neq 0 \end{cases}$$

Listing 9.15

```
/*  Program to compute GCD of two natural numbers m and n */
int gcd( int m, int n );    /*  function prototype */
void main()
{
    int m, n;
    printf( "\nEnter value of m : " );
    scanf( "%d", &m );
    printf( "\nEnter value of n : " );
    scanf( "%d", &n );
    k = gcd( m, n );
    printf( "\nValue of GCD(%d,%d) = %d\n", m, n, k );
}
int gcd( int m, int n )
{
    if ( m % n == 0 )
        return n;
    else
        return gcd( n, m % n );
}
```

Test Run

```
Enter value of m: 35
Enter value of n: 125
Value of GCD(35,125) = 5
```

9.10 COMMAND-LINE ARGUMENTS

To run a program, the name of the program is entered at the operating system's prompt. In addition to entering the name of the program, we can also enter other items, such as the name of the file the program may be using. For example, you can start Turbo C's IDE in the following ways:

1. By just typing TC at DOS prompt as

   `C:\TC>TC↵`

   In this case, the opening screen will appear, and you can load any file into the environment by using *File* option.

2. By typing TC and the name of the program file, to work on, as

   `C:\TC>TC test↵`

   In this case, if program file *test.c* already exists, it will be loaded into the environment, and the editing session will start. If program file *test.c* does not exist, then the environment will start the edit session to create a new file.

In the later case, the string *test* is used as a *command-line argument — an argument that appears on the command line following the operating system prompt*. We can give multiple arguments on the command line. For example, the COPY command needs two arguments, where first argument is the name of the source file and the second argument is the name of the destination file.

The question here is — *How can these arguments be used in our program*? As you know that the execution of the program always begins with the *main()* function. Accordingly, these arguments can be accessed via arguments in the *main()* function. The C system automatically adds the capability to read these arguments to all C programs. The following program illustrates this.

Listing 9.16

```
/* Program to demonstrate the use of command line arguments
 *  filename : cmdline.c
 */
# include <stdio.h>
int main ( int argc, char *argv[] )
{
   int i;
   printf( "\nCommand = %s\n", argv[0] );
   printf( "\nNumber of arguments = %d", argc-1 );
   printf( "\n\nThese are\n\n" );
   for ( i = 1; i < argc; i++ ) {
        printf( "Argument number %d = %s\n", i, argv[i] );
   }
   return 0;
}
```

Test Run

```
>cmdline first second third fourth↵
Command = C:\TC\CMDLINE.EXE
Number of arguments = 4
These are
Argument number 1 = first
Argument number 2 = second
Argument number 3 = third
Argument number 4 = fourth
```

Two arguments used in the definition of function *main()* are:

- First argument is *argc* that is of type integer and holds total number of entries on command line, including the program name.

- Second argument is **argv[]*, an array of pointers to strings which holds individual arguments. Argument *argv[0]* holds the name of the program, *argv[1]* holds first argument, *argv[2]* holds second argument and so on. These arguments can also be accessed using pointer notation as **(argv+0)*, **(argv+1)*, **(argv+2)* and so on.

If we have written a program that accepts a specified number of command-line arguments, then we must check in the program whether the correct number of arguments is passed to it or not.

For example, suppose we have written a program that emulates DOS COPY command, and is given a name KOPY. Then we can add the following segment in very beginning of the program.

```
if ( argc < 3 )
{
    printf("\nCorrect Usage:>" );
    printf("\nKOPY <sourceFile><destinationFile>\n" );
    return;
}
```

to ensure that the correct number of command-line arguments as expected by the program are specified.

Note that it is not mandatory that names of formal arguments for the main function be *argc* and *argv*, only thing you have to ensure that is — first argument is of type *int* and second argument is array of pointers of type *char*.

9.11 STORAGE CLASSES REVISITED

Every variable in C language possesses a characteristic called its *storage class.* The storage class defines two main characteristics of the variable, apart from the place of storage and the initial value. These are

- Lifetime
- Visibility or scope

Lifetime — *Lifetime* of a variable is the length of time it retains a particular value.

Visibility — *Visibility* or *scope* of a variable refers to those parts of a program that will be able to recognize it.

At first instance, a question arises: *why are storage classes necessary*? The answer is that by using variables with the appropriate lifetime and visibility, we can write programs that use memory more efficiently, run faster, and are less prone to programming errors. The correct use of storage class is especially important in large programs.

The storage class precedes the variable's declaration and tells the compiler regarding where the variable should be stored, what should be its initial value, what will be its scope, and for how much time it will be available for use.

The C language supports following four storage classes:

- Auto
- Static
- register
- extern

If the storage class in the variable declaration is omitted, it is assumed of type *auto*, *i.e.*, storage class is *auto* by default. Let us examine these storage classes one by one.

9.11.1 The *auto* Storage Class

The variables declared with *auto* storage class, which is the storage class by default, have following features:

- Stored in memory.
- If not initialized in the declaration statement, their initial value is unpredictable, which is often called *garbage value.*
- Local (visible) to the block in which the variable is declared. If the variable is declared within a function, then it is only visible to that function. But if the variable is declared outside the function, then it is visible to all the functions following its declaration.
- It retains its value till the control remains in the block in which the variable is declared. For example, if the variable is declared within a function, it will retain its value till the control is within that function. As the execution of the function terminates, it is cleared/destroyed.

The following examples illustrate these features.

Listing 9.17

```
/* Program to illustrate the auto storage class   */
#include <stdio.h>
int main()
{
   auto int i, j = 10;
   printf( "Value of i = %d\n", i );
   printf( "Value of j = %d\n", j );
   return 0;
}
```

This example program shows how an automatic storage class is declared, and the fact that if the variable is not initialized, it contains a garbage value.

Test Run

```
Value of i = 1250
Value of j = 10
```

Next time you run this program you may get different value of *i*. The reason is that we have not initialized variable *i* prior to outputting it, so it can be assigned any garbage value depending on the storage location set aside for it. Most of the times, for un-initialized variables, we will be getting value zero. I would suggest our readers to check this fact.

Consider the following program.

Listing 9.18

```
/*  Program to illustrate the auto storage class
 *  when variable is declared outside all the blocks
 */
#include<stdio.h>
int main()
{
  auto int j = 10;
  {
     {
        {
           printf( "Value of auto variable j in" );
           printf( " innermost block = %d\n", j );
        }
        printf( "Value of auto variable j in" );
        printf( " next innermost block = %d\n", j );
     }
     printf( "Value of auto variable j in" );
     printf( " outermost block = %d\n", j );
  }
  return 0;
}
```

Test Run

```
Value of auto variable j in innermost block = 10
Value of auto variable j in next innermost block = 10
Value of auto variable j in outermost block = 10
```

In this program, all the *printf()* statements output value 10, the reason being that all the *printf()* statements occur within the outermost block in which the variable *j* is declared.

Consider the following program.

Listing 9.19

```
/*  Program to illustrate the auto storage class
 *  when variable is declared in different blocks    */
#include <stdio.h>
int main()
{
   auto int j = 10;
   {
      {
         auto int j = 50;
         printf( "Value of auto variable j in" );
         printf( " innermost block = %d\n", j );
      }
      printf( "Value of auto variable j in" );
      printf( " outer block = %d\n", j );
   }
   return 0;
}
```

Test Run

```
Value of auto variable j in inner block = 50
Value of auto variable j in outer block = 10
```

Note that the compiler treats these *j*s as different variables, since they are defined in different blocks, although, within the same function (main function in this case). As soon as the execution of the innermost block is complete, the variable *j* loses its value 50. And in the outer loop the variable *j* refers to value 10. And finally, when the execution of the function terminates, the variable *j* also loses its value.

9.11.2 The *static* Storage Class

The variables declared with *static* storage class have the following features:

- Stored in memory.
- If not initialized in the declaration statement, their initial value is zero.
- Local (visible) to the block in which the variable is declared. If the variable is declared within a function, then it is only visible to that function. But if the variable is declared outside the function, then it is visible to all the functions following its declaration.
- It retains its value between different function calls. That is, when for the first time a function is called, a static variable is created with initial value zero, and in subsequent calls it retains its present value.

Consider the following program.

Listing 9.20

```
/*
 *   Program to illustrate the auto storage class
 *   when variable is declared in different blocks
 */
#include <stdio.h>
int sumOfDigits(int n);    /*   function prototype    */
int main()
{
   int sumOfDigits(int);
   int a, b;
   a = sumOfDigits(125);
   b = sumOfDigits(125);
   printf("%d, %d", a, b);
   return 0;
}
int sumOfDigits(int n)
{
  static int sum = 0;
  while ( n > 0 )
  {
     sum = sum + n % 10;
     n = n / 10;
  }
  return sum;
}
```

Test Run

```
8, 16
```

You may be expecting answer "8, 8" since we are attempting to find the sum of digits of the same number 125.

The local variable *sum* is declared with *static* storage class. Further, lifetime of a static variable is program execution time, hence a static variable is able to retain its existing value between different calls. Therefore, on the first call, the *sum* variable is initialized to 0 and at return time has value 8. In the second call, the initial value of *sum* is 8, and further 8 (sum of digits of 125) is added thus giving value 16.

9.11.3 The *register* Storage Class

Register storage class is like *auto* in the sense that the visibility and lifetime of the variable is limited to the block in which it is declared. The difference is that the compiler, if possible, will assign a register variable to one of the processor's register instead of storing it in the memory. Value stored in register can be accessed much faster than the value stored in memory. But do remember that there is limited number of registers in a processor, and secondly not all variables can be register variables. Most of the compilers (especially Turbo C compiler) only guarantee registers storage of type *int* and pointer types.

Now the question arises, *what variables should be made register variables*? The obvious candidates are those variables on which the program spends most of its time to process them such as index variables used of *for loop*.

The variables declared with *register* storage class have the following features:

- Stored in processor registers, if a register is available. If no register is available, then variable is stored in memory, and works as if its storage class is *auto*. As soon as register becomes available, storage class of that variable is converted to register type.
- If not initialized in declaration statement, their initial value is unpredictable, *i.e.*, *garbage*.
- Local (visible) to the block in which the variable is declared. If the variable is declared within a function, then it is only visible to that function. But if the variable is declared outside the function, then it is visible to all the functions following its declaration.
- It retains its value till the control remains in the block in which the variable is declared. For example, if the variable is declared within a function, it will retain its value till the control is within that function. As the execution of the function terminates, it is cleared/destroyed.

9.11.4 The *extern* Storage Class

The scope of external variables is global. External variables are declared outside all functions, *i.e.*, in the beginning of the program file. As all the function will be defined after this variable declaration, so these variables are visible to all functions in a program file. However, if we have written our functions in separate program files, then we have to declare that variable again with *extern* storage class to tell the compiler that this variable or these variables are external variables and are defined in other source file(s).

The variables declared outside all the functions (if all functions are in single file) or declared with *extern* storage class (if you have multi-file programs) have the following features:

- Stored in memory.
- If not initialized in the declaration statement, their initial value is zero.
- Global, *i.e.*, visible to all functions.
- Retains its values till the program execution terminates.

Consider the following programs stored in separate files as shown:

Listing 9.21

```
/* file : extern1.c
 *  Program to illustrate the extern storage class.
 *  A global variable is defined here.  */
#include <stdio.h>
int commonVariable = 10;     /* definition of global variable */
int main()
{
   void sample();                    /* prototype of function */
   printf("\nValue of global variable in main function=");
   printf("%d\n", commonVariable);
   sample();
   return 0;
}
```

Listing 9.22

```
/* file : extern2.cpp
 *  Function that uses global variable declared in other file.
 *  A global variable is defined here.   */
#include <stdio.h>
extern int commonVariable; /* declaration of global variable */
void sample()
{
   printf("\nValue of global variable in sample function=");
   printf("%d\n", commonVariable);
}
```

Test Run

```
Value of global variable in main function = 10
Value of global variable in sample function = 10
```

Table 9.3 summarizes lifetime and visibility of various types of variables. The first three types are local; those defined within a function. Their visibility is confined to that function, and their lifetime — except for static variables — is also the same as that of the functions. These first three types of variables can also be defined within a block (a pair of braces); in this case, they will only be visible within block.

Table 9.3: Lifetime and visibility of various types of variables

Place of Declaration	Storage class	Lifetime	Visibility
Function	auto	Function	function
function	register	function	function
function	static	program	function
file	static	program	one file
file	not specified	program	one file
file	extern	program	multi-file

The next three types are defined in the beginning of the file, *i.e.*, external to any function. In the first case, the variable is confined to a single file. The *extern* keyword cannot be used to reference it from other files. We use this storage class when we wish to hide the variables in one file.

In second case, which is most common, the variable is defined externally but is not declared in any other files. The visibility of such variables is confined to a single file. In this case the variable is visible only from the place in the file where it is defined to the end of the file. It is invisible to functions occurring before it even in the same program file.

In third case, a variable is defined in one file, but also declared in other files. In this case, its visibility will extend to the appropriate parts of all the files in which it is declared.

9.11.5 Advantages of Storage Classes

By using the appropriate storage class for a variable, the programmer can economize on memory requirements by restricting the lifetime of the variables. It will also help to maintain a modular, well-structured program by allowing access to variables only to those files and functions that need it. The number of ways in which the visibility and lifetime of variables can be restricted is an important reason for C's popularity with developers of large programs and systems.

ILLUSTRATIVE EXAMPLES

Examples 9.5: Write a function, say *int smallestDigit(int x)*, that returns the smallest digit in number *x*. Use minimum number of variables and minimum number of statements. Don't use array at all. Use this function in a program to demonstrate its use.

The procedure to find the smallest digit is as follows: We know that the largest digit is 9 and the smallest digit is 0. We start with taking 9 as the current smallest digit. In each iteration, we separate a digit and compare it with current smallest digit. If digit under consideration is less than current smallest digit, we take digit under consideration as current smallest digit. This process is repeated till all digits in a number have been examined.

Listing 9.23

```
/*
 * Program to find smallest digit in a positive integer number
*/
#include <stdio.h>
int smallestDigit(int); /* prototype   */
int main()
{
    long int number;
    printf( "\nEnter positive integer number : " );
    scanf( "%ld", &number);
```

```
    printf( "\nSmallest digit in %ld is %d\n", number,
            smallestDigit(number));
    return 0;
}
/* function that returns smallest digit in a +ve integer number  */
int smallestDigit(int k)
{
    int smallDigit = 9;
    int digit;
    while ( k > 0 )
    {
       digit = k % 10;
       k /= 10;
       if ( digit < smallDigit )
           smallDigit = digit;
    }
    return smallDigit;
}
```

Test Run

```
Enter positive integer number : 23145
Smallest digit in 23145 is 1
```

Examples 9.6: Write a program that takes a list of numbers as its command line arguments, and prints the sum of these numbers.

The procedure to be followed is as follows: The command line arguments are received by the *main()* function as string. Since the command line arguments are just numbers, therefore the string are made up of only letters 0-9 and decimal point '.'. We use here a library function *atof()* to convert a string of representing a number to numeric value. Then we add these numeric values one by one after conversion. We need to include *stdlib.h* header file in order to use *atof()* library function.

Listing 9.24

```
/*  Program to add list of numbers given as command line arguments
 *  filename : sum.c        */
# include <stdio.h>
# include <stdlib.h>
void main ( int argc, char *argv[] )
{
   int i;
   float sum = 0;
   for ( i = 1; i < argc; i++ ) {
      sum = sum + atof( argv[i] );
   }
   printf("\nSum of numbers on command line = %.2f\n", sum);
}
```

Test Run

```
>sum 1 2 3 4 5 6 7 8 9 10
Sum of numbers on command line = 55.00
```

MULTIPLE CHOICE QUESTIONS

1. Which of the following statement about functions is false?
   (*a*) More than one function is allowed in a program unit.
   (*b*) A function can call another function.
   (*c*) A function can call itself.
   (*d*) Constants can appear in the formal argument list.
2. Which of the following is not a valid reason for using functions?
   (*a*) They use less memory than repeating the same code.
   (*b*) They keep different program activities separate.
   (*c*) They run faster.
   (*d*) They keep variables safe from other parts of the program
3. What is the default return type of a function?
   (*a*) void (*b*) int
   (*c*) float (*d*) char
4. The program execution starts from the
   (*a*) *main()* function (*b*) function which is defined first
   (*c*) function that is defined last (*d*) None of above
5. The C language allow arguments to be passed
   (*a*) only call by value (*b*) only call by reference
   (*c*) both *a* & *b* (*d*) None of above
6. Consider the following program

```
void main()
{
      int i = 5, j = 10;
      swap(i, j);
      printf( "\n%d, %d", i, j );
}
swap(int i, int j) {
    i = i + j; j = i - j; i = i - j;
}
```

   What will happen when we attempt to compile & run the program?
   (*a*) Program will fail to compile because same variable names cannot be used in *main()* and *swap()* functions.
   (*b*) Program will fail to compile because return type of the swap function is not specified.
   (*c*) Program will compile and execute giving output as 5, 10.
   (*d*) Program will compile and execute giving output as 10, 5.

7. Identify the correct statement
   (*a*) A function can be defined more than once in a program.
   (*b*) One function cannot be defined within another function definition.
   (*c*) All functions must be in the same file.
   (*d*) Function should appear in the order they are called in the main function.
8. Consider the following function definition, and identify the correct statement

```
myFunction( a + b, c, d, 5 )
int a, b, c, d;
{
      int sum;
      sum = a + b +  c + d + 5;
      return sum;
}
```

   (*a*) Function will fail to compile because expression is not permitted as formal argument.
   (*b*) Function will fail to compile because constant is not permitted as formal argument.
   (*c*) Function will fail to compile because return type of the function is omitted.
   (*d*) Both (*a*) and (*b*).
9. The process of dividing a program into functions which in turn are further divided into function until they consists of elementary operations and cannot be further divided is known as
   (*a*) Structuring (*b*) Charting (*c*) Factoring (*d*) Splitting
10. Which of the following is not a valid definitions for the *main()* function so that it can compile and execute successfully?
   (*a*) `main(void)  { /* . . .  body . . .   */ }`
   (*b*) `void main(int count, char *args[] )`
   `{ /* . . .  body  . . .  */  }`
   (*c*) `void main(void)  {  /*  . . .  body  . . .  */  }`
   (*d*) `int main(float count, char *argv[])`
   `{  /* . . .  body  . . .  */  }`
11. Which of the following is a part of the function header?
   (*a*) Function name (*b*) Return type
   (*c*) Argument list (*d*) All
12. Consider the following declaration statement

```
extern int a;
```

   What information does it convey?
   (*a*) Variable *a* is declared in some other source file.
   (*b*) Variable *a* has its initial value as 0.
   (*c*) Variable a can retain its value during different function calls.
   (*d*) None of above.

13. Which of the following statement is true about *register* storage class?
   (*a*) Variable always stored in processor registers.
   (*b*) Can only be used with integral data types and pointers.
   (*c*) Variable takes initial value 0 if not initialized.
   (*d*) None of above.
14. Find the output of the following program.

```
void main()
{
    f();
    f();
    f();
}
void f(void)
{
    static int x =0;
    printf("%d ", ++x);
}
```

   (*a*) 0 1 2 (*b*) 1 1 1 (*c*) 1 2 3 (*d*) 3 3 3
15. Find the output of the following program.

```
void main()
{
    static int i =0;
    i++;
    if ( i <= 5 ) {
       printf("%d ",i);
       main();
    } else {
       return;
    }
}
```

   (*a*) 5 4 3 2 1 (*b*) 1 2 3 4 5 (*c*) 1 1 1 1 1 (*d*) None of above

REVIEW EXERCISE

1. What is the difference between a user-defined function and library function?
2. When is the execution of the function terminated?
3. How many arguments can be passed to a function?
4. What are the rules regarding the relationship between formal arguments and actual arguments?
5. How is a function invoked?
6. How many times can a function be called?
7. What is wrong with the following function definition?

```
testFunction( int k ) {
    float temp = 5.25;
    temp = k / 2.0;
    return temp;
}
```

8. Find the errors, if any, in the following function definition

```
void fun( int x, int y ) {
    int z;
    /*  some statement  */
    return z;
}
```

9. Find the errors, if any, in the following function definition

```
void fun( int x, y )
{
    int z;
    /*  some statement  */
    return;
}
```

10. Find the errors, if any, in the following function definition

```
int fun1( int x, int y ) {
    int z;
    /*  some statement  */
    int fun2( int t ) {
        return (t-2);
    }
    /*  some more statements  */
    return z;
}
```

12. What will be the output of the following program?

```
void main() {
    float r = 5.0, c;
    float func( float t );      /* function prototype   */
    c = func( r );
    printf( "\nValue returned by function" );
    printf( " = %d\n", c);
}
float func( float t ) {
    float temp;
    temp = t * t / 2;
    return 5.0;
}
```

13. Find the errors, if any, in the following function declarations:

(*a*) `int diff(int x, y);`

(*b*) `void fun(void, void);`

(*c*) `float diff(float, float);`

(*d*) `int test(float x, int y)`

PROGRAMMING EXERCISE

1. Write a function, say *int sumOfDigits(int number)*, that returns the sum of digits of a positive integer number.
2. Write a function, say *float absValue(float value)*, that returns the absolute value of a number of type *float*.
3. Write a function, say *int round(float x)*, that returns rounded value of x to nearest integer value.
4. Write a function, say *int sumOfN(int n, int m)*, which computes the sum of n integers starting with mth integer, *i.e.*, $m + (m + 1) + (m + 2) + . . . + (m + n - 1)$.

5. The combination function *C(n, r)* can be defined recursively as *C(n–1,r) + C(n–1,r–1)*. What is the base case? Write a recursive program to compute *C(n, r)*.
6. Write a function, say *int isPrime(int n)*, that returns 1 if number *n* is prime else returns 0.
7. A number is a palindrome if it is the same number when read forward or backward. For example, numbers 1991, 1001 and 1221 palindrome numbers. Write a function, say *int isPalindrome(unsigned int k)*, that returns value 1 if the number *k* is palindrome else returns value 0.
8. A positive integer number IJK is said to be *well-ordered* if I < J < K. For example, number 138 is called *well-ordered* because the digits in the number (1, 3, 8) increase from left to right, *i.e.*, 1 < 3 < 8. Number 365 is not *well-ordered* because 6 is larger than 5. Write a function, say *int isWellOrdered(unsigned int k)*, that returns value 1 if the *k* is a *well-ordered* number else returns 0.
9. Write a function, say *int largestDigit(int x)*, that returns the largest digit in number *x*.
10. Write a function, say *int unitDigit(int n)*, that returns the unit digit of a number represented in argument *n*.
11. Write a function, say *int isLeapYear(int year)*, that returns value 1 if the argument *year* represents a leap year else return value 0.
12. Write a recursive function, say *int sumOfDigits(int n)*, that returns sum of first *n* natural numbers. Using this function, write a program to find the sum of first *n* natural numbers.
13. Write a recursive function, say *int decToBinay(int n)*, that prints binary number equivalent to given decimal number *n*. Using this function, write a program to convert a decimal number to binary number.
14. Write a function, say *int isValidDate(int dd, int mm, int yyyy)*, that returns value 1 if the date is valid else returns value 0.
15. Write a function, say *int dayNumber(int dd, int mm, int yyyy)*, that returns day of week as number (0..6) for the given date.
16. Write a function, say *int isAlternateBits(int n)*, that returns value 1 if the binary representation of *n* comprises of alternate bits else returns value 0.
17. Write a function, say *int isTrianglePossible(int a, int b, int c)*, that returns value 1 if the triangle can be passed using *a*, *b*, and *c* as its sides else returns value 0.
18. Write a recursive function, say *void printInReverse(int n)*, that prints number *n* in reverse order.
19. Write a function that takes month and year as its arguments, and returns the number of days in the given month.
20. Write a function, say *int countDigits(int n)*, that return the number of digits in *n*, *i.e.*, size of number *n*.

❑ ❑ ❑

Chapter **10**

Arrays and Strings

Learning Outcomes

After reading this chapter, students will be able to

- explain the concept of arrays
- explain the usefulness of arrays in handling class of related problems
- declare, initialize and do input/output of arrays
- explain the way arrays are passed as arguments to functions
- write efficient programs using arrays to solve variety of problems
- comprehend the handling of string as array of characters
- declare, initialize and perform I/O of string data
- explain the use of various library function for manipulating strings
- explain the way strings are passed to functions as arguments
- write efficient programs for manipulating strings as a part of overall problem

10.1 INTRODUCTION

The variables used so for are also called *scalar* variables or *non-subscripted* variables. Each such variable represents a memory cell/location in which a single value can be stored. However, there are many situations in which it is necessary to process a collection of identical/homogeneous values. To illustrate this, consider that, for a class of 50 students, we have with us score of each student in final exam, and we are interested in having the following information:

- maximum score,
- average score of the class, and
- number of students whose score is greater than class's average score

One way of solving the above problem is

- use 50 variables, where each variable represents individual score,
- input these 50 scores,
- use about 50 *if* statements to find the maximum of these scores,
- add these 50 variables scores to find the total score of the class
- divide the total score by 50 to get the average score
- use another set of about 50 *if* statements to compare the individual scores with the average score to find the scores greater than average score and count such scores
- output the results

If at later stage, we want to process the scores of students of another class with number of students less or more than 50, the above program have to be modified as the number of variables will vary in the resulting program. This change in number of variables will affect the other parts of the program. The readers must find out themselves those parts that are affected.

Thus the above approach of processing a collection of homogeneous data is too cumbersome and not flexible enough to handle the collection of data of varying size.

The C language, like other high-level languages, provides a convenient way of processing such collections of homogeneous data. This is using *arrays* or *subscripted* variables.

In this chapter, we describe the processing of arrays and strings. Strings, which basically are array of characters with null character ('\0') as the last character, are also described.

10.2 ARRAYS

An *array* is a collection of homogeneous data elements (i.e. of same data type) described by a single name, and each individual element of an array is referenced by a *subscripted* variable, formed by affixing to the array name a *subscript* or *index* enclosed in brackets.

The term subscript has the same meaning as in the mathematical notation. If single subscript is required to refer to an element of an array, the array is known as *one-dimensional* or *linear array*, and if two subscripts are required to refer to an element of an array, the array is known as *two-dimensional* array, and so on. Analogously, the arrays whose elements are referred by two or more subscripts are called *multi-dimensional* arrays.

10.2.1 One-Dimensional Arrays

A list of items can be given a common name and the individual items can be referenced using only one subscript. Such variables are referred to a *single-subscripted* variable or a *one-dimensional* array.

In mathematics, we often deal with variables that are single-subscripted. For instance, we use the equation

$$\overline{x} = \frac{\sum_{i=1}^{n} x_i}{n}$$

to calculate the average of n values of a list named x. The subscripted variable x_i refers to the ith element of x.

In C language, single subscripted variable is referred as $x[i]$.

For illustration, let us consider a list of 10 scores. The array named *scores* can be visualized as shown in Figure 10.1.

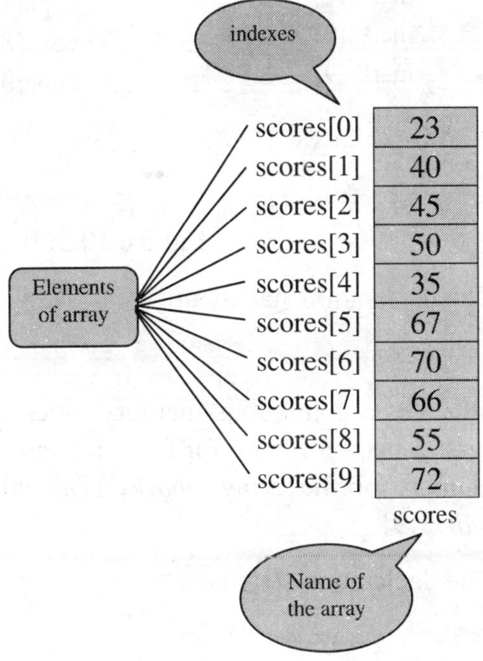

Figure 10.1: The Scores Array

10.2.1.1 Declaration and Definition

An array must be declared and defined before it can be used. Array declaration and definition tell the compiler *the name of the array*, *the type of its elements*, and *the size or number of elements in the array*. The size of the array is a constant and must have a value at compilation time.

The syntax for declaration of one-dimensional array is

```
type arrayName[arraySize];
```

Figure 10.2 shows three different array declarations, one for integers, one for floating-point numbers, and one for characters.

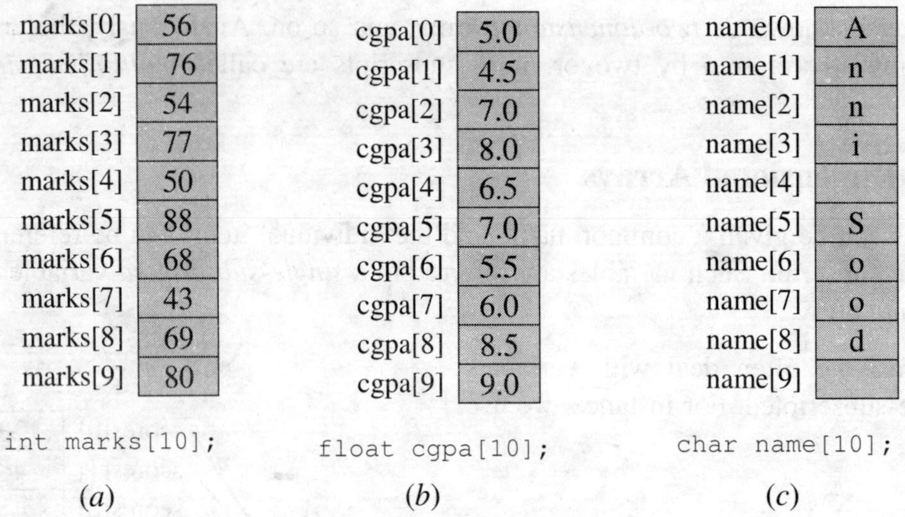

marks[0]	56
marks[1]	76
marks[2]	54
marks[3]	77
marks[4]	50
marks[5]	88
marks[6]	68
marks[7]	43
marks[8]	69
marks[9]	80

`int marks[10];`

(*a*)

cgpa[0]	5.0
cgpa[1]	4.5
cgpa[2]	7.0
cgpa[3]	8.0
cgpa[4]	6.5
cgpa[5]	7.0
cgpa[6]	5.5
cgpa[7]	6.0
cgpa[8]	8.5
cgpa[9]	9.0

`float cgpa[10];`

(*b*)

name[0]	A
name[1]	n
name[2]	n
name[3]	i
name[4]	
name[5]	S
name[6]	o
name[7]	o
name[8]	d
name[9]	

`char name[10];`

(*c*)

Figure 10.2: Declaring and defining arrays

The declaration statement

```
int marks[10];
```

allocates a contiguous memory block of 20 bytes, 2-bytes for each element of type *int*. First 2-bytes are used for first element of the array (*marks*[0]), next 2-bytes for the second element of the array (*marks*[1]), and so on. The last 2-bytes are used for element *marks*[9].

The declaration statement

```
float cgpa[10];
```

allocates a contiguous memory block of 40 bytes, 4-bytes for each element of type *float*. First 4-bytes are used for first element of the array (*cgpa*[0]), next 4-bytes for the second element of the array (*cgpa*[1]), and so on. The last 4-bytes are used for element *cgpa*[9].

The declaration statement

```
char name[10];
```

allocates a contiguous memory block of 10 bytes, 1-byte for each element of type *char*. First byte is used for first element of the array (*name*[0]), next byte for the second element of the array (*name*[1]), and so on. The last byte is used for element *name*[9].

10.2.1.2 Accessing Elements

The C language uses single *index* to access individual elements in a one-dimensional array. The index must be an integral value or an expression that evaluates to an integral value.

For example, given the *marks* array in Figure 10.2 (*a*), we would access the first element as

```
marks[0]
```

To process all the elements in *marks*, a loop similar to the following code is used:

```
for (i = 0; i < 10; i++ )
    process ( marks[i] );
```

One very interesting point that you may be keen to know is that how the C system knows where an individual element is located in memory. In *marks* one-dimensional array, there are ten elements. *How does it find just one of them*? The answer is simple – the array's name is a symbolic reference for the address to the first byte of the block of memory allocated for the array. The address of the first byte for the array is also known as *base address* of the array. Whenever we use the array's name, we actually refer to the first byte of the array. The index represents an offset from the beginning of the array to the element being referenced. With these two facts of information, the C system can calculate the address of any element in the array using the following simple formula:

```
element address = base address + ( sizeof(element) * index )
```

or simply as

```
address(marks[i]) = base address + ( sizeof(element) * i )
```

For example, assume that the base address for array *marks* is 1000. Since the elements of the array *marks* are of type *int*, the size of one element is the size of *int*, *i.e.*, 2 bytes. The address of an element at index 3 is

```
address(marks[3]) = 1000 + ( 2 * 3 ) = 1006
```

In C language, there is no built-in facility to ensure that while accessing array elements, the subscript is within the permissible range. For example, the declaration statement is *int a[10]*, and even if we attempt to access the 16th element as *a*[15], which of course is not a valid array element, in our program, the C system will not flag any error at any stage. It will just pick up a garbage value from the accessed memory location. The programmer must take care of this problem. Further, *i*th element of above array can also be accessed as *i*[*a*], in addition to normal way as *a*[*i*].

10.2.1.3 Storing Values

Declaration and definition only reserve space for the elements of the array. No values are stored in the array at this moment. To store values in the array, we must initialize the elements, read values from the keyboard, or assign values to each individual element.

10.2.1.3.1 Initialization

Just as simple variables can be initialized along with declaration and definition, the elements of the arrays can also be initialized. For each element in the array, we provide a value. The only difference is that the values must be enclosed in braces, and if there is more than one, separated by comma. The following examples provide all possible ways of initializing one-dimensional array. The following statement

```
int b[10]={12, 0, 14, -4, 7, 8, 10, 11, 9, 15};
```

initializes element *b*[0] to 12, *b*[1] to 0, *b*[2] to 14, *b*[3] to –4, *b*[4] to 7, *b*[5] to 8, *b*[6] to 10, *b*[7] to 11, *b*[8] to 9, *b*[9] to 15.

If the array is initialized like this

```
int b[] = {12, 0, 14, -4, 7};
```

Since size of the array is not specified, the compiler counts the number of elements in the initialization list and fixes that as the array size.

10.2.1.3.2 Inputting Values

The second way to fill the array with values is to read the values from the keyboard or a data file. This method of inputting values is done using a loop. The following segment shows the way data is entered into one-dimensional array.

```
for ( i = 0; i < 10; i++ )
{
   scanf( "%d", &marks[i] );
}
```

```
for ( i = 0; i <= 9; i++ )
{
   scanf( "%d", &marks[i] );
}
```

Observe the following things:

1. We start the index, *i*, at 0. Since the array has 10 elements, we must store the values from index location 0 through 9. The limit test, therefore, is set at $i < 10$ or $i <= 9$, which conveniently is the number of elements in an array.
2. Even though, we are dealing with array elements, the *address of* operator (&) is still necessary in the *scanf()* function call.

10.2.1.3.3 Assigning Values

The third way to fill the array with values is to assign the values to the individual elements using the assignment operator. Any value that is compatible can be assigned to the individual array element.

A simple assignment statement for *marks* is given below:

```
marks[3] = 45;
```

We can also assign the elements of one array to second array provided they match fully in type and size. Suppose that two arrays named *first* and *second* are compatible, and size of each array is 25, and we want to copy the elements of array *first* to *second*, the following loop will serve the purpose:

```
for ( i = 0; i < 25; i++ ) {
  second[i] = first[i];
}
```

If the values of elements of an array follow a certain pattern, we can use a loop to assign values. For example, following loop assigns a value to the element that is twice of its index:

```
for ( i = 0; i < 10; i++ ) {
  a[i] = 2 * i;
}
```

Following loop assigns value 0 to elements at even index and value 1 to elements at odd index:

```
for ( i = 0; i < 10; i++ )
{
   if ( i % 2 == 0 )
      a[i] = 0;
   else
      a[i] = 1;
}
```

One array cannot be copied to another using assignment. For instance, if *a* and *b* are two arrays that match fully in type and size, the statement "a = b;" is not acceptable.

10.2.1.4 Outputting Values

To output values of elements of an array, we use the loop as follows:

```
for ( i = 0; i < n; i++ ) {
   printf( "%d  ", a[i] );
}
printf( "\n" );
```

All the values are displayed on one line. However, if all the values cannot be displayed in one line either because of their magnitude of values or number of values, they are displayed in subsequent lines. After, the *for* loop completes, a final call to *printf()* function advances the cursor to the next line. The following *for* loop display one value per line:

```
for ( i = 0; i < n; i++ ) {
   printf( "\n%d", a[i] );
}
printf( "\n" );
```

However, if the values are more than 25, these values will scroll up. However, we can set up some of counter to control the display so that we able to view all the values.

Having understood the various aspects of the one-dimensional arrays, let us write some C functions and complete programs to demonstrate the power of one-dimensional arrays to handle complex problems.

ILLUSTRATIVE EXAMPLES

Example 10.1: Given an array, named *x*, of size (≤ 50) whose elements are of type *int.* Write a program that reads the actual size 'n' of the array, assigns array elements with value that is square of its index, and finally prints the elements of array such that maximum of 5 elements are printed on one line.

Listing 10.1

```
/* Program that assigns elements of the array with square of its
   index, and then prints the array such that maximum of 5 elements
   on one line
*/
#include<stdio.h>
int main()
{
   int x[50], i, n, counter = 0;
   printf("\nEnter actual size of array (<=50) : ");
   scanf("%d", &n);
   for ( i = 0; i < n; i++ )
      x[i] = i  * i;
   for ( i = 0; i < n; i++ )
   {
      printf("%d  ", x[i]);
      counter++;
      if ( counter == 5 ) {
         printf("\n");
         counter = 0;
      }
   }
   return 0;
}
```

Test Run

```
Enter actual size of array (<=50) : 18
  0    1    4    9   16
 25   36   49   64   81
100  121  144  169  196
255  256  289
```

Example 10.2: Let us consider that for a class of *n* students, we have with us score of each student in final exam, and we are interested in generating the following information

- maximum score of the class,
- average score of the class, and
- number of students whose score is greater than class's average score

The following program accomplishes this task.

Listing 10.2

```
/* Program to find the maximum score, average score, and the
   number of students whose score is greater than class average  */
#include<stdio.h>
int main()
{
   float score[50], maximumScore, averageScore, float totalScore;
   int i, n, noOfStudents = 0;
   printf("\nEnter number of students in class: ");
   scanf( "%d", &n );
   printf( "\nEnter %d scores\n\n", n );
   for ( i = 0; i < n; i++ )
      scanf( "%f", &score[i] );
   totalScore = 0;
   for ( i = 0; i < n; i++ )
      totalScore += score[i];
   averageScore = totalScore / n;
   maximumScore = score[0];
   for ( i = 1; i < n; i++ ) {
      if ( score[i] > maximumScore )
         maximumScore = score[i];
   }
   for ( i = 0; i < n; i++ ) {
      if ( score[i] > averageScore )
         noOfStudents++;
   }
   printf( "\nMaximum score of class = %.2f", maximumScore );
   printf( "\nAverage score of class = %.2f", averageScore );
   printf( "\nNo. of students with score > %.2f = %d\n",
            averageScore, noOfStudents );
   return 0;
}
```

Test Run

```
Enter number of students in class: 10
Enter 10 scores
75.25
80
65.5
59.5
72.25
35.5
55.75
69
43
85
Maximum score of class = 85.00
Average score of class = 63.58
No. of students with score > 63.58 = 6
```

Example 10.3: Program to find the range of a set of value collected from some experiment or survey.

The *range* is defined as difference between the smallest and largest element of the given set of values. In order to find the range, first we have to find the largest and smallest element of the set. The following program accomplishes this task.

Listing 10.3

```
/* Program to find the range from a set of values    */
#include<stdio.h>
int main()
{
   float a[50], smallest, largest, range;
   int i, n;
   printf( "\nEnter size of array: " );
   scanf( "%d", &n );
   printf("Enter %d elements of the array\n",n);
   for ( i = 0; i < n; i++ )
     scanf( "%f", &a[i] );
   smallest = largest = a[0];
   for ( i = 1; i < n; i++ )
   {
      if ( a[i] > largest ) largest = a[i];
      if ( a[i] < smallest ) smallest = a[i];
   }
   range = largest - smallest;
   printf( "\nRange of set of value = %.2f\n", range );
   return 0;
}
```

Test Run

```
Enter size of array: 10
Enter 10 elements of the array
75.25
80
65.5
59.5
72.25
35.5
55.75
69
43
85

Range of set of value = 49.50
```

Example 10.4: Program to find standard deviation (σ) for a given set of data, say X (x_0, x_1, x_2,..., x_{n-1}). Standard deviation is a valuable statistics that measures the dispersion of data, *i.e.*, how the data is dispersed (scattered) about the mean ($\bar{x}$) of data.

The formula for computing standard deviation is

$$\text{standard deviation } (\sigma) = \sqrt{\frac{\sum_{i=0}^{n-1}(x_i - \bar{x})^2}{n}}, \text{ where mean } (\bar{x}) = \frac{\sum_{i=0}^{n-1} x_i}{n}$$

Listing 10.4

```
/*  Program to find standard deviation for a given set of data  */
#include<stdio.h>
#include<math.h>
int main()
{
   float x[50], average, sum, stdDeviation, temp;
   int i, n;
   printf("\nProgram to compute standard deviation for" );
   printf(" a given set of data\n\n");
   printf("\nEnter number of elements in data set (n): ");
   scanf( "%d", &n );
   printf("\n\n");   /*  insert two blank lines */
   for ( i = 0; i < n; i++ ) {
      printf( "Enter data element #%d : ", i+1 );
      scanf( "%f", &x[i] );
   }
   for ( sum = 0.0, i = 0; i < n; i++ )
      sum += x[i];
   average = sum / n;
   for (sum = 0.0, i = 0; i < n; i++ ) {
      temp = x[i] - average;
      sum += temp * temp;
   }
```

```
    stdDeviation = sqrt( (double) sum / n );
    printf( "\nStandard deviation = %.3f", stdDeviation );
    printf( " (rounded to three decimal digits)\n" );
    return 0;
}
```

Test Run

```
Program to compute standard deviation for a given set of data
Enter number of elements in data set (n): 5
Enter data element #1: 1
Enter data element #2: 2
Enter data element #3: 3
Enter data element #4: 4
Enter data element #5: 5
Standard deviation = 1.414 (rounded to three decimal digits)
```

Example 10.5: Program to sort a linear array *a* with *n* elements in ascending order. The elements of *a* are indexed from 0 to (*n*–1).

In order to sort the above array, we will use one very important method known as Bubble sort, which is described in *Example 2.21*.

Following listing shows the implementation of bubble sort method to sort an array in ascending order.

Listing 10.5

```
/* Program to sort an array of integers in ascending order
   using Bubble sort method   */
*/
#include<stdio.h>
int main()
{
    int a[50], i, n, k, temp;
    printf( "\nEnter size of array: " );
    scanf( "%d", &n );
    printf("\nEnter %d elements of the array\n",n);
    for ( i = 0; i < n; i++ )
       scanf ( "%d", &a[i] );
    for ( k = 0;  k < n-1;  k++ ) {
       for ( i = 0; i < n-k-1; i++ ) {
           if ( a[i] > a[i+1] ) {
               temp = a[i];
               a[i] = a[i+1];
               a[i+1] = temp;
           }
       }
    }
    printf( "\nArray elements after sorting\n" );
    for ( i = 0; i < n; i++ )
```

```
        printf ( "%d  ", a[i] );
    printf( "\n" );
    return 0;
}
```

Test Run

```
Enter size of array: 10
Enter 10 elements of the array
75 80 65 59 72 35 55 69 43 85

Array elements after sorting
35  43  55  59  65  69  70  72  80  85
```

10.2.1.5 Passing One-Dimensional Array as Argument

Note that instead of passing values of all elements of an array to a function, only address of the array is passed. As you know, in C, the name of an array is a primary expression whose value is the address of the first element of the array.

```
/*  main function      */
#include <stdio.h>
void fun( int x[], int m );
void main()
{
    int a[10], n;
    /*  other local declarations */

    func(a, n); /* function call */
    /*  other statements         */
}
```

```
/*  function definition    */
void fun( int x[], int m )
{
    /*  local declarations */
    /*  other statements   */
}
```

Figure 10.3: Passing one-dimensional array as an argument to a function

Since, indexed references are simply calculated addresses, all we need to refer to any of the elements of the array is the address of the array, *i.e.*, address of the first element.

Because the name of the array is in fact its address, therefore, passing the array name allows called function to refer to the array back in the calling function.

Listing 10.6

```
/* Program to demonstrate the passing of an array as an argument  */
#include <stdio.h>
int largest( int x[], int size );  /*  function prototype */
int main(void)
{
    int a[50]={12,15,20,17,25,50,11,10,8,13};
    int i;
    printf( "Largest of the array element is %d\n", largest(a,50));
    return 0;
}
/* function that returns largest element of the array   */
```

```
int largest( const int x[], int size )
{
    int big, i;
    big = x[0];
    for ( i = 1; i < size; i++ ) {
        if ( x[i] > big )
            big = x[i];
    }
    return big;
}
```

Test Run

```
Largest of the array element is 50
```

10.2.2 Two-Dimensional Arrays

The array we have discussed for far are known as one-dimensional arrays because the data are organized linearly only in one direction (dimension). Many applications require that data be organized in more than one dimension. One common example is a matrix (a table), which is an array that consists of rows and columns.

Figure 10.4 shows a matrix (table) with 4 rows and 4 columns which is commonly used as a two-dimensional array.

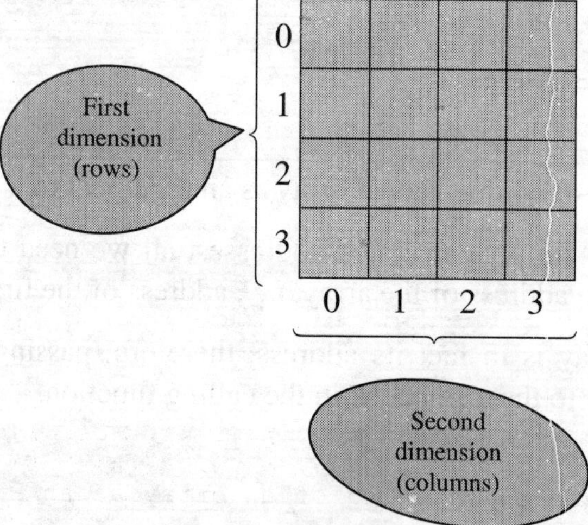

Figure 10.4: Two-dimensional array

Although two-dimensional array is exactly what is shown in Figure 10.4, C language looks at it in a different way. It looks at the two-dimensional array as am array of arrays. In other words, a two-dimensional array in C language is an array of one-dimensional arrays. This concept is shown in Figure 10.5.

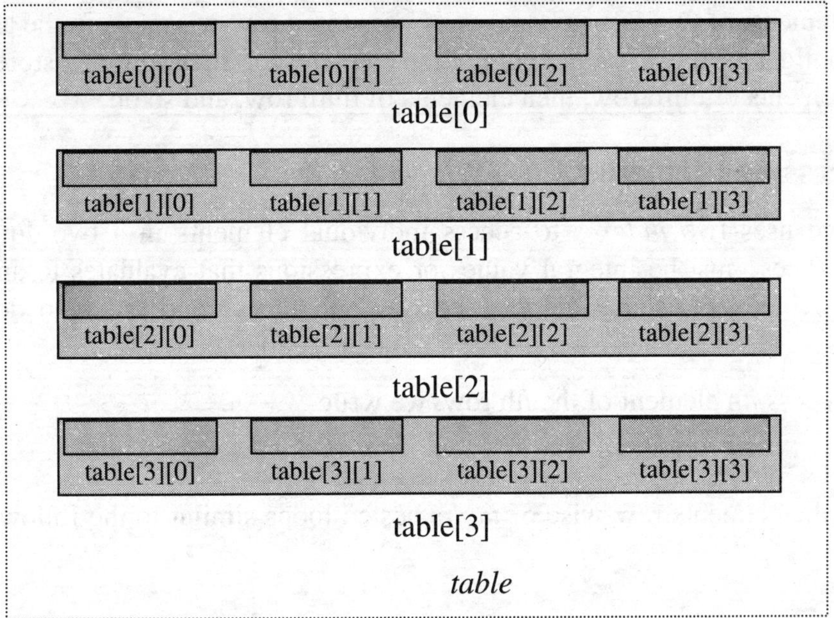

Figure 10.5: Array of arrays

10.2.2.1 Declaration and Definition

Two-dimensional arrays, like one-dimensional arrays, must be declared before being used. Declaration tells the compiler *the name of the array*, *the type of its elements*, and *the size of each dimension*. The syntax for declaration of two-dimensional array is

```
type arrayName[sizeOfDimension1][sizeOfDimension2];
```

By convention, the first dimension specifies the number of *rows* in the array while the second dimension specifies the number of *columns* in each row.

The declaration statement, declares the array shown in Figure 10.9

```
int table[4][4];
```

This declarations statement allocates a contiguous memory block of 32 bytes, 8-bytes for each row. First set of 8-bytes are used to store elements of first row, second set of 8-bytes are used to store elements of second row, third set of 8-bytes are used to store elements of third row and the last set of 8-bytes are used to store elements of the fourth (last) row. Within in each row, first set of 2-bytes are used to store element of the first column, next set of 2-bytes are used to store element of second column, next set of 2-bytes are used to store elements of the third column, and final set of 2-bytes are used to store element of the fourth (last) column.

Elements of two-dimensional array are stored row-wise, *i.e.*, in the allocated contiguous block of memory, first elements of first row are stored, then elements of third row, then elements of third row, and so on.

10.2.2.2 Accessing Elements

The C language uses two *indexes* to access individual elements in a two-dimensional array. These indices must be integral values or expressions that evaluates to an integral value. The first index is used to denote the row and the second index to denote the column.

Therefore, to access *j*th element of the *i*th row, we write

```
table[i][j]
```

To process all the elements row-wise of *table*, nested loops similar to the following code will be used:

```
for (i = 0; i < 4; i++ )
{
    for (j = 0; j < 4; j++ )
    {
        process ( table[i][j] );
    }
}
```

10.2.2.3 Storing Values

Like one-dimensional array, the declaration and definition of two-dimensional array also only reserve space for the elements of the array. No values are stored in the array at this moment. To store values in the array, we must initialize the elements, read values from keyboard, or assign values to each individual element.

10.2.2.3.1 Initialization

One way to initialize two-dimensional arrays is as shown below:

```
int table[4][4] = { 0, 0, 0, 0, 1, 1, 1, 1, 2, 2, 2, 2, 3, 3, 3, 3 };
```

It will initialize all elements of first row with value 0, elements of second row with value 1, elements of third row with 2 and elements of fourth row with value 3.

Though, the above initialization has no problems, it highly recommended that nest braces be used to show the exact nature of the two-dimensional array.

The array *table* is better initialized as shown below:

```
int table[4][4] = { { 0, 0, 0, 0 },
                    { 1, 1, 1, 1 },
                    { 2, 2, 2, 2 },
                    { 3, 3, 3, 3 },
                  };
```

In above way of initialization, we initialize each row as a one-dimensional array of four elements enclosed in braces. The array of four rows also has its set of braces. Note that we use commas between the elements in the columns and also commas between the rows.

10.2.2.3.2 Inputting Values

The second way to fill the array with values is to read the values from the keyboard or a data file. This method of inputting values is done using nested loops. Following segment shows the way elements can be read row-wise using nested for loops:

```
for (i = 0; i < 4; i++ )   {
    for (j = 0; j < 4; j++ ) {
        scanf( "%d", &table[i][j] );
    }
}
```

In general, if the one-dimensional array is of size $m \times n$, the first loop varies from 0 to (m-1) and the second loop varies from 0 to (n-1).

10.2.2.3.3 Assigning Values

The third way to fill the array with values is to assign the values to the individual elements using the assignment operator. Any value that is compatible can be assigned to the individual array element. A simple assignment statement for *table* is given below:

```
table[1][2] = 40;
```

We can also assign the elements of one array to second array provided they match fully in type and size. Suppose that two arrays named a and b are compatible, and size of each array is 4×4, and we want to copy the elements of array a to b, the following loop will serve the purpose:

```
for (i = 0; i < 4; i++ )
{
     for (j = 0; j < 4; j++ )
     {
          b[i][j] = a[i][j];
     }
}
```

If the values of elements of an array follow a certain pattern, we can use nested loops to assign values. For example, if we want to initialize a 4×4 two-dimensional array to represent an identity matrix, this can be done using following nested loops as shown:

```
for (i = 0; i < 4; i++ )   {
    for (j = 0; j < 4; j++ )   {
       if ( i == j )
          a[][] = 1;
       else
          a[i][j] = 0
    }
}
```

$$\begin{pmatrix} 1 & 0 & 0 & 0 \\ 0 & 1 & 0 & 0 \\ 0 & 0 & 1 & 0 \\ 0 & 0 & 0 & 1 \end{pmatrix}$$

Figure10.6: Identity matrix

10.2.2.4 Outputting Values

To output values of elements of two-dimensional array, we use nested loops as follows:

```
for ( i = 0; i < 4; i++ ) {
    for ( j = 0; j < 4; j++ ) {
       printf( "%d", a[i][j] );
    }
    printf( "\n" );
}
```

The elements of the two-dimensional array will be output row-wise by the above segment. The first loop controls the rows while the second loop controls the columns.

ILLUSTRATIVE EXAMPLES

Example 10.6: Program to compute the transpose of a matrix *A* of order *m*×*n*.

We know that if we interchange the rows with columns, we get transpose of a given matrix, say B of order *n*×*m*, *i.e.*,

$$b_{ji} = a_{ij}$$

Listing 10.7

```
/*  Program to compute transpose of matrix A of size mxn   */
#include<stdio.h>
int main()
{
    int a[10][10], b[10][10];
    int n, m, i, j;
    printf( "Enter size of matrix A as m,n: " );
    scanf( "%d,%d", &m, &n);
    printf("\nEnter %d elements of matrix A row-wise\n",m*n);
    for ( i = 0 ; i < m; i++ ) {
        for ( j = 0; j < n; j++ ) {
           scanf( "%d", &a[i][j] );
        }
    }
```

```
    for ( i = 0 ; i < m ; i++ ) {
        for ( j = 0 ; j < n ; j++ ) {
            b[j][i] = a[i][j];
        }
    }
    printf( "\nTranspose is\n\n" );
    for ( i = 0 ; i < n; i++ ) {
        for ( j = 0; j < m; j++ ) {
            printf( "%5d", b[i][j] );
        }
        printf ( "\n" );
    }
    return 0;
}
```

Test Run

```
Enter size of matrix A as m,n: 3,3
Enter 9 elements of matrix A row-wise
1 2 3
4 5 6
7 8 9
Transpose is
    1    4    7
    2    5    8
    3    6    9
```

Example 10.7: Program to add two matrices *A* and *B*, each of order *m*×*n*.

We know that if to add to matrices, we add corresponding elements to get the element of a resultant matrix, i.e.,

$$c_{ij} = a_{ij} + b_{ij}$$

Listing 10.8

```
/*  Program to add matrix A and B, each of order mxn   */
#include<stdio.h>
int main()
{
    float a[10][10], b[10][10], c[10][10];
    int n, m, i, j;
    printf( "Enter size of matrices as m,n: " );
    scanf( "%d,%d", &m, &n);
    printf( "Enter %d elements of matrix A row-wise\n", m*n );
    for ( i = 0 ; i < m; i++ ) {
        for ( j = 0; j < n; j++ ){
            scanf( "%f", &a[i][j] );
        }
    }
    printf( "Enter %d elements of matrix B row-wise\n", m*n );
    for ( i = 0 ; i < m; i++ ) {
        for ( j = 0; j < n; j++ ) {
            scanf( "%f", &b[i][j] );
```

```
        }
    }
    for ( i = 0 ; i < m ; i++ ) {
        for ( j = 0 ; j < n ; j++ ) {
            c[i][j] = a[i][j] + b[i][j];
        }
    }
    printf( "\nSum A+B is\n\n" );
    for ( i = 0 ; i < m; i++ )
    {
         for ( j = 0; j < n; j++ )
         {
            printf( "%7.2f", c[i][j] );
         }
         printf ( "\n" );
    }
    return 0;
}
```

Test Run

```
Enter size of matrices as m,n: 3,3
Enter 9 elements of matrix A row-wise
1 1 1
1 1 1
1 1 1
Enter 9 elements of matrix B row-wise
2 2 2
2 2 2
2 2 2

Sum A+B is

  3.00   3.00   3.00
  3.00   3.00   3.00
  3.00   3.00   3.00
```

Example 10.8: Program to multiply two matrices *A* and *B*. The matrix *A* is of order $m\times n$ matrix and the matrix *B* is of order $p\times q$ matrix, provided $n = p$.

As we want to multiply two matrices *A* and *B*, where the matrix *A* is of order $m\times n$ and the matrix *B* is of order $p\times q$, the product matrix, say matrix *C*, then will be of order $m\times p$ with entry c_{ij}, which appears in the *i*th row and *j*th column is given as:

c_{ij} = sum of the products of the entries in the *i*th row of matrix *A* with the corresponding entries in the *j*th column of matrix *B*

$$= a_{i0}\times b_{0j} + a_{i1}\times b_{1j} + a_{i2}\times b_{2j} + \ldots + a_{i(n-1)}\times b_{(n-1)j}$$

$$= \sum_{k=0}^{n-1} a_{ik} b_{kj}$$

Listing 10.9

```
/*
*  Program to multiply matrix A of size mxn by matrix B of
*  size pxq. Program also checks for feasibility of product.
*/
#include<stdio.h>

int main()
{
   float a[10][10], b[10][10], c[10][10];
   int n, m, p, q, i, j, k;
   printf( "Enter size of matrix A as m,n: " );
   scanf( "%d,%d", &m, &n);
   printf( "Enter size of matrix B as p,q: " );
   scanf( "%d,%d", &p, &q);
   if ( n != p ) {
      printf( "\nMatrix Product AxB is not feasible\n" );
      return;
   }
   printf("\nEnter %d elements of matrix A row-wise\n", m*n);
   for ( i = 0 ; i < m; i++ )
   {
      for ( j = 0; j < n; j++ )
      {
         scanf( "%f", &a[i][j] );
      }
   }
   printf("\nEnter %d elements of matrix B row-wise\n", p*q);
   for ( i = 0 ; i< p; i++ )
   {
      for ( j = 0; j < q; j++ )
      {
         scanf( "%f", &b[i][j] );
      }
   }
   for ( i = 0 ; i < m ; i++ )
   {
      for ( j = 0 ; j < q ; j++ )
      {
         c[i][j] = 0;
         for ( k = 0; k < n; k++ )
         {
            c[i][j] += a[i][k] * b[k][j];
         }
      }
   }
   printf( "\nProduct AxB is\n\n" );
   for ( i = 0 ; i < m; i++ )
   {
      for ( j = 0; j < q; j++ )
      {
         printf( "%7.2f\t", c[i][j] );
      }
```

```
        printf ( "\n" );
    }
    return 0;
}
```

Test Run

```
Enter size of matrix A as m,n: 3,3
Enter size of matrix B as p,q: 3,3
Enter 9 elements of matrix A row-wise
1 1 1
1 1 1
1 1 1
Enter 9 elements of matrix B row-wise
2 2 2
2 2 2
2 2 2
Product AxB is
 6.00  6.00  6.00
 6.00  6.00  6.00
 6.00  6.00  6.00
```

10.2.2.5 Passing Two-Dimensional Array as Arguments

When we pass two-dimensional array to a function, we use the array name as we did with one-dimensional array. The formal argument in the argument list in the called function header must indicate the array has two dimensions. This is done by including two sets of brackets, one for each dimension. During function call, only the base address of the array is passed.

```
/*  main function    */
#include <stdio.h>
void fun(int x[][], int m, int n);
void main()
{
   int a[4][4];
   /*  other local declarations  */

   func(a,m,n); /* function call */
   /*  other statements          */
}
```

```
/*  function definition     */
void fun(int x[][], int m, int n)
{
   /*  local declarations  */
   /*  other statement     */
}
```

Figure 10.7: Passing entire two-dimensional array

Listing 10.10

```
/* Program to multiply matrix A of size mxn by matrix B of size
*  pxq. Program also checks for feasibility of product. This
*  program uses function to read, multiply and display matrices   */
#include<stdio.h>
/*    function declarations     */
void readMatrix(float a[][], int m, int n);
void displayMatrix(float a[][], int m, int n);
void multiplyMatrix(float a[][],b[][],c[][],int m,int n,int q);
int main()
{
    float a[10][10], b[10][10], c[10][10];
    int n, m, p, q;
    printf( "Enter size of matrix A as m,n: " );
    scanf( "%d,%d", &m, &n);
    printf( "Enter size of matrix B as p,q: " );
    scanf( "%d,%d", &p, &q);
    if ( n != p ) {
        printf( "\nMatrix Product AxB is not feasible\n" );
        return 1;
    }
    readMatrix(a,m,n);              /* input matrix 'a'  */
    readMatrix(b,p,q);              /* input matrix 'b'  */
    multiplyMatrix(a,b,c,m,n,q); /* multiply matrix a by matrix 'b'*/
    printf( "\nProduct AxB is\n\n" );
    displayMatrix(c,m,q);            /* output matrix 'c' */
    return 0;
}
/*  function that reads a matrix 'a' of order mxn  */
void readMatrix(float a[][], int m, int n)
{
   int i, j;
   printf("\nEnter %d elements of matrix A row-wise\n", m*n );
   for ( i = 0 ; i < m; i++ ) {
      for ( j = 0; j < n; j++ ) {
         scanf( "%f", &a[i][j] );
      }
   }
}
/*  function that displays a matrix 'a' of order mxn  */
void displayMatrix(float a[][], int m, int n)
{
    int i, j;
    for ( i = 0 ; i < m; i++ ) {
       for ( j = 0; j < n; j++ ) {
           printf( "%7.3f\t", c[i][j] );
       }
       printf ( "\n" );
    }
}
/* function that multiplies matrices 'a' and 'b' and
 *  stores the product in matrix 'c'  */
void multiplyMatrix(float a[][],b[][],c[][],int m,int n,int q)
```

```
{
    int i, j, k;
    float temp;
    for ( i = 0 ; i < m ; i++ ) {
        for ( j = 0 ; j < q ; j++ ) {
            c[i][j] = 0;
            for ( k = 0; k < n; k++ )
                c[i][j] = c[i][j] + a[i][k] * b[k][j];
        }
    }
}
```

10.3 CHARACTER ARRAYS AND STRINGS

Each of the arrays illustrated so far contained numeric elements. We can also have array of characters. Using array of characters, we can store string data. But do remember that as such strings are not directly supported in C language, though sometimes loosely we refer to array of characters as strings.

A string in C language is a variable length array of characters that is delimited by null character ('\0'). Generally, characters comprising a string are selected only from printable characters; however, C language does permit the use of any character except the null character. In fact, it is common to use formatting characters, such as tabs, format specifiers, etc., in strings.

Here we will discuss the various aspects related to handling of strings in C language.

The C language uses variable length, delimited strings. The null character '\0' is used as delimiting character. Though the null character '\0' looks like a sequence of two characters, '\' and '0', but in C language it is treated as single character.

10.3.1 Storing Strings

In C language, a string is stored in an array of characters. It is terminated by the null ('\0') character. Figure 10.8 shows how a string "Hello" is stored in memory. Because a string is stored in an array, the name of the array, and hence string, is a pointer to the beginning of the string.

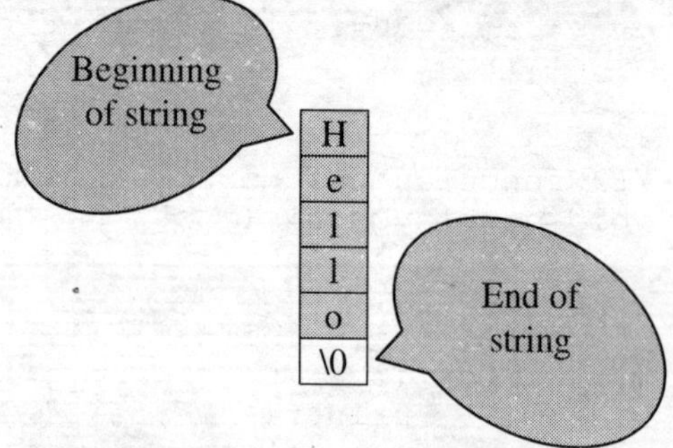

Figure 10.8: Storing a string in memory

Figure 10.9 shows the difference between the storage of a character and a one character string. The character required only one storage location (of 1-byte) whereas one character string requires two storage location (of 1-byte each); one for the character and one for the delimiter. The figure also shows how an empty string is stored. The empty string consists of null character only.

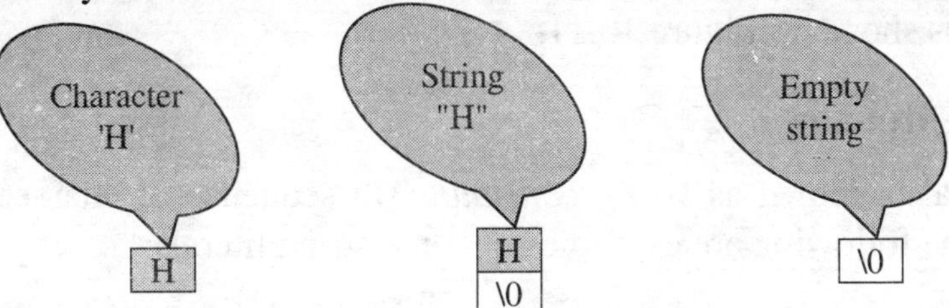

Figure 10.9: Difference in storage of characters and strings

10.3.2 Need of String Delimiter

The question – *Why actually we need a null character to terminate a string*? - may be coming in your mind. The answer to this question is that a string is not a data type; it is data structure – *an array*. Therefore, implementation of string is logical not physical. The physical structure is the array in which string is stored. Since, strings are of varying length, there is a need to identify the logical end of data within physical structure.

Let look at the need of string delimiter from another point of view. If the data is of fixed length, then we don't need string data structure to store them rather it can easily be stored in an array. When data is stored in an array, the end of the data is always indicated by the last element of the array. But, if the data is not of fixed length (*i.e.*, varying length), then we need some other way to determine the end of the data. The null character is used to mark the end of the string data.

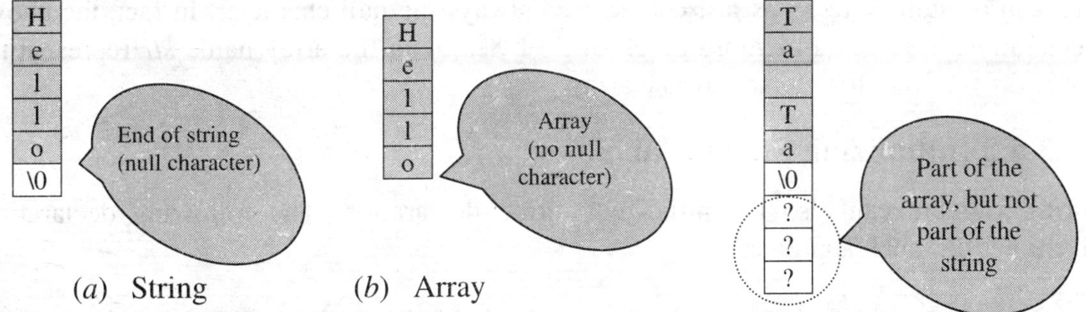

Figure 10.10: Difference between a string and an array of characters

Figure 10.11: String stored in part of array

Because strings are variable-length structures, enough space for the maximum-length string that we have to store plus one for the null character (delimiter) be reserved.

In many situations it may be possible that the entire array is not filled and we may have null character in the middle of the array. In that case, the array from the beginning to the null character is treated as string and remaining elements of the array are ignored. This means that a part of an array of characters can be treated as a string provided it ends with a null character, as shown in Figure 10.11.

10.3.3 String Literals

A *string literal*, also known as *string constant*, is a sequence of characters enclosed in double quotes. The following are examples of some string literals:

```
"Hello! you are wel come"
"Hari\'s Book"
""
"A"
```

Note that embedded spaces are significant. And if quotes, single (') or double ("), are part of the string, they must be used with the escape sequence.

10.3.4 String Variables

A string variable is actually an array of characters.

10.3.4.1 Declaring String Variables

String variables are declared as shown in the following statement

```
char str[20];
```

This statement declares string variable *str* of length 20. In fact, the maximum characters that can be stored are 19 as last character is always the null character. In fact, the above declarations reserves a memory block of 20 bytes, with the array name *str* representing the address of the first element of the array.

10.3.4.2 Initializing String Variables

String variable can also be initialized during declaration. The following declaration illustrates one such approach.

```
char mess[] = {'Y', 'o', 'u', ' ', 'a', 'r', 'e', ' ',
               'w', 'e', 'l', ' ', 'c', 'o', 'm', 'e', '\0'};
```

Here string variable *mess* is initialized as sequence of character constants. In this style it is programmer's job to specify the null character as the last character. In addition to this, as you can see, the size of the array is not specified. *Can you guess what will be the size of mess*? In fact it is the number of character constants specified, including the null character.

Since strings will be used very often, the C language provides a short and efficient approach for initializing strings.

For example, the task of above statement can also be accomplished as

```
char mess[] = "You are wel come";
```

Note that in this approach, the compiler automatically appends the null character.

10.3.5 Input/Output of Strings

A string can be read and written. The C language provides library function *gets()* and *puts()* for input and output of string data.

Listing 10.11

```
/* Program to perform I/O of string data with string I/O functions */
#include<stdio.h>
#include<string.h>
void main()
{
    char str[30];
    printf( "\nEnter string of length <= 29" );
    printf( "\nand terminate with ENTER key\n\n" );
    gets( str );
    printf( "\nYou have entered\n\n" );
    puts( str );
}
```

Test Run

```
Enter string of length <= 29
and terminate with ENTER key
There is no substitute of hard work.↵
You have entered
There is no substitute of hard work.
```

10.3.6 String Manipulation Functions

Almost every C Compiler provides a large number of library functions for manipulating strings. Table 10.1 lists some frequently used string functions.

Table 10.1: Frequently Used String Functions

String Function	Operation Performed
`strlen`	Finds length of a string.
`strcat`	Appends the characters of second argument at the end of first string variable.
`strcpy`	Copies the character of source string constant or variable to destination string variable.
`strcmp`	Compares two strings (comparison is case sensitive).

Let us discuss some of these string functions one by one.

10.3.6.1 The *strlen()* Function

This function takes one argument that can be a string constant or a variable. It counts the number of character present in the string. Do remember that null character '\0' is not a part of the character; it is merely used to mark the end of the string, so it is not counted.

Listing 10.12

```
/*  Program to illustrate the use of strlen()function   */
#include <stdio.h>
#include <string.h>
int main()
{
    char str[30];
    int len;
    printf( "\nEnter a string of length <= 29" );
    printf( "\nand terminate with ENTER key" );
    gets (str);
    len = strlen ( str );
    printf( "Number of characters in %s = %d\n", str, len );
    return 0;
}
```

Test Run

```
Enter string of length <= 29
and terminate with ENTER key
Turbo↵
Number of characters in Turbo = 5
```

10.3.6.2 The *strcat()* Function

This function takes two arguments, of which first is a string variable and second can be a string constant or a variable. It appends the character(s) of the second arguments at the end of the first argument. There must be sufficient space available to accommodate the incoming characters from destination string, otherwise overflow occurs.

Listing 10.13

```
/*  Program to illustrate the use of strcat()function   */
#include <stdio.h>
#include <string.h>
void main()
{
    char str1[30] = "Wel", str2[30] = " Come";
    printf( "\nFirst string is %s\n", str1 );
    printf( "Second string is %s\n", str2 );
    strcat( str1, str2 );
```

```
    printf( "First string after appending\n" );
    printf( "second one: %s\n", str1 );
}
```

Test Run

```
First string is Wel
Second string is Come
First string after appending
second one: Wel Come
```

10.3.6.3 The *strcpy()* Function

This function takes two arguments, of which first is a string variable and second can be a string constant or a variable. It copies the character(s) of the second arguments to the first argument.

The following program illustrates its use.

Listing 10.14

```
/* Program to illustrate the use of strcpy()function */
#include <stdio.h>
#include <string.h>
void main()
{
    char str1[30] = "Wel", str2[30] = " Come";
    printf( "\nFirst string is %s\n", str1 );
    printf( "Second string is %s\n", str2 );
    strcpy( str1, str2 );
    printf( "First string after copying\n" );
    printf( "second one: %s\n", str1 );
}
```

Test Run

```
First string is Wel
Second string is Come
First string after copying
second one: Come
```

10.3.6.4 The *strcmp()* Function

This function takes two arguments, of which both can be string variables or constants. It compares the characters of each but one at a time and returns a value indicating the result of the comparison. For example, in the statement

```
strcmp ( str1, str2 );
```

value returned will have the meanings explained overleaf.

Table 10.2: Interpretation of value returned *by strcmp()* function

Value Returned	Meaning
Less than zero	*str1* is less than *str2*, *i.e.*, *str1* comes first *str2* in dictionary order.
Zero	*str1* is identical to *str2*.
Greater than zero	*str1* is greater *str2* i.e. *str1* comes after *str2* in dictionary order.

Listing 10.15

```
/*  Program to illustrate the use of strcmp()function  */
#include <stdio.h>
#include <string.h>
void main()
{
    char str1[31], str2[31];
    int value;
    printf( "Enter first  string: " );
    gets(str1);
    printf( "\nEnter second string: " );
    gets( str2 );
    value = strcmp( str1, str2 );
    if ( value > 0 ) {
       printf("\n%s comes after %s", str1, str2);
       printf(" in dictionary order\n" );
    } else if ( value < 0 ) {
       printf("\n%s comes before %s", str1, str2);
       printf(" in dictionary order\n" );
    } else
       printf("\nBoth strings are same\n"  );
}
```

Test Run

```
Enter first string: software
Enter first string: hardware
hardware comes before software in dictionary order
```

10.3.7 Passing String to a Function

A string can be passed as an argument to a function using subscripted notation for arrays. In a function definition, the formal parameter is declared as an array of characters. The following program illustrates the passing of a string to a function.

Listing 10.16

```
/* Program to illustrates passing of a string to a function  */
#include<stdio.h>
#include<string.h>
void func( char [] );   /*  function prototype  */
void main()
{
   char str1[] = "Sample string";
```

```
    func( str1 );
}
void func( char temp[] )
{
    printf( "String passed to it: " );
    puts( temp );
    printf( "\nand its length is: %d\n", strlen( temp ) );
}
```

Test Run

```
String passed to it: Sample string
and its length is: 13
```

10.3.8 Array of Strings

In the last section, we discussed string variables that can store one string value at a time. Suppose we want to handle list (array) of such strings, then *what is the solution*? The answer to this question can be found by using two-dimensional array of characters. In two-dimensional arrays, the first row stores the first string, the second row stores the second string, and so on.

Suppose, we want to handle a list of names of 4 persons, where the name of each person can be maximum of 30 characters long, the following will be the required declaration

```
char names[4][31];
```

The following statement shows the way list of strings can be initialized.

```
char names[4][31] = { "Ram Parkash",
                      "Inder Mohan Singh Sidhu",
                      "Amanpreet",
                      "Rajan"
                    };
```

Listing 10.17

```
/* Program to illustrates the input/output of array of strings
*  represented as two-dimensional array of characters  */
#include <stdio.h>
#include <string.h>
int main()
{
    char names[50][31];
    int i, n;
    printf( "Enter number of strings: " );
    scanf( "%d", &n );
    for ( i = 0; i < n; ++i ) {
        printf( "Enter string number %d: ", i+1 );
        gets( names[i] );
```

```
    }
    printf( "\nList of given strings\n\n" );
    for ( i = 0; i < n; ++i )
       puts ( names[i] );
    return 0;
}
```

But this scheme of representing list of strings has one serious problem. The problem is of memory wastage. As in the above example, we assumed that the name could be up to 30 characters long, so we have declared the string length equal to 30. In fact, some names may be of few characters long while others roughly half of the declared length. So, on average, half of the allocated space is not utilized at all. Using array of pointers to strings can solve this problem of memory wastage. We will discuss this aspect in *Chapter 12*.

ILLUSTRATIVE EXAMPLES

Example 10.9: Write a function that takes an array of *n* integer numbers as its argument, and reverses the order of its elements without using any other array. For example, if elements of the array are

2 4 1 6 7 9 15 12

The function should rearrange the array as

12 15 9 7 6 1 4 2

Solution: To accomplish the above task, we use two index variables *i* and *j*, where index variable *i* is initialized with the index of the first element, and index variable *j* is initialized with the index of the last element. Now elements at index *i* and at index *j* are swapped. Then index variable *i* is incremented and index variable *j* is decremented, and this is continued till $i \leq j$. The function *reverseElements()* implements the above logic. Following is the complete listing for the program to accomplish above task.

Listing 10.18

```
/* Program to reverse the order of elements of the array  */
#include <stdio.h>
void reverseElements(int a[], int n);  /* function prototype */
int main()
{
    int a[100], n, i;
    printf("\nEnter even value for n as size of array : ");
    scanf("%d", &n);
    printf("Enter %d elements of array A\n", n);
    for ( i = 0; i < n; i++ )
        scanf("%d", &a[i]);
    reverseElements(a,n);       /*  call function   */
    printf("\nElements of array after reversing the order of ");
    printf(" elements of a given array\n\n");
    for ( i = 0; i < n; i++ )
        printf("%d  ", a[i]);
    return 0;
```

```
}
void reverseElements(int a[], int n)
{
   int i, j, temp;
   i = 0; j = n-1;
   while ( i <= j )
   {
      temp = a[i];
      a[i] = a[j];
      a[j] = temp;
      i++;
      j--;
   }
}
```

Example 10.10: Write a function that takes an array of *n* (where *n* is even) integer numbers as its argument, and rearrange its elements in such a way that values of alternate locations of the array are exchanged. For example, if the elements of the array are

2 4 1 6 7 9 15 12

The function should rearrange the array as

4 2 6 1 9 7 12 15

Solution: To accomplish the above task, we use index variable *i* and vary it from 0 to (*n*-1) in steps of 2. In each iteration, element at index *i* and at index (*i*+1) are swapped. The function *rearrangeElements()* implements the above logic.

Listing 10.19

```
/*
*  Program to rearrange the elements of array in such a way
*  that values of alternate locations of the array are exchanged.
*/

#include <stdio.h>
void rearrangeElements(int a[],int n); /* function prototype */
int main()
{
    int a[100], n, i;
    printf("\nEnter even value for n as size of array : ");
       scanf("%d", &n);
    printf("Enter %d elements of array A\n", n);
    for ( i = 0; i < n; i++ )
       scanf("%d", &a[i]);
    rearrangeElements(a,n);    /*  call function    */
    printf("\nElements of array after swapping the alternate");
    printf(" elements of a given array\n\n");
    for ( i = 0; i < n; i++ )
       printf("%d   ", a[i]);
    return 0;
}
```

```
void rearrangeElements(int a[], int n)
{
    int i, temp;
    for ( i = 0; i < n; i += 2 )
    {
       temp = a[i];
       a[i] = a[i+1];
       a[i+1] = temp;
    }
}
```

Example 10.11: Write a program that read a matrix of size *m*×*n*, computes sum of its rows and columns and prints the output in the following format.

```
12   9  10  13  | 44
 7  11  22   5  | 45  <——— row sums
 6  25  21  15  | 67
20   8  33  16  | 77
----------------------
45  53  86  49  <——— column sums
```

Solution: To accomplish the above task, we create two one-dimensional arrays named *colSum* and *rowSum* to store the sum of elements of respective rows and columns. Once these sums are computed, we print the elements of the array *rowSum* as the last column after printing all elements of a given row of the given matrix. After printing all the rows of the given matrix, we print the elements of the array *colSum* as the last row.

The following program implements the details.

Listing 10.20

```
/*  Program to find row sums and column sums of the given matrix  */
#include<stdio.h>
int main()
{
    int a[10][10], rowSum[10] = {0}, colSum[10] = {0};
    int i, j, n, m;
    printf( "Enter size of matrix A as m,n: " );
    scanf( "%d,%d", &m, &n);
    for ( i = 0 ; i < m; i++ ) {
       for ( j = 0; j < n; j++ ) {
          scanf( "%d", &a[i][j] );
       }
    }
    /*   compute the sum of elements of rows   */
    for ( i = 0 ; i < m; i++ ) {
       for ( j = 0; j < n; j++ ) {
          rowSum[i] += a[i][j];
       }
```

```
    }
    /*   compute the sum of elements of columns   */
    for ( j = 0 ; j < n; j++ ) {
      for ( i = 0; i < m; i++ ) {
        colSum[j] += a[i][j];
      }
    }
    /*   print the output    */
    for ( i = 0 ; i < m; i++ ) {
      for ( j = 0; j < n; j++ ) {
        printf("%5d", a[i][j]);
      }
      printf("  |%5d\n", rowSum[i]);
    }
    for ( j = 0; j < n; j++ )
       printf("-----");
    for ( j = 0; j < n; j++ ) {
      printf("%5d", colSum[j]);
    }
    printf("\n");
    return 0;
}
```

MULTIPLE CHOICE QUESTIONS

1. Elements of two-dimensional array are stored in
   (*a*) column major order (*b*) row major order
   (*c*) random order (*d*) None
2. The value within [] in an array declaration specifies
   (*a*) size of the array (*b*) largest permitted subscript value
   (*c*) both (*a*) & (*b*) (*d*) None
3. Consider the following program in C language

```
void main()
{
   int i;
   int a[] = { 5, 7, 9, 11, 13 };
   for ( i = 0; i <=4; i++ ) {
     if ( i )
       continue;
     else
       printf( "%5d", i[a] );
   }
}
```

   What will be result of attempting to compile and run the above program?
   (*a*) Program will fail to compile because accessing *i*th element of array *a* as *i*[*a*] is not permitted in C.
   (*b*) The program will output: 5 7 9 11 13
   (*c*) The program will output: 5
   (*d*) Program will fail to compile because in array declaration, size of the array is not specified.

4. Given the declaration "int a[10];" Identify the which of the following is wrong?
   (*a*) *a*[−1] (*b*) *a*[0]
   (*c*) *a*[10] (*d*) ++a
5. Consider the following program in C language.

```
void main()
{
    int i;
    int a[] = { 5, 8, 9, 10, 13 };
    for ( i = 0; i <=4; i++ ) {
        if ( a[i] % 2 != 0 )
            a[i] /= 2;
        else
            a[i] *= 2;
        if ( a[i] % 2 == 0 )
            a[i] *= 2;
        else
            a[i] /= 2;
        printf( "   %d", a[i] );
    }
}
```

   What will be its output?

   (*a*) 5 8 9 10 13 (*b*) 10 16 18 20 36
   (*c*) 4 32 8 40 12 (*d*) 2 4 4 5 6

6. When we should use an array?
   (*a*) When we need to hold constants.
   (*b*) When we need to hold data of same type.
   (*c*) When we need to hold data of different type.
   (*d*) When we need to obtain automatic memory cleanup functionality.
7. The use of arrays makes a program __________.
   (*a*) general and capable of handling class of problems.
   (*b*) difficult to understand.
   (*c*) compact and efficient.
   (*d*) both *a* & *c*.
8. The maximum number of dimensions an array can have is __________.
   (*a*) 3 (*b*) 4
   (*c*) user-defined (*d*) compiler dependent
9. What is the difference between the 5's in the below expressions?

```
int num[5];
num[5]=10;
```

   (*a*) First specifies the array size, the second one specifies individual element
   (*b*) First specifies the individual element, the second one specifies the size
   (*c*) Both specify array size
   (*d*) None of above

10. What would be the output of the following program if the array begins at 65486 address location?

```
void main()
{
    int a[6]={ 12,14,15,17,19 };
    printf("%u, %u ", arr, &arr );
}
```

(*a*) 65486, 65488 (*b*) 65488, 65486
(*c*) 65486, 65486 (*d*) 65488, 65488

11. What would be the output of the program, if the array begins at address 1200?

```
void main()
{
    int a[5]={ 2,4,6,8,10 };
    printf("%u, %d ", a, sizeof(a) );
}
```

(*a*) 1202, 10 (*b*) 10, 1202
(*c*) 10, 1200 (*d*) 1200, 10

12. A two dimensional array is called as a ____________ .

(*a*) Matrix (*b*) Vector
(*c*) Stack (*d*) queue

13. The number of passes required in Bubble sort for an array of size *n* is ______ .

(*a*) *n* (*b*) *n*-1
(*c*) > *n*-1 (*d*) < *n*-1

14. What will be the output of the following program?

```
void main()
{
    void fun(int x[]);
    int a[] = {1, 2, 3, 4, 5 };
    fun(a);
    printf("\n%d,%d", a[0], a[1]);
    printf("%d,%d,%d\n", a[2], a[3], a[4]);
}
void fun(int x[])
{
    x[2] = x[2] + 2;
    x[4] = x[4] + 4;
}
```

(*a*) 5,6,7,8,9 (*b*) 2,4,6,8,10
(*c*) 3,4,5,6,7 (*d*) None

15. Which would be the output of the following program?

```
void main()
{
    int a[5] = { 1, 2 };
    printf("\n%d,%d,%d\n", a[2], a[3], a[4]);
}
```

(*a*) 0, 0, 0 (*b*) 2, 2, 2
(*c*) 1, 1, 1 (*d*) Garbage

16. What will happen if in a C program, you assign a value to an array element whose subscript exceeds the size of the array?
    (*a*) Nothing will happen, and the element will get a given value.
    (*b*) Compiler will flag a syntax error.
    (*c*) Memory location immediately after the last memory location of the array will be overwritten, and this may cause system to crash.
    (*d*) The array size will automatically be increased to accommodate the new value.
17. Which of the following is a correct output of the program?

```
void main() {
    int i, j, m;
    int a[5] = { 5, 1, 15, 20, 25 };
    i = ++a[1];
    j = a[1]++;
    m = a[i++];
    printf("\n%d, %d, %d\n", i, j, m);
}
```

    (*a*) 2, 1, 15 (*b*) 3, 2, 15
    (*c*) 1, 15, 20 (*d*) 1, 2, 15
18. When an array is passed as an argument, what actually get passed?
    (*a*) All the elements.
    (*b*) Only the first element.
    (*c*) The base address of the array, *i.e.*, address of the first element.
    (*d*) Addresses of all the elements.
19. In which of the following cases mentioning the name of array does not yield the base address?
    (*a*) When the array name is passed to *scanf()* function.
    (*b*) When the array name is used with the *sizeof* operator.
    (*c*) When the array name is operand of the *addressof* (&) operator.
    (*d*) All of above.
20. Which of the following is a correct output of the program?

```
void main() {
    int a[] = { 11 };
    printf("\n%d", 0[a]);
}
```

    (*a*) 0 (*b*) 1
    (*c*) 11 (*d*) Syntax error
21. If the declaration is `ar sample[80];`", what is the length of string that can be correctly represented y *sample*?
    (*a*) 80 (*b*) 79
    (*c*) 81 (*d*) None
22. Which of the following statement is true about strings in C?
    (*a*) Every string must be terminated by null character ('\0').
    (*b*) String is a primitive data type in C.
    (*c*) String can be assigned to another string using assignment operator '='.
    (*d*) All of above.

23. Which of the following function is not supported in C on Unix/Linux operating system?
   (*a*) strlen (*b*) strcat
   (*c*) strcmp (*d*) strrev()
24. What will be the output of the following program?
```
void main()
{
    printf(5+"Fascimile");
}
```
   (*a*) Error (*b*) Fascimile
   (*c*) mile (*d*) Fasci
25. What will be the output of the following program?
```
void main()
{
    char str1[] = "Hello", str2[] = "Hello";
    if ( str1 == str2 )
        printf("Equal");
    else
        printf("Unequal");
}
```
   (*a*) Equal (*b*) Unequal
   (*c*) Error (*d*) None
26. What will be the output of the following program?
```
void main() {
    char a[] = "Visual C++", *b = "Visual C++";
    printf("%d,%d", sizeof(a), sizeof(b));
}
```
   (*a*) 11, 2 (*b*) 2, 11
   (*c*) 11, 11 (*d*) None
27. Which of the following is not a correct way of creating and initializing a *char* array named *vowels* with string value "aeiou"?
   (*a*) `char vowels[] = { 'a', 'e', 'i', 'o', 'u', '\0' };`
   (*b*) `char vowels[] = { 'a', 'e', 'i', 'o', 'u', '\n' };`
   (*c*) `char vowels[] = "aeiou";`
   (*d*) None of above
28. Consider the following statements
```
char *ptr = "abba";
putchar(*ptr);
```
   What will be printed?
   (*a*) first letter (*b*) entire string
   (*c*) it is a syntax error (*d*) last letter
29. If two strings *str1* and *str2* are identical, then the value *y* in the following code segment will be ______.
```
int y;
y = strcmp(str, str2);
```
   (*a*) 1 (*b*) –1
   (*c*) 0 (*d*) None of above

30. Which of the following is a correct output of the following program?

```
void main() {
    char str[] = { "\0" };
    if ( printf("%s", str) )
        printf("\nString is not empty\n");
    else
        printf("\nString is empty\n");
}
```

(*a*) String is empty (*b*) String is not empty
(*c*) Syntax error (*d*) None of above

31. Which of the following is a correct output of the following program?

```
void main() {
    printf("Touch me, ""Touch me!");
}
```

(*a*) Touch me, (*b*) Touch me!
(*c*) Touch me, Touch me! (*d*) Compile time error

32. Which of the following is a correct output of the following program?

```
#include <string.h>
void main() {
    char str[] = "Spider\0man\0";
    printf("%d", strlen(str));
}
```

(*a*) 6 (*b*) 10
(*c*) 11 (*d*) 13

33. Which of the following is a correct output of the following program?

```
void main()
{
    char str[] = "Sales\0man";
    puts(str);
}
```

(*a*) Sales (*b*) man
(*c*) Salesman (*d*) Error

REVIEW EXERCISE

1. What is an array?
2. Is it possible to declare and initialize an array in C simultaneously? If yes, how?
3. What are the rules for naming an array?
4. What is the subscript of the first element of the array?
5. How a real array named *data* with 100 elements will be declared? What will be the subscript of the last element?
6. Is it possible to declare more than one array in the same declaration statement?
10. Is the following array declaration is correct? If not, why?

```
int a(50);
```

11. If the declaration segment is

```
#define ROWS  100
float height[ROWS];
```

Is the following segment to enter data in array is correct?

```
for ( i = 0; i <= ROWS; i++ )
    scanf ( "%f", height[i] );
```

12. When an array is passed as an argument to a function, what is actually passed?
13. When an entire array is passed as an argument to a function, the function can alter the values of the elements of the array. We want to prevent the function from doing so, how this task can be accomplished?
14. Is any thing wrong with the following declaration? Explain.
```
int a[3][] = { { 1, 2, 3 }, { 3, 2, 1 }, { 4, 5, 2 } };
```
15. How a string is stored in C language?
16. What will be the declaration for array named *str* in which string "*C is just like sea*" is to be stored?
17. What is the difference between "A" and 'A'?
18. Consider the following declaration
```
char name[20];
```
What is the maximum length of string which can be stored correctly in name?

20. Consider the following declaration
```
char name[20];
```
Is following the correct way of assigning string "*Rajan Mehta*" to *name*?
```
name = "Rajan Mehta";
```

PROGRAMMING EXERCISE

1. Write a program that accepts an array, interchanges first element with last element, second element with second last element, and so on, and finally prints the new array.
2. Write a function, say *void transpose(float a[][], int n)*, that computes the transpose of a square matrix *A* of size $n \times n$. Note that the transpose of the matrix is to be stored in the original matrix itself not in a separate matrix.
3. Write a program to read a set of *n* integers between 1 and 10 and count the number of times each integer appears in the set. Print the integers in order of frequency, the number occurring the highest number of times first.
4. A department store chain has *m* (≤10) stores, and each store has the same *n* (≤15) departments. The weekly sales of the chain are stored in $m \times n$ array *SALES* (say). Write down the program which
   - Prints the total weekly sales of each store.
   - Prints the total weekly sales of each department.
   - Prints the total weekly sales of the chain.
5. Write a program which inserts an element *d* in *k*th position of an array *a* of size *n*.
6. Write a program which removes an element from the *k*th position of an array *a* of size *n*.
7. Write a program to remove duplicate elements from an array.
8. Given a matrix *A* of order $m \times n$, write a program to find the row having the maximum number of elements which are even.
9. Suppose you are given a sample of *n* (≤100) values between 1 and 18. Write a program that builds the frequency array and prints it in the form of a histogram.

10. A square matrix *A* of size *nxn* is said to be symmetric if element $a_{ij} = a_{ji}$ for all values of *i* and *j*. Write a function, say *int isSymmetric(int a[][], int n)* that returns value 1 if the input matrix is symmetric else returns value 0.
11. Given a matrix *A* of order *m*×*n*, write a program to sort the even rows in ascending order and odd rows in descending order
12. Write a program that asks the user to enter a sentence and then splits out the words in the sentence and puts them in a table.
13. Write a program that asks the user for a list of words and prints the list in reverse order.
14. Suppose you translate a number into a string of digits spelled out, one word for each digit, followed by a single space. For example, the number 407 become the digit string: "*Four Zero Seven*". If the number 407.8, then its digit string is "*Four Zero Seven Point Eight*". Write a program to accept a positive number and print out its digit string.
15. Write a program that accepts an amount in figures and prints that in words. For example, for an amount of Rs. 12, 500.25 it should output the string *Rs. Twelve Thousand Five Hundred and paise Twenty Five only.*
16. Given an array 'a' of size 'n'. Write a program to find all the triplets such that sum of any two elements is equal to the third element.
17. Write a program to find third largest element in an unsorted array
18. Given an array 'a' of size 'n' of non-negative integers. Write a program to reverse every sub-array of size 'k', where 'n' is multiple of 'k'.
19. An element is a leader if it is greater than all the elements to its right side, i.e., following it. And the rightmost element is always a leader. For example in the array {16, 17, 4, 3, 5, 2}, leaders are 17, 5 and 2. Write a program to print all the leaders in the array.
20. A majority element in an array of size 'n' is an element that appears more than 'n/2' times.
    (*a*) If there exist a majority element, then display that element.
    (*b*) If such element doesn't exists, then display "No Majority Element exists".
    Write a program that prints the majority element in an array.
21. Write a program to find the missing integer among the array of 'n' integers that are in the range 1 to (n+1).
22. Given an unsorted array *a* of size *n* of non-negative integers, find a continuous sub-array which adds to a given number *sum*.
23. Equilibrium position in an array is a position such that the sum of elements before of it is equal to the sum of elements after it.
24. Given an array *a* of size *n* of non-negative integers, find the index in the array where equilibrium occurs first in the array.
25. A *left rotation* operation on an array of size *n* shifts each of the array's elements 1 unit to the left. Given an integer, *d*, rotate the array that many steps left.
26. Given an array *a* of *n* integers, a triplet (*a*[*i*], *a*[*j*], *a*[*k*]) is called Magic Triplet if *a*[*i*]<*a*[*j*]< *a*[*k*] and $i < j < k$. Write a program to count number of magic triplets in a given array.

❑ ❑ ❑

Chapter 11

User-Defined Data Types: *Structure, Union & Enumeration*

Learning Outcome

After reading this chapter, students will be able to

- explain the concept of structure
- explain how a structure is different from an array
- declare, initialize and do input/output of structures
- explain the way array of structures are handled
- explain the way structures are passed as arguments to functions
- explain the concept of union
- declare and initialize unions
- explain the concept of enumeration
- describe various aspects of enumerations
- write efficient programs using user-defined data types

11.1 INTRODUCTION

This chapter describes tree useful C constructs, the *structure*, *union,* and *enumeration.*

You can think of a *structure* as an array of closely related items. However, unlike an array, a structure permits the data items to be of different data types. These data items occupy contiguous memory locations.

Union, which are in some way similar to a structure, but in other ways are completely different, allow you to store data items of different data types. But these data items are stored in same memory, and store one value of given type at one time, and another value at different time. The size of memory allocated depends on the data item whose memory requirements are more.

Enumeration is based on the standard integer data type. In an enumeration, each integer value is given an identifier called *enumeration constant*. We can thus use the enumerated constants as symbolic names, which make our programs much more readable.

Unions, which are in some way similar to a structure, but in other ways are completely different, allow you to store data items of different data types. But these data items are stored in same memory, and store one value of given type at one time, and another value at different time. The size of memory allocated depends on the data item whose memory requirements are more.

A *structure*, *union*, or *enumerations* are types defined by the users as per their requirements. This way a user creates a new data type. These newly created data types are known as *user-defined data types*.

This chapter describes aspects associated with these user-defined data types.

11.2 STRUCTURES

You can think of a structure as an array of closely related items. However, unlike arrays, a structure is a collection of related data items that can belong to different data types. The data items that make a structure occupy contiguous memory locations, the same way elements of an array are stored in contiguous memory locations.

Structures can be used to organize complex data in a more meaningful way.

Suppose we want to store information of employees in an organization. This information may include employee's name (character array), department code (integer number), salary (floating point number), and so forth. And we want our program to deal with them as elements of an array.

One possible approach is to use several arrays — a character array for names, an integer array for department code, a floating point array for salary, and so forth. And using common index we can access the information for a particular employee, but this approach obscures the fact that we are dealing with data items related to a single entity, *an employee* in our example.

To solve this sort of problem, C language provides a special data type: *the structure*. As already mentioned, you can think of a structure as group of data items of different data types held together in a single unit. In our example, structure would consist of employee's name, department number, salary, and so forth.

Here are few more examples of structures

Name of Structure	Probable Constituent Data Items
Date	day, month, year
Time	hours, minutes, seconds
Book	title, author, publisher, price
Address	name, house number, locality, city, state, pincode
Student	roll number, name, father's name, grade
Customer	name, address, mobile no
Inventory	item code, description, quantity, unit price
Employee	employee id, name, address, mobile no

In this section, we learn about various aspects associated with structures.

11.2.1 Defining a Structure

The C language has two ways to define a structure: *tagged structure* and *type-defined structure*.

Tagged Structure

The first way to define a structure is to use tagged structure. A tagged structure can be used to declare variables, arguments, and return types. The syntax for declaring a tagged structure is

```
struct STUDENT
{
    . . .
    Field list
    . . .
};
```

(*a*) Format

```
struct STUDENT
{
    int rollNo;
    char name[20];
    char fname[20];
    char grade;
};
```

(*b*) Example

Figure 11.1: Defining a tagged structure

The tagged structure starts with the keyword *struct*. The next element in the definition is *tag*. The tag is the identifier for the structure, and will be used to declare variable, arguments of the functions, and return types for the functions.

If we conclude the structure definition with semicolon after the closing brace, no variables are declared. In this case, the structure is simply a type template with no associated storage.

Type-defined Structure

The more powerful way to define a structure is by using type definition *typedef*. The syntax for declaring a type-defined structure is

```
typedef struct
{
   . . .
   Field list
   . . .
} TYPE;
```

(*a*) Format

```
typedef struct
{
   int rollNo;
   char name[20];
   char fname[20];
   char grade;
} STUDENT;
```

(*b*) Example

Figure 11.2: Declaring a type-defined structure

Type-defined structure differs from tagged structure in following aspects:

1. The keyword *typedef* is used before the *struct* keyword.
2. Identifier after the *struct* keyword is not mandatory as is the case with tagged structure.
3. Identifier is required after the closing brace and before the semicolon.
4. As we will see later, the variable can be declared with the declaration of tagged structure whereas it cannot be with type-defined structures.

A type definition, *typedef*, gives a name to a data type by creating a new type that can be used anywhere a type is permitted. Syntax for type definition is

```
typedef dataType IDENTIFIER;
```

where *dataType* is either built-in data type or user-defined data type, and *IDENTIFIER*, usually in uppercase, is the new and convenient name for the *dataType*. The *typedef* keyword tells the compiler to recognize the *IDENTIFIER* as synonymous of *dataType*.

Following are few more examples of definitions of structures:

```
struct Date
{
  int day;
  int month;
  int year;
};
```

```
struct Time
{
  int hours;
  int minutes;
  int seconds;
};
```

```
struct Book
{
  char title[25];
  char author[25];
  char publisher[15];
  float price;
};
```

```
struct Customer
{
  char name[25];
  char address[80];
  long mobile_no;
};
```

```
struct Inventory
{
  char item_code[25];
  char desc[25];
  int qty;
  float unit_price;
};
```

```
struct Address
{
  char name[25];
  int house_no;
  char locality[25];
  char city[15];
  char state[15];
  long pincode;
};
```

11.2.2 Declaring Structure Variables

Once a structure is defined, we can declare variables using these declarations. Usually, the definition is placed in the global area of the program, *i.e.*, before the *main()* function, so that it is visible to all the functions in the program. The variables, on the other hand, are usually declared in the local area of the functions or in the argument list in the functions headers. Following statements shows the way structure variables are declared.

For tagged structure, the variable is declared as

```
struct STUDENT aStudent;
```

For type-defined structure, the variable is declared as

```
STUDENT bStudent;
```

By comparing both the declarations, you will find the type-defined structure is handier in the declarations of the variable because you don't need to use the *struct* keyword. This can be quite time and effort saving if the variable declarations are required at many placed in the program.

Further note that, in case of tagged structure, variable declaration can be combined with its definition as shown below. In fact any number of variables can be declared.

```
struct EMPLOYEE {
  int   code;
  char  name[25];
  char  department[15];
  float salary;
} aEmployee, bEmployee;
```

Usually, this approach of declaring variables is not recommended, because of the following reasons:

1. Since structure definition is placed in the global area, and so the variables will also be global.
2. In case we want to declare local variables using this approach, we may have to place the structure definition in local scope, which will make definition of structure invisible to other functions.

11.2.3 Initializing Structures

Like simple variables and arrays, structure variables can also be initialized at declaration time. The format used is quite similar to that used to initialize arrays.

The following segment illustrates this

```
typedef struct
{
  int    rollno;
  char   name[25];
  int    age;
  float height;
} STUDENT;

STUDENT aStudent = { 1000, "Surbhi", 18, 5.6 };
```

Like partially initialized arrays, if a structure variable is partially initialized, the remaining numeric fields will be initialized to zero and the string fields will be initialized to null.

11.2.4 Accessing Structure Elements

Individual structure elements of a structure variables can be accessed using the *dot operator* '.' This dot operator is also referred to as *membership operator*.

The syntax for accessing a structure element is as follows

```
  sname.vname
```

where *sname* is the name of structure variable, and vname is the name of the data field.

Consider the following statements

```
struct Student {
  int  rollno;
  char name[25];
  char fname[25];
  char grade;
};

Student aStudent;
```

The individual elements of structure variable *aStudent* will be accessed as shown below:

```
aStudent.rollno      // reference to element rollno
aStudent.name        // reference to element name
aStudent.fname       // reference to element fname
aStudent.grade       // reference to element grade
```

11.2.5 Reading/Writing Structures

We can also read into and write data from a structure element in the same manner as we do with simple variables. For example, the elements of *aStudent* structure, we can read data, interactively, from the keyboard and place in the *aStudent* structure using following statements

```
printf("\nEnter roll number of the student : ");
scanf("%d", &aStudent.rollno);
printf("\nEnter name of the student : ");
scanf("%d", aStudent.name);
printf("\nEnter age of the student : ");
scanf("%d", &aStudent.age);
printf("\nEnter height of the student : ");
scanf("%d", &aStudent.height);
```

Note that in the *scanf()* function, the expression

```
&aStudent.rollno
```

is interpreted by the compiler as

```
&(aStudent.rollno)
```

because of higher precedence of direct selection operator over *addressof* operator (&), and that is what exactly is required.

Listing 11.1

```
/* Program to illustrate way data is entered into structures */

#include<stdio.h>
#include<sting.h>

struct EMPLOYEE
{
    int code;
    char name[25];
    char department[15];
    float salary;
};

int main()
{
```

```
    struct EMPLOYEE aEmployee;
    printf("Enter employee's code: " );
    scanf( "%d", &aEmployee.code );
    fflush(stdin);
    printf("Enter employee's name: " );
    gets( aEmployee.name );
    printf("Enter employee's department: " );
    gets( aEmployee.department );
    printf("Enter employee's salary: " );
    scanf("%f", &aEmployee.salary );
    printf("\n\nParticulars of employee as" );
    printf(" entered by user\n" );
    printf("\nEmployee's code: %d", aEmployee.code );
    printf("\nEmployee's name: %s", aEmployee.name );
    printf("\nEmployee's department: %s",aEmployee.department);
    printf("\nEmployee's salary: %.2f\n", aEmployee.salary );
    return 0;
}
```

Test Run

```
Enter employee's code: 826
Enter employee's name: Rajesh Kumar
Enter employee's department: Computer Science
Enter employee's salary: 20000
Particulars of employee as entered by user
Employee's code: 826
Employee's name: Rajesh Kumar
Employee's department: Computer Science
Employee's salary: 20000.00
```

Following is a program that uses structures and direct selection operator. It multiplies two fractions and prints the result. For example, if the fractions are 2/5 and 3/5, it prints the result as 6/25.

Listing 11.2

```
/*   Program to multiply two fractions      */
#include<stdio.h>
struct FRACTION
{
  int nom;        /*   nominator    */
  int denom;      /*   denominator */
};
int main()
{
    struct FRACTION aFraction, bFraction, cFraction;
    printf( "Enter first fraction as x/y : " );
    scanf( "%d%*c%d",&aFraction.nom, &aFraction.denom);
```

```
    printf( "Enter second fraction as x/y : " );
    scanf( "%d%*c%d",&bFraction.nom, &bFraction.denom);
    cFraction.nom = aFraction.nom * bFraction.nom;
    cFraction.denom = aFraction.denom * bFraction.denom;
    printf( "\nResult of (%d/%d) x (%d/%d) = (%d/%d)\n",
            aFraction.nom, aFraction.denom,
            bFraction.nom, bFraction.denom,
            cFraction.nom, cFraction.denom );
    return 0;
}
```

Test Run

```
Enter first fraction as x/y : 2/5
Enter second fraction also as x/y : 3/5
Result of (2/5) x (3/5) = (6/25)
```

11.2.6 Assigning of Structures

A structure is an entity that can be treated as a whole. There is only one operation that can be performed on the structure as a whole is that a structure can be assigned/copied into another structure of the same type using the assignment operator.

Again consider the following statements

```
typedef struct
{
   int   rollno;
   char  name[25];
   int   age;
   float height;
} STUDENT;
STUDENT aStudent, bStudent;
```

The following assignment statement

```
aStudent = bStudent;
```

assigns the contents of structure *bStudent* to *aStudent*.

It is a unique feature of structures as compared to arrays, where in order to copy one array to another array of same type, we have to set up a loop to copy element-by-element.

This is rather an amazing capability when you think about it. When you assign one structure to another, all the values in the structure are assigned to corresponding structure elements. Simple assignment statement cannot be used this way for arrays. That is, to assign one array to another, we have to assign element by element.

11.2.7 Arrays of Structures

In many practical situations, we need to create an array of structures. As an example, we need to create an array of students to work with a group of students in a class. By putting the data of students in an array of structures, we can quickly and efficiently process the student's data.

An array of structures to store the data of students will be declared as

```
STUDENT students[50];
```

11.2.7.1 Accessing Elements of Array of Structures

Individual elements of a structure in an array of structures are accessed by referring to structure variable name, followed by index, followed by direct selection operator, and ending with the structure element desired.

In our example of students of a class, the name of the ith student can be accessed as follows:

```
students[i].name;
```

11.2.7.2 Initializing Array of Structures

Like any other array, an array of structures can also be initialized. Since each element of array structures is a different structure, we need to include elements of each structure in a separate set of braces as shown:

```
STUDENT students[50] = { { 1000, "Monika",{56,76,85,69}, 'A' },
                         { 1001, "Ram", {50,66,70,60}, 'B' },
                                :
                         { 1049, "Raju", {65,62,68,59}, 'B' },
                       };
```

Following program illustrates input, processing and output of array of structures.

Listing 11.3

```
/*
*   Program to illustrate processing of array of structures by
*   storing list of students in memory, and computes their grades
*/
#include <stdio.h>

typedef struct {
   int rollno;
   char sname[25];
   int  marks[4];
   char grade;
} STUDENT;
```

```
int main()
{
   STUDENT students[50];
   int totalMarks, i, j, n;
   printf( "\n\nEnter number of students in a class : " );
   scanf( "%d", &n );
   for ( i = 0; i < n; i++ ) {
      printf("\nParticulars of student #%d\n\n", i+1);
      printf( "Enter students' roll number : " );
      scanf("%d", &students[i].rollno );
      printf( "\nEnter students' name : " );
      gets( students[i].sname );
      printf( "\nEnter students marks in four subjects\n\n" );
      for ( j = 0; j < 4; j++ )
         scanf("%d", &students[i].marks[j] );
   }
   for ( i = 0; i < n; i++ ) {
      totalMarks = 0;
      for ( j = 0; j < 4; j++ )
         totalMarks += students[i].marks[j];
      if ( totalMarks > 75 )
         students[i].grade = 'A';
      else if ( totalMarks > 60 )
         students[i].grade = 'B';
      else if ( totalMarks > 50 )
         students[i].grade = 'C';
      else if ( totalMarks > 40 )
         students[i].grade = 'D';
      else
         students[i].grade = 'F';
   }
   printf("\nPerformance of students\n\n");
   printf("%-10s", "Roll No.");
   printf("%-30s", "Name");
   printf("%6s\n", "Grade");
   for ( i = 0; i < n; i++ ) {
      printf("%-10d", students[i].rollno);
      printf("%-30s", students[i].sname);
      printf("%4c\n", students[i].grade);
   }
   return 0;
}
```

Test Run

```
Enter number of students in a class : 3

Particulars of student #1

Enter students' roll number : 1000
Enter students' name : Ram Kumar
Enter students marks in four subjects
98 99 89 88
```

```
Particulars of student #2

Enter students' roll number : 1001
Enter students' name : Mohit Kumar
Enter students marks in four subjects
65 55 45 60

Particulars of student #3

Enter students' roll number : 1002
Enter students' name : Jaspreet Singh
Enter students marks in four subjects
41 44 45 50

Performance of students

SNo.          Student's Name      Grade
  1                    Sonu         A
  2                  bhollu         C
  3                jaspreet         D
```

11.2.8 Structures and Functions

In order to be fully useful, we must be able to pass structures to functions as arguments and returning them from functions.

```
typedef struct {
   int day;
   int month;
   int year;
} DATE;
void printDate(DATE);
void main()
{
   DATE aDate = { 10, 6, 2008 };
   int x;
   printDate(aDate);
   /*  more statements    */
}
```

```
/*  function definition  */
void printDate(DATE tDate )
{

   /*  local declarations */
   /*  other statements   */

}
```

'igure 11.3: Passing structure by value

Structure can be passeu oy value as well as by reference. In this section, we will consider passing a structure by value only. Passing structure by reference, we will consider in the *Chapter 12*.

Listing 11.4

```
/*  Program to illustrate passing of a structure by value  */
#include <stdio.h>
typedef struct
{
   int day;
```

```
   int month;
   int year;
} DATE;
void printDate( DATE aDate );      /*  function declaration  */
void main()
{
   DATE tDate = { 10, 6, 2008 }; /* initialize a structure  */
   printDate( tDate );            /*  function call          */
}
void printDate( DATE aDate )
{
  printf( "\nDate in format dd/mm/yy: %02d/%02d/%2d\n",
          aDate.day, aDate.month, aDate.year );
}
```

Test Run

```
Date in format dd/mm/yy: 10/06/2008
```

A function can also return a value of structure type as well via *return* statement as illustrated below.

```
typedef struct
{
   int day;
   int month;
   int year;
} DATE;
DATE getDate(void);
void main()
{
   DATE aDate;
   int x;
   aDate = getDate();
   /*  more statement    */
}
```

```
/*  function definition    */
DATE getDate(void)
{
   DATE tDate;
   /*  local declarations */
   /*  more statements    */
   return tDate;
}
```

Figure 11.4: Function returning a structure

Listing 11.5

```
/* Program to illustrate function returning a structure  */
#include <stdio.h>
typedef struct
{
   int day;
   int month;
   int year;
} DATE;
DATE getDate( void );          /*  function declaration    */
```

```
void main()
{
    DATE tDate;
    printf( "Enter date in format dd/mm/yyyy : " );
    tDate = getDate();
    printf( "\nDate entered by you: %02d/%02d/%2d\n",
            tDate.day, tDate.month, tDate.year );
}
DATE getDate( void )
{
    DATE xDate;
    scanf( "%d/%d/%d", &xDate.day, &xDate.month, &xDate.year );
    return xDate;
}
```

Test Run

```
Enter date in format dd/mm/yy: 10/6/2008
Date entered by you: 10/06/2008
```

11.3 UNIONS

As mentioned in the very beginning of this chapter, both structures and unions are used to group a number of variables of different type in single unit. While a structure enables us to treat this unit as a number of different variables stored at different places in memory, a union enables us to treat the same space in memory as a number of different variables. That is, a union permits a section of memory to be treated as a variable of one type on one occasion, and as a different variable of a different type on another occasion.

If two or more variables are mutually exclusive, unions provides an opportunity to the programmer to economize the usage of memory by sharing memory among those variables by declaring them members of a union.

The syntax for declaring a union, declaring variables of union type, accessing elements of a union, is identical to that of structures, except for storage of members of union and their initialization.

11.3.1 Defining a Union

Like structures, the C language also provides two ways to declare a union: *tagged union* and *type-defined union*.

Tagged Union

The first way to declare a union is to use tagged union. A tagged union can be used to define variables, arguments, and return types.

The syntax for declaring a tagged union is

```
union MIXTYPES
{
    . . .
    Field list
    . . .
};
```

(*a*) Format

```
union MIXTYPES
{
    char   cVar;
    int    iVar;
    flota  fVar;
    double dVar;
};
```

(*b*) Example

Figure 11.5: Declaring a tagged union

The tagged union starts with the keyword *union*. The next element in the declaration is *tag*. The tag is the identifier for the union, and will be used to declare variable, arguments of the functions, and return types for the functions.

If we conclude the union declaration with semicolon after the closing brace, no variables are defined. In this case, the union is simply a type template with no associated storage.

Type-defined Union

The more powerful way to declare a union is by using type definition *typedef*. The syntax for declaring a type-defined union is

```
typedef union
{
    . . .
    Field list
    . . .
} TYPE;
```

(*a*) Format

```
typedef union
{
    char   cVar;
    int    iVar;
    flota  fVar;
    double dVar;
} MIXTYPES;
```

(*b*) Example

Figure 11.6: Declaring a type-defined union

Type-defined union differs from tagged union in following aspects:

1. The keyword *typedef* is used before the *union* keyword.
2. Identifier after *union* keyword is not mandatory as is the case with tagged union.
3. Identifier is required after the closing brace and before the semicolon.
4. As we will see later, the variable can be declared with the declaration of tagged union whereas it cannot be with type-defined unions.

To understand how the elements of a union variable are stored in memory, consider the following declarations.

```
typedef union {
    char   cVar;
    int    iVar;
    float  fVar;
    double dVar;
} EXAMPLE;
EXAMPLE bExample;
```

11.3.2 Declaring Union Variables

Once a union is defined, we can declare variables using these definitions. Usually, the definition is placed in the global area of the program, *i.e.*, before the *main()* function, so that it is visible to all the functions in the program. The variables, on the other hand, are usually declared in the local area of the functions or in the argument list in the functions headers. Following statements shows the way union variables are declared.

For tagged union, the variable is declared as

```
union MIXTYPES aUnion;
```

For type-defined structure, the variable is declared as

```
MIXTYPES bUnion;
```

11.3.4 Initializing a Union

Like structures, the unions cannot be initialized at the declaration time. Consider the following type-defined union declaration

```
typedef union  {
   char   cVar;
   int    iVar;
   float  fVar;
} EXAMPLE;
```

then the declaration statement

```
EXAMPLE bExample = { 'A', 300, 250.25 };
```

will not be acceptable.

11.3.4 Accessing Union Elements

Like structures, the elements of a union can be accessed using direct selection operator. Consider again the following declarations

```
typedef union
{
     char    cVar;
     int     iVar;
     float   fVar;
} EXAMPLE;
EXAMPLE bExample = { 'A', 300, 250.25 };
```

We can refer to the individual elements of the union *bExample* as shown below:

```
bExample.cVar
bExample.iVar
bExample.fVar
```

Listing 11.6

```
/*  Program to illustrate the use of a union  */
#include <stdio.h>
typedef union
{
   char    cVar;
   int     iVar;
   float   fVar;
} EXAMPLE;
void main()
{
   EXAMPLE bExample;
   printf( "\nSize of EXAMPLE = %d\n", sizeof(EXAMPLE) );
   bExample.cVar = 'A';
   printf( "\nbExample.cVar = %c\n", bExample.cVar );
   bExample.iVar = 300;
   printf( "\nbExample.iVar = %d\n", bExample.iVar );
   bExample.fVar = 250.25;
   printf( "\nbExample.fVar = %.2f\n", bExample.fVar );
}
```

Test Run

```
Size of EXAMPLE = 4
bExample.cVar = A
bExample.iVar = 300
bExample.fVar = 250.25
```

11.4 UNIONS AND STRUCTURES

Just as there can be arrays within structures, structures within structures, array of structures, there can be structures within unions, and unions within structures. In this section, we will consider a case of union of structures.

Consider the following structure declarations

```
typedef struct
{
   int code;
   char name[25];
   char department[15];
   float salary;
} EMPLOYEE;
```

```
typedef struct
{
   int  rollno;
   char name[25];
   int  MARKS[4];
   char grade;
} STUDENT;
```

Union of these two structures will be declared as

```
typedef union  {
   EMPLOYEE aEmployee;
   STUDENT  aStudent;
} COMMON;
COMMON example;
```

The elements of this union of structures are accessed using direct selection operator twice as we do with nested structures. For example

```
example.aStudent.rollno
```

refer to element *rollno* in the structure *aStudent* in union *ex*, where as

```
example.aEmployee.code
```

refer to element *code* in the structure *aEmployee* in union *ex*.

11.5 ENUMERATIONS

An *enumeration* is a user-defined data type based on the standard integer data type. In an enumerated data type, each integer value is given an identifier called *enumeration constant*. We can thus use the enumerated constants as symbolic names, which make our programs much more readable.

To make the usefulness of enumerated of data type more clear, let us consider a practical application dealing with personnel database of an organization. In an organization, there is large variety of designations (categories of employees). In absence of enumerated data type, the one of the possible options left is to use numeric codes such as 0, 1, 2, . . ., for various category of employees.

For example, code 0 for Chief Managing Director (CMD), 1 for Managing Director (MD), 2 for Executive Director (ED), 3 for Finance and Development Officer (FDO), 4 for Chief Manager (CM), 5 for Deputy Chief Manager (DCM), and so on. This list will be very long.

The main problems with this coding scheme are:

1. It is very difficult to memorize this large number of codes.
2. It makes code programs more cryptic, and thus reduces its readability & understandability.

The use of enumerated data type helps to overcomes all these problems, by associating symbolic names with these categories of employees.

11.5.1 Defining an Enumeration

To declare an enumerated data type, we must declare its identifier and its values. Since enumerated data type id derived from the integer data type, its operations are the same as for integers. The syntax for declaring an enumerated data type is

```
enum typeName { identifier list };
```

The keyword *enum* is followed by an identifier which is followed by an identifier list enclosed in a set of braces, and finally terminated by semicolon. The identifier list contains enumeration identifiers (symbolic names) separated by commas.

Each enumeration identifier is assigned an integer value. If we do not explicitly assign the values, the compiler assigns value 0 to the first enumeration identifier, value 1 to the second enumeration identifier, and so on until all of the enumeration identifiers have a value.

For example, consider an enumerated data type for *Boolean* constants as defined in the following statement

```
enum Boolean { FALSE, TRUE };
```

Note that, as convention, we use uppercase alphabetic character for enumeration identifiers. This convention also helps to distinguish an enumeration identifier from normal identifiers.

The *Boolean* type has two and only two possible values, 0 and 1 with the FALSE enumeration identifier representing the value 0 and TRUE enumeration identifier representing the value 1. Similarly, for the following declaration statement

```
enum Color { RED, BLUE, GREEN, WHITE };
```

the *Color* type has four possible values, the range being 0..3, with the RED enumeration identifier representing the value 0 and BLUE enumeration identifier representing the value 1, GREEN enumeration identifier representing the value 2, and finally the WHITE enumeration identifier representing the value 3.

11.5.2 Declaring Enumeration Variables

To declare an enumerated data type, we must declare its identifier and its values. Since Once the enumerated data type is declared, we can create variable from it just as we can create variables from the standard data types.

Following example, defined three variable of *Color* type The Boolean type has two and only two possible values, 0 and 1 with the FALSE enumeration identifier representing value 0 and TRUE enumeration identifier representing value 1.

```
enum Color flowerColor, skyColor, bodyColor;
```

Since, the enumerated data type is internally represented as integer, for each variable of enumerated data type, compiler reserves 2-bytes of memory.

10.5.3 Initialization of Enumerations

While the compiler automatically assigns values to the enumerated types starting with 0, we can override it and assign our own values. Consider following enumerated data type declarations

```
enum Days { SUN, MON, TUE, WED, THU, FRI, SAT };
```

The compiler assign value 0 to SUN, 1 to MON, 2 to TUE, 3 to WED, 4 to THU, 5 to FRI and 6 to SAT. This assignment of values can be confusing, and to avoid this confusion, we would like to start with the value 1 for SUN. This can be done using the following declaration statement

```
enum Days { SUN = 1, MON, TUE, WED, THU, FRI, SAT };
```

Note that you don't have to assign value to each enumeration identifier. If we don't specify the value for each of the enumeration identifier, the compiler automatically takes the next value.

Therefore, compiler assign value 2 to MON, 3 to TUE, 4 to WED, 5 to THU, 6 to FRI and 7 to SAT. This assignment of values can be confusing, and to avoid this confusion, we would like to start with the value 1 for SUN.

10.5.4 Assigning Values to Enumerated Variables

Once we have declared an enumerated variable, we can assign values to it. Remember that only those values that are specified during declaration of enumeration data type can be assigned. If we attempt to assign any other values, it will result in an error.

Consider the following variable declaration statement

```
enum Color xColor, yColor, zColor;
```

The following statements are valid assignment statement

```
xColor = RED;
yColor = BLUE;
```

However, the following statement is invalid assignment statement

```
zColor = PINK;
```

since there is no PINK enumeration identifier in the identifier list.

Once a variable is declared, defined and assigned a value, we can use it further for any computation just like an integer value; even the value of one enumerated variable can be assigned to another variable of same enumerated data type.

The following assignment statements are valid

```
xColor = yColor;
zColor = yColor;
```

10.5.5 Enumeration Data Type Conversion

Enumerated data types can be implicitly and explicitly casted. When we assign an enumerated type to an integral type, the compiler automatically cast it, *i.e.*, assigns its integer value to integral type. However, when we assign is an integral type to an enumerated type, we either gets a warning or an error depending on the compiler.

Following example demonstrate implicit casts:

```
int x;
enum Days day1;
x = WED;          /*  valid, x is assigned value 4  */
day1 = 5;         /*  warning or error  */
```

We can explicitly cast an integer to an enumerated type as shown below:

```
enum Days day1;
day1 = (Days) 5;    /*  valid, day1 is assigned       */
                    /*  enumeration identifier WED    */
```

11.5.6 Input/Output of Enumerated Data

Because the enumerated data type is derived data type, it cannot be read and written using the formatted input/output functions. When we read an enumerated type, we must read it as integer. When we write an enumerated type, it displays an integer. The following example shows how we can read and write two days:

```
enum Days { SUN = 1, MON, TUE, WED, THU, FRI, SAT };
enum Days day;
scanf("%d", &day);                  /*  valid input is from 1-7  */
printf("day = %d", day);            /*  outputs 1-7   */
```

The above method of reading and writing enumerated types is very cumbersome, and makes the input/output handling of enumerated types impractical.

However, we can output the enumerated type indirectly as shown below:

```
enum Days day;
if ( day == MON )
   printf( "\nMonday\n" );
```

Similarly, input of enumerated data can be performed indirectly as shown below:

```
enum Days day;
printf( "\nDay codes are as follows\n" );
printf( "\n1 - Sunday\n2 - Monday\n3 - Tuesday\n4 - Wednesday" );
printf( "\n5 - Thurssay\n6 - Friday\n7 - Saturday\n" );
printf( "\nEnter day code : " );
scanf( "%d", &day );
if ( day == 1 )
   day = SUN;
else if ( day == 2 )
   day = MON
     :
```

11.5.7 Operations on Enumerations

In general, any operation defined on integers can be performed on enumerated data types. We can add, subtract, multiply, and divide enumerated data types provided the resulting values lies in the range of integral values assigned to enumeration identifiers. In case, the resulting value after operation is not in the defined set, it may cause the program to fail. It is important to note that the C language does not ensure that the values stored in enumerated variables are valid.

We can also pass then to library and user-defined functions.

Enumerated data types can be compared using relational operators, as shown below:

```
if ( xColor == yColor )
    . . .
if ( xColor == BLUE )
    . . .
```

Another common use of the enumerated data type is with the *switch* statement. Because the enumerated data types are derived from the integer types, they can be used in the case expressions as shown in the following segment of code

```
enum Days { SUN, MON, TUE, WED, THU, FRI, SAT };
enum Days dayOfWeek;
switch(dayOfWeek)
{
   case SUN: /* statements-block-1 */
             break;
   case MON: /* statements-block-2 */
             break;
   case TUE: /* statements-block-3 */
             break;
   case WED: /* statements-block-4 */
             break;
   case THU: /* statements-block-5 */
             break;
   case FRI: /* statements-block-6 */
             break;
   case SAT: /* statements-block-7 */
             break;
   default:  /* default-statements-block */
}
```

ILLUSTRATIVE EXAMPLES

Example 11.1: Given the following declaration

```
struct STUDENT
{
  int  rollNo;
  char name[31];
  int   marks;
};
```

to represent the information of a student.

Using this structure, write a function that takes list of students, in the form of an array, as its argument, and sorts them in the descending order of the marks secured by the students.

Using the above structure and function, write a program that reads the data for *n* students and displays them in descending order of marks secure by the students.

Listing 11.7

```
/* Program to the given information of students in
*  descending order of the marks secured by them    */

typedef struct {
  int  rollno;
  char name[20];
  int  marks;
} STUDENT;

void arrangeBubble(STUDENT students[], int n);

int main()
{
  STUDENT students[50];
  int i, n;
  printf( "\n\nEnter number of students : " );
  scanf( "%d", &n );

  for ( i = 0; i < n; i++ ) {
     printf("\nParticulars of student #%d\n\n", i+1);
     printf( "Enter students' roll number : " );
     scanf("%d", &students[i].rollno );
     printf( "Enter students' name : " );
     gets(students[i].name );
     printf( "Enter students' marks : " );
     scanf("%d", &students[i].marks );
  }

  arrangeBubble(students,n);    /*  function call  */
  printf("\nStudent's Information after sorting\n\n");
  printf("%-10s", "Roll No.");
  printf("%-20s", "Name");
  printf("%6s\n\n", "Marks");

  for ( i = 0; i < n; i++ ) {
     printf("%-10d", students[i].rollno);
     printf("%-20s", students[i].name);
     printf("%4d\n", students[i].marks);
  }
  return 0;
}
void arrangeBubble(STUDENT st[], int n)
{
   int i, j, k;
   STUDENT temp;
   for ( i = 1; i < n; i++ )
   {
      k = n-i;
      for ( j = 0; j < k; j++ )
      {
         if ( st[j].marks < st[j+1].marks )
         {
            temp = students[j];
            st[j] = st[j+1];
            st[j+1] = temp;
```

```
            }
        }
    }
}
```

Test Run

```
Enter number of students : 5
Particulars of student #1
Enter students' roll number : 1000
Enter students' name : Ravi
Enter students' marks : 80
Particulars of student #2
Enter students' roll number : 1001
Enter students' name : Shankar
Enter students' marks : 79
Particulars of student #3
Enter students' roll number : 1002
Enter students' name : Ankit
Enter students' marks : 65
Particulars of student #4
Enter students' roll number : 1003
Enter students' name : Rupinder
Enter students' marks : 75
Particulars of student #5
Enter students' roll number : 1004
Enter students' name : Mehak
Enter students' marks : 88
Student's Information after sorting
Roll No.  Name              Marks
 1004     Mehak              88
 1000     Ravi               80
 1001     Shankar            79
 1003     Rupinder           75
 1002     Ankita             65
```

Example 11.2: Use a structure to model a complex number. Using this structure write functions to accomplish the following tasks:

- Function that returns the structure read from the keyboard.
- Function that takes two structures using call be value mechanism as its arguments and return their sum.
- Function that takes single structure using call be value mechanism as its arguments and prints it.

Use these functions in the program to demonstrate their working.

Listing 11.8

```
/*
 *   Program to demonstrate passing structure to a function using
 *   call by value mechanism and function returning a structure
 */
#include <stdio.h>
typedef struct
{
   float real;
   float imag;
} COMPLEX;
COMPLEX input(void);          /*  function declarations    */
void output(COMPLEX);
COMPLEX add(COMPLEX, COMPLEX);
COMPLEX multiply(COMPLEX, COMPLEX);
int main()
{
   COMPLEX n1, n2, n3, n4;
   printf("\nProgram to add two complex numbers\n\n");
   printf("\nEnter first complex number\n");
   n1 = input();
   printf("\nEnter second complex number\n");
   n2 = input();
   n3 = add(n1,n2);
   printf("\nSum of given complex numbers\n\n");
   output(n3);
   n4 = multiply(n1,n2);
   printf("\nProduct of given complex numbers\n\n");
   output(n4);
   return 0;
}
COMPLEX input(void)
{
   COMPLEX t;
   printf("\nEnter real part : ");
   scanf("%f", &t.real);
   printf("Enter imaginary part : ");
   scanf("%f", &t.imag);
   return t;
}
void output(COMPLEX t)
{
   printf("Real part : %.2f", t.real);
   printf("\nImaginary part : %.2f", t.imag);
}
COMPLEX add(COMPLEX a, COMPLEX b)
{
   COMPLEX t;
   t.real = a.real + b.real;
   t.imag = a.imag + b.imag;
   return t;
```

```
}
COMPLEX multiply(COMPLEX a, COMPLEX b)
{
    COMPLEX t;
    t.real = a.real * b.real - a.imag * b.imag;
    t.imag = a.real * b.imag + a.imag * b.real;
    return t;
}
```

Test Run

```
Program to add two complex numbers
Enter first complex number
Enter real part : 1.0
Enter imaginary part : 2.5
Enter second complex number
Enter real part : 2.5
Enter imaginary part : -1.5
Sum of given complex numbers
Real part : 3.50
Imaginary part : 1.00
Product of given complex numbers
Real part : 6.25
Imaginary part : 4.75
```

MULTIPLE CHOICE QUESTIONS

1. A structure is
   (*a*) Scalar data type (*b*) Derived data type
   (*c*) Primitive data type (*d*) None of above
2. Which of the following statement is true about structures?
   (*a*) The elements of the structure must be of different type.
   (*b*) Structure tag is mandatory in all cases.
   (*c*) While declaring a variable of structure type, you need to prefix keyword *struct* before tag name.
   (*d*) Structure element are accessed using star (*) operator.
3. A structure is a data type in which
   (*a*) Each element must have the same type.
   (*b*) Elements can be of different type.
   (*c*) Each element must be of pointer type.
   (*d*) None

4. Consider the following declaration

```
struct ex {
    char cVvar;
    int iVar;
    long lVar;
};
```

What value will be returned by the *sizeof* operator called on *ex*?

(*a*) 4 (*b*) 7
(*c*) 1 (*d*) 6

5. Keyword used to create a structured data type is

(*a*) structure (*b*) structr
(*c*) struct (*d*) struc

6. The operator used to access the structure member is

(*a*) * (*b*) .
(*c*) & (*d*) []

7. If one or more members of a structure are other structures, then structure is known as

(*a*) Nested structure (*b*) Structured structure
(*c*) Invalid structure (*d*) Self-referential structure

8. Identify the wrong statement

(*a*) Structure variable can be passed as an argument.
(*b*) Member of a structure can be passed as an argument.
(*c*) More than one structure variable can be declared in a single declaration statement.
(*d*) None of above.

9. Consider the statement: The size of a *struct* is always equal to the sum of its members.

(*a*) Valid (*b*) Can't say
(*c*) Invalid (*d*) None

10. Consider the following statement

```
honda.engine.bolt = 20;
```

Which of the following statement is incorrect?

(*a*) Structure *bolt* is nested within structure *engine*.
(*b*) Structure *engine* is nested within structure *honda*.
(*c*) Structure *honda* is nested within structure *engine*.
(*d*) Structure *bolt* is nested within structure *honda*.

11. Consider the following code segment

```
strcut Date {
    int day, month, year;
};
struct Date dt, *pdt;
pdt = &dt;
```

Which of the following statement assigns value 11 to month?

(*a*) pdt.month = 11;. (*b*) &pdt.month = 11;

(*c*) pdt->month = 11; (*d*) None of above

12. Enumeration (or enum) is a ______ data type in C?

(*a*) built-in (*b*) pre-defined

(*c*) user-defined (*d*) derived

13. What is the correct syntax of enum?

(*a*) `enum flag{constant1, constant2, constant3, ...};`

(*b*) `enum flag(constant1, constant2, constant3, ...};`

(*c*) `enum flag[constant1, constant2, constant3, ...];`

(*d*) `enumflag{constant1, constant2, constant3, ...};`

14. If we do not explicitly assign values to *enum* names, the compiler by default assigns value ___ .

(*a*) NULL (*b*) 1

(*c*) 0 (*d*) –1

15. A user defined data type, which is used to assign names to integral constants, is

(*a*) array (*b*) structure

(*c*) enum (*d*) typedef

16. What will be the output of the following C code?

```
enum color { blue, red, yellow };
main()
{
      enum color c;
      c=yellow;
      printf("%d",c);
}
```

(*a*) 1 (*b*) 2

(*c*) 0 (*d*) Error

17. What will be the output of the following C code?

```
enum sample { a = 2, b = 2.5 };
main()
{
      printf("%d,%f",a,b);
}
```

(*a*) 2,2.5 (*b*) 0,1

(*c*) 2,3 (*d*) Error

REVIEW EXERCISE

1. What is the main reason for using structures?
2. How a structure is different from an array?
3. What is the structure tag and what is its purpose?
4. What are different ways of declaring a structure variable?

5. Is any there thing wrong with the following?

```
struct Time
{
    int hrs;
    int mts;
    int secs;
}
time t1;
```

6. What is wrong with the following, if any?

```
struct Time
{
    int hrs =10;
    int mts = 20;
    int secs = 35;
} t1, t2, t3;
```

7. Identify the errors, if any, in the following

```
struct {
    int hrs;
    int mts;
    int secs;
} time;
time t1, t2, t3;
```

8. Give the output of the following program:

```
#include <stdio.h>
struct MyBox {
    int Length, Breadth, Height;
};
void Dimension( MyBox M )
{
    printf("\n%d x %d x %d\n",M.Length,M.Breadth,M.Height);
}
void main()
{
    MyBox B1 = { 12, 20, 8 }, B2, B3;
    ++B1.Height;
    Dimension(B1);
    B3 = B1;
    ++B3.Length;
    B3.Breadth++;
    Dimension(B3);
    B2 = B3;
    B2.Height += 5;
    B2.Length--;
    Dimension(B2);
}
```

PROGRAMMING EXERCISE

1. Design a structure named *distance* to store a length in *yards*, *feets* and *inches*. Using this structure, write a program that accepts two measurements from the user as represented by these structures and prints the absolute difference between these measurements.

2. Design a structure named *student* to store data about a student which contains following elements — *rollno* of type *int*, *name* an array of type *char* of size 20, *college* an array of type *char* of size 40, and *score* of type *float*. Assume that there are not more than 100 students. Write a program to input the data about students, and output the stored data according to the merit of the students.
3. Consider the following declaration:
```
struct EMPLOYEE
{
    int code;
    char name[31];
    float salary;
};
```
Given that there are *n* (≤50) employees in an organization. Write a program that output the details of the employee having highest salary.
4. Create a structure named *Time* to model a time in 12 hour system having three fields as *hours*, *minutes* and *seconds*. Write a function called *elapsedTime* that takes two arguments, the start time and the end time. The function should return a *Time* structure containing the time elapsed between the two arguments. Write a program to test the above function.
5. A point in plane in Cartesian system can be represented by its coordinates *x* and *y*. We can create a structure to represent a point in plane as shown below:
```
struct POINT
{
    int x;
    int y;
};
```
Write a function that accepts the structure representing a point and returns an integer value 1, 2, 3 or 4 indicating the quadrant in which the point is located. Write a program to test the above function.
6. A straight line is a geometric object connecting two points, therefore, it can be represented by a nested structure having two structures of POINT type
```
struct LINE
{
    POINT beg;
    POINT end;
};
```
Write a function that accepts two argument of POINT type and returns a structure of type LINE representing the line connecting the two points Write a program to test the above function.
7. Write a function that accepts a structure type LINE as its argument and return an integer (1,2,3), where 1 means line is vertical, 2 means line is horizontal, and 3 means line is oblique. Write a program to test the above function.
[**Hint:** A vertical line is line whose *x*-coordinates are same. Horizontal line is a line whose *y*-coordinates are same. An oblique line is a line which is neither vertical nor horizontal.]

8. Define a structure that contains three members – name of country, name of the capital, and per capita income. Using this structure, write a program to list the countries along with their capitals in the decreasing order of their per capita income.
9. Consider the following structure definition:

```
struct BOX
{
    char make[25];
    float length, breadth, height;
};
```

Write a function that takes the address of the *BOX* structure and returns its volume.

10. Consider the following declaration:

```
struct EMPLOYEE
{
    int code;
    char name[31];
    float salary;
};
```

Given that there are *n* (≤50) employees in an organization. Write a program, using pointer notation to access the data of an employee that output the details of the employee having highest salary.

❑ ❑ ❑

Chapter 12

Pointers & Dynamic Memory Management

Learning Outcomes

After reading this chapter, students will be able to

- comprehend the world of pointers, one of the richest feature of C language
- explain how main memory is organized and addresses are assigned to each memory location
- declare , initialize and access through pointer
- explain the usefulness of pointers in passing arguments to a function
- explain the permissible operations on pointers
- explain the concept of memory allocation
- distinguish between static allocation and dynamic allocation
- comprehend and use various memory management functions for dynamic memory management
- explain what possible problems can occur while working with pointers
- write efficient programs using pointers to solve variety of problems

12.1 INTRODUCTION

Pointers are one of the most important features of C language. Pointers, by most people, are regarded as one of the most difficult topics in C. Let me tell you frankly, it is more a myth rather a fact. I would prefer to say that pointers are one of the most delicate aspects of C language, and you know delicate things need very careful handling. Therefore, pointers need very careful handling.

There are many reasons for using pointers; some of them are enumerated below:

- A pointer allows a function or a program to access a variable (better we say memory location) outside the preview of function or a program. In fact, using a pointer, your program can access any memory location in the computer's memory.
- Since using *return* statement, a function can only pass back a single value to the calling function, pointers allows a function to pass back more than one value by writing them into memory locations that are accessible to the calling function.
- Using pointers, arrays and structures can be handled in more efficient way.
- Without pointers, it will be impossible to create complex data structures, such as *linked lists*, *trees* and *graphs*.
- To communicate with operating system about memory. For example, the operator *new* returns the location of free memory block by using a pointer and the operator *delete* returns the memory block pointed to by a pointer to the operating system.

In totality, we can say that the real power of C language lies in the judicious use of pointers. Therefore, every reader of C must learn and master the art of using pointers.

In this chapter our goal is to explain the pointers as clearly as possible. In an attempt to achieve our goal, we ensure that the groundwork is laid carefully before we move on to use pointers in more advanced situations.

12.2 UNDERSTANDING POINTERS

Computer's memory is organized as a linear collection of bytes or words. For Intel processors, memory is byte-organized whereas for most of the non-Intel processors, memory is word-organized. In byte-organized memory, each memory location is of one byte, whereas in word-organized, each memory location is of one word. Each memory location is assigned a unique number called *location number* or *address*. Figure 12.1 shows the schematic of memory organization.

We know that one byte is equal to 8 bits, whereas one word may be made up of two or more bytes. A word is a measure of the capacity of processor in terms of the size of operands that it can process (*i.e.* add, subtract, multiply, divide, compare, etc.) in single operation. When we say that a processor is a 16-bit processor, it means that its word size is 16-bits (2 bytes) and it can process 16-bit operands in single operations. For large size operands, it needs more operations.

Memory locations Addresses

0
1
2
3
4
5
6
7
.
.
.
MAX

Figure 12.1: Memory Organization

A *memory cell* is either of one byte/word or more contiguous bytes/words. Each cell can store one value at a given time. The address of a cell is always the address of the first byte/word.

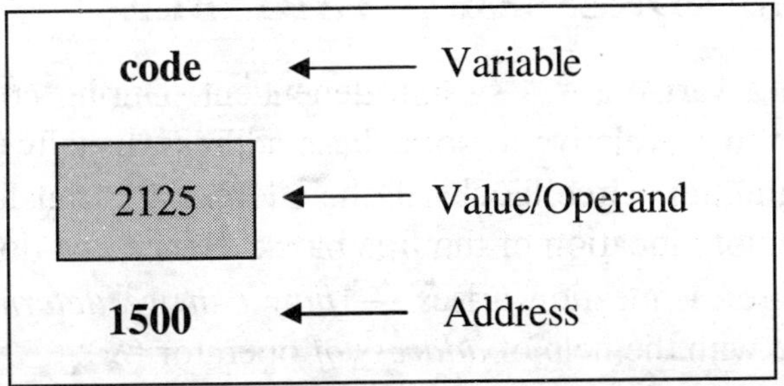

Figure 12.2: Representation of a Variable in Memory

When you give a command to run a program, the operating system first finds a free block of requisite size from computer (main) memory, and then loads the program into that block. Even in that block, usually, the program's code is loaded in one part and the program's data in another part of the block. Each of the data operand is stored in a cell and the system associates each of the variables with these addresses. This is illustrated in Figure 12.2.

In order to access the data operand either we use either variable name or we can use address of the memory cell. Since these addresses are simply positive integer numbers, they can also be assigned to some variables and stored in memory, like any other variable. Such variables that hold addresses are known as a *pointer*. A pointer is, therefore, nothing but a variable that contains address of another variable in memory.

Since pointer is a variable, its value is also stored in memory in another address. Suppose we assign the address of variable *code* to a variable *ptr*. The link between the variable *ptr* and *code* can be visualized as shown in Figure 12.3.

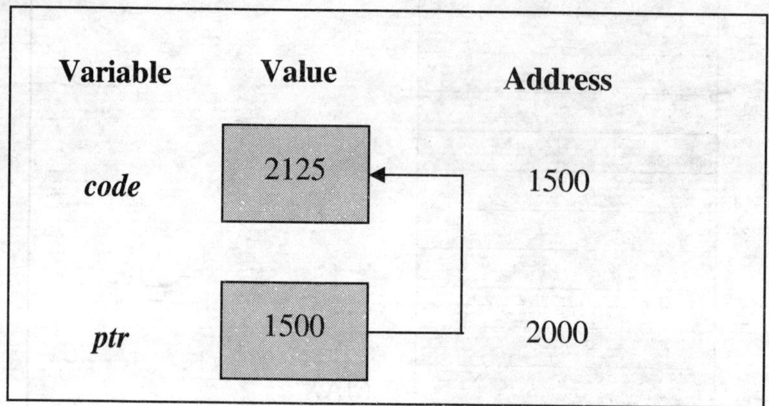

Figure 12.3: Pointer as a Variable

Since the value of the variable *ptr* is the address of the variable *code*, we can access the value of variable *code* by using the value of variable *ptr* and therefore, we say that the variable *ptr* points to the variable *code*, and hence *ptr* gets the name pointer.

12.3 ACCESSING ADDRESS OF A VARIABLE

The actual address of a variable is a system dependent. During compilation and linking, addresses are assumed to be relative to some base address, usually 0. When the operating system loads the program in a free block, all the address are transformed with relative to address of the first memory location of the free block. Hence, we do not know the address of a variable. Therefore, a question arises — *how can we determine the address of a variable*? This is done with the help of *address of* operator (&).

Consider the following statement

```
ptr = &code;
```

It will assign the number 1500 (address of variable *code*) to the variable *ptr*.

The following restrictions must be observed carefully:

1. The address of operator can only be used with variables or an array element. The following are illegal use of *address of* operator:

   (*a*) `&1200` (pointing at constant expression)
   (*b*) `&(a/b)` (pointing at expression)

2. Pointer variable can be initialized only with the use of *address of* operator or by assigning the value of another compatible pointer variable. The following way of initializing a pointer variable *ptr* is illegal:

   ```
   ptr  = 1500;
   ```

   Though, the number 1500 happen to be the address of variable code.

12.4 DECLARING AND INITIALIZING POINTERS

Like other variables, the pointer variables are to be declared first in the declaration block to tell the compiler that to which kind of values these variables will be pointing. The syntax for declaring pointer variable is

```
type *ptr_name;
```

where *type* is a *pre-defined* or *user-defined* data types and indicates that the pointer will point to the variables of that specific data type, and character '*' indicates that variable is a pointer variable. For example, the statement

```
int *ptr1;
```

declares a pointer variable *ptr1* that can point to an integer type variable.

The pointer variable(s) declaration can also be combined with the declaration of other variables, provided they are of same type. For example, the statement

```
int x, *ptrX;
```

declares an integer variable *x* and a pointer variable *ptrX*.

Once a pointer has been declared, it must be initialized prior to its use. It is important to note here that like other variables, a pointer variable will take a garbage value, which can be an address of any storage location. Therefore, if not properly initialized, a pointer may point to any location in memory including those locations where operating system is running, and your system may hang-up i.e. stop responding. Such un-initialized pointers are sometimes referred to as *dangling/wild pointers*.

12.5 ACCESSING A VARIABLE THROUGH ITS POINTER

Once a pointer has been assigned the address of a variable, the question remains as how to access the value of the variable using the pointer. This is done using indirection operator (*). Consider the following statements

```
int *ptrX, x = 10, y;
ptrX = &x;
y = *ptrX;
printf("Value of variable x = %d\n", x );
printf("Value of variable pointed to by ptrX = %d\n",*ptrX );
printf("Address of variable x = %u\n", ptrX );
printf("Value of variable y  = %d\n",  y );
```

The declaration statement tells the compiler that *ptrX* is pointer variable that will point to an integer value, *x* is an integer variable and it also initializes the variable *x* with value 10, and *y* is another integer variable.

And suppose that during the execution of the above segment, a memory location with address 1098 is set aside for variable *x*.

The output of the above segment will look like:

```
Value of variable x = 10
Value of variable pointed to by ptrX = 10
Address of variable x = 1098
Value of variable y = 10
```

It is important to note that assignment of pointers and addresses is always done symbolically, by means of symbolic names. You cannot access the value stored at address 1098 by writing *1098. It will not work.

Listing 12.1

```
/* Program to illustrate use of addressof and indirection operators */
#include<stdio.h>
int main()
{
   int *ptrX, x = 10, y;
   ptrX = &x;
   y = *ptrX
   printf( "Value of variable x = %d\n", x );
   printf( "Value of variable pointed to by ptrX = %d\n", *ptrX );
   printf( "Address of variable x = %u\n", ptrX );
   printf( "Value of variable y  = %d\n",  y );
   return 0;
}
```

12.6 POINTER ARITHMETIC

The C language is very consistent in its approach to pointer arithmetic. The integration of pointers, arrays, and pointer arithmetic is one of the strengths of the C language. In this section, we will briefly describe the various operations permissible on pointers.

The following operations are permitted on pointers.

1. **Addition of a number to a pointer variable.**

   Suppose *p* is a pointer variable pointing to an element of integer type, then the statement

   ```
   p++;        or    ++p;
   ```

   increments the value of *p* by a factor of 2, so that it points to the next location that holds another value of integer type. This increment factor will be 1 for *character*, 4 for *long integer* and *float*, and 8 for *long float*. The statement

   ```
   p += i;
   ```

   where *i* is either a positive integer constant or an integer variable having positive value, increments *p* such that it points to the *i*th location beyond the location to which it is currently pointing.

2. **Subtraction of a number from a pointer variable.**

   Suppose *p* is a pointer variable pointing to an element of integer type, then the statement

   ```
   p--;        or    --p;
   ```

   decrements the value of *p* by a factor of 2, so that it now points to the location preceding the current location. The statement

   ```
   p -= i;
   ```

   where *i* is either a positive integer constant or an integer variable having positive value, decrements *p* such that it points to the *i*th location before the location to which it is currently pointing.

3. **Subtraction of one pointer variable from another.**

   One pointer variable can be subtracted from another provided both point to the same data type. The difference of the two indicates the number of bytes separating the corresponding elements.

In addition to these arithmetic operations, pointer variables can be compared with one another, provided they are compatible, *i.e.*, point to variables of same data type.

The following arithmetic operations on pointers are not permitted.

- Addition of two pointer variables.
- Multiplication of a pointer variable by a number.
- Division of a pointer variable by a number.

12.7 MEMORY ALLOCATION

It is important to note that anything that needs to be executed (programs) or processed (data) must be loaded into memory before it can be executed or processed. Therefore, for every program to be executed or data to be processed, some memory must be allocated for them. The memory allocation can be done in two ways – *statically* and *dynamically*. Let us look at both the ways.

12.7.1 Static Memory Allocation

When you give a command to run your program, the operating system allocates a block of memory for your program, loads it from the secondary storage (Floppy disk, hard disk, CD or DVD, etc.) into that block and then initiates its execution. When the program finishes its execution, the memory occupied by the program is automatically released. It means that the memory requirements are determined prior to execution of your program, and thereafter your program neither can acquire more memory nor can release the free memory, if not required.

The above situations depicts a situation where the amount of memory required is known before hand, and the compiler then can determine the memory requirements for the data as well as the instructions. In such a situation, memory allocation is known as *static memory allocation*.

12.7.2 Dynamic Memory Allocation

Further, it is important to note that the memory requirements for the instructions are always fixed. However, there may be situation when the exact memory requirements for the data may not be known in advance or the data requirements may vary from one program execution to another program execution, *i.e.*, memory requirements for the data are dynamic.

Therefore, when the amount of memory is not known before hand for a particular data item(s), then it is allocated at the execution time, *i.e.*, when the program is running. In such a situation, memory allocation is known as *dynamic memory allocation*.

12.7.3 Heap/Free Store

As mentioned earlier, every program is provided with a pool of unallocated memory that it can utilize during execution. This pool of unallocated memory is known as *heap* or *free store*. Therefore, whenever the memory of requisite size (amount) is required by the program, it is taken from the free store, and when the previously allocated memory is not required further, it is returned back to the free store.

The memory requirements for the instructions (program code) are always fixed, whereas the memory requirements for data may change when your program is dealing with dynamic structures.

12.8 DYNAMIC MEMORY MANAGEMENT

As mentioned earlier, when you give a command to run your program, the operating system allocates a block of memory for your program, loads it from the disk into that block and then initiates its execution. When the program finishes its execution, the memory occupied by the program is automatically released. It means that the memory requirements are determined prior to execution of your program, and there after your program can neither acquire more memory nor can release the free memory, if not required.

However, C language provides library functions (listed in Table 12.1), called *dynamic memory management functions*, to allocate and de-allocate memory at execution time, *i.e.*, dynamically.

By allocation, we mean that your program can obtain as much memory as required by your program even during execution of your program.

By de-allocation, we mean that the memory acquired dynamically can be released at any time during your program execution.

It is important to note that memory allocated dynamically, must be de-allocated before your program finishes its execution. Otherwise, even if your program terminates, memory allocated dynamically is never released automatically, and from operating system point of view that memory is still in use.

Therefore, if you run the same program many times, many users are using your program concurrently, the operating system may run out of memory.

Table 12.1: Memory Management Functions

Function Name	Description
`malloc`	Allocates memory from heap.
`calloc`	Allocates memory from heap and initializes the allocated memory to zeroes.
`realloc`	Readjusts the existing block and copies the contents to the new location.
`free`	Deallocates a block allocated by *malloc*, *calloc* and *realloc* functions.
`coreleft`	Returns a measure of unused memory. (*supported by Turbo C Only*)

12.8.1 Dynamic Memory Allocation With *malloc()* Function

It takes one argument that specifies the size of the block in bytes. The function allocates the memory of requisite size from heap and returns a pointer to the allocated memory on success or a NULL pointer (0) in case of failure. In case of failure, the program should make a safe exit.

The pointer returned is of type *void* which need to be casted into a pointer of desired type.

The memory allocated is left un-initialized, i.e., all locations get the garbage value.

The syntax for use of *malloc()* function is as shown below

```
datatype *ptr
ptr = (datatype *) malloc(n * size_of_data_bytes);
```

where *n* is the number of elements.

For example,

```
int *ptr
ptr = (int *) malloc( 10 * sizeof(int) );
if ( ptr == NULL ) {
   printf("\nMemory not allocated\n");
   exit(1);
}
```

Since the size of *int* is 2-bytes for 16-bit C Compiler and 4-bytes for 16-bit C Compiler, therefore, this statement will allocate block of 20/40 bytes of memory. And, the pointer *ptr* holds the address of the first byte in the allocated memory.

12.8.2 Dynamic Memory Allocation With *calloc()* Function

It takes two arguments that specify the number of items for which memory is to be allocated and the data type of the item. The function allocates the memory of requisite number of element from heap and returns a pointer to the allocated memory on success or a NULL pointer (0) in case of failure. In case of failure, the program should make a safe exit.

The pointer returned is also of type *void* which need to be casted into a pointer of desired type.

The memory allocated is initialized with 0, i.e., all locations get the 0 value.

The syntax for use of *calloc()* function is as shown below

```
datatype *ptr
ptr = (datatype *) calloc(n, datatype);
```

where *n* is the number of elements.

For example,

```
int *ptr
ptr = (int *) malloc( 10, sizeof(int) );
if ( ptr == NULL ) {
   printf("\nMemory not allocated\n");
   exit(1);
}
```

Since the size of *int* is 2-bytes for 16-bit C Compiler and 4-bytes for 16-bit C Compiler, therefore, number of items is 10, therefore the *calloc()* function will allocate block of 20/40 bytes of memory. And, the pointer *ptr* holds the address of the first byte in the allocated memory.

12.8.3 Resizing the Allocated Block With *realloc()* Function

There may be circumstances, where the allocated memory falls short of the retirements during the course of execution of the program. Hence, there is need to resize the memory so that more free memory can be appended in the exiting allocated block.

This task can be accomplished by the *realloc()* function. The *realloc()* function allocates a block of memory (which can be of larger or smaller in size than the original) and copies the contents of the old block to the new block of memory.

Internally, *realloc()* function allocates the new block of requite memory, copies the contents of the existing block, releases/reallocates the exiting block, and at the end returns the address of the new block.

The *realloc()* function takes two arguments — first argument is the pointer to the original block, and the second argument is the size of the new block.

It also returns NULL pointer (0) in case of failure.

The syntax for use of *realloc()* function is as shown below

```
datatype *ptr1, *ptr2
ptr2 = (datatype *) realloc(ptr1, size_in_bytes);
```

where *n* is the number of elements.

For example,

```
int *ptr1, *ptr2;

ptr1 = (int *) malloc( 10 * sizeof(int) );
if ( ptr1 == NULL ) {
   printf("\nMemory not allocated\n");
   exit(1);
}
```

It allocates memory for 10 integers.

Let us suppose, later on there is need to store 20 more integers, the task for storage of 30 integers can now be accomplished as shown below:

```
ptr2 = (int *) realloc( ptr1, 20 * sizeof(int) );
if ( ptr2 == NULL ) {
   printf("\nUnable to expand/inrease the Memory\n");
   exit(1);
}
```

12.8.4 Releasing Dynamically Allocated Memory

The *free()* function releases/deallocates a dynamically allocated block of memory. The syntax for use of *free()* function is as shown below

```
free(ptr);
```

It takes one argument that specifies the pointer to the allocated block. It is important to note that only the allocated block is deallocated, the pointer variable is not deleted.

12.8.5 Querying Existing Memory

During the execution of the program, if it is required at any stage to know the free memory available in heap, the task can be accomplished sing the *coreleft()* function. The syntax for use of *coreleft()* function is as shown below

```
unsigned long free_mem;
free_mem = coreleft();
```

Listing 12.2

```
/* Example program to demonstrate the use of malloc() function. */
#include <stdio.h>
#include <stdlib.h>
void main()
{
   int *ptr, n, i;
   printf("\nEnter number of integers to handled (n) : ");
   scanf("%d", &n);
   ptr = (int *)malloc(n*sizeof(int));
   if (ptr == NULL) {
      printf("Could not allocate required memory\n");
      exit(1);
   }
   for ( i =0; i <n; i++ )
      ptr[i] = i+1;
   printf("\nElements of the array are\n\n");
   for ( i =0; i <n; i++ )
      printf("%d  ", ptr[i]);
   free(ptr);
}
```

Test Run

```
Enter number of integers to handled (n) : 10

Initial contents of memory after allocation

1992528  0  1966416  0  1634300513  1129270272  1346456641  1413563472
977485121  1702057308

Elements of the array, after loading data in allocated memory, are

1  2  3  4  5  6  7  8  9  10
```

Listing 12.3

```
/* Example program to demonstrate the use of calloc() function. */
#include <stdio.h>
#include <stdlib.h>
void main()
{
   int *ptr, n, i;
   printf("\nEnter number of integers to handled (n) : ");
   scanf("%d", &n);
   ptr = (int *)calloc(n, sizeof(int));
   if (ptr == NULL) {
      printf("Could not allocate required memory\n");
      exit(1);
   }
   for ( i =0; i <n; i++ )
      ptr[i] = i+1;
```

```
    printf("\nElements of the array are\n\n");
    for ( i =0; i <n; i++ )
        printf("%d  ", ptr[i]);
    free(ptr);
}
```

Test Run

```
Enter number of integers to handled (n) : 10
Initial contents of memory after allocation
0  0  0  0  0  0  0  0  0  0
Elements of the array, after loading data in allocated memory, are
1  2  3  4  5  6  7  8  9  10
```

Listing 12.4

```
/* Example program to demonstrate the use of realloc() function. */
#include <stdio.h>
#include <stdlib.h>
#include <string.h>
void main()
{
    char *ptr1, *ptr2;
    char *str1 = "Initial string";
    char *str2 = ", appended afterward.";
    int length1 = strlen(str1), length2 = strlen(str2);
    ptr1 = (char *)malloc(length1);
    if (ptr1 == NULL) {
        printf("Could not allocate required memory\n");
        exit(1);
    }
    strcpy(ptr1, str1);
    printf("Initial string: %s\n", ptr1);
    ptr2 = (char *)realloc(ptr1, length2);
    if (ptr2 == NULL) {
        printf("Could not expand memory\n");
        exit(1);
    }
    strcat(ptr2, str2);
    printf("Final string: %s\n", ptr2);
    free(ptr2);
}
```

Test Run

```
Initial string: Initial string
Final string: Initial string, appended afterward.
```

Listing 12.5

```
/* Example program to demonstrate the use of coreleft() function.
   Only on Turbo C/C++ Compiler  */
#include <stdio.h>
#include <stdlib.h>
void main()
{
  unsigned long free_mem;
  free_mem = coreleft();
  printf( "\nFree memory available = %lu bytes.\n", free_mem );
}
```

Test Run (on Turbo C/C++ 3.0 for Windows)

```
Free memory available = 429496264 bytes.
```

12.9 POINTERS AND ARRAYS

When we declare an array, the compiler allocates a block of memory of appropriate size so that all the elements of an array can be stored in contiguous memory locations. The address of the first memory location (also the address of the first element of the array) of allocated block is known as the *base address*, and is assigned to the array name. Thus, array name behaves as a pointer variable.

12.9.1 Pointers and One-dimensional Arrays

Consider the following declaration statement

```
float a[9] = { 0, 1, 2, 3, 4, 5, 6, 7, 8 };
```

This statement set aside a contiguous block of 9 memory locations for the array named *a*. Here the array named *a* holds the address of the first element of the array i.e. base address of the array. The memory map of the array *a* is shown in Figure 12.4. By adding offset (index of the element) to this base address, the other elements of the array can also be accessed. For example, using pointer notation, elements of array *a* can be accessed as

```
*(a+0),  *(a+1),  *(a+2),  ...,  *(a+8)
```

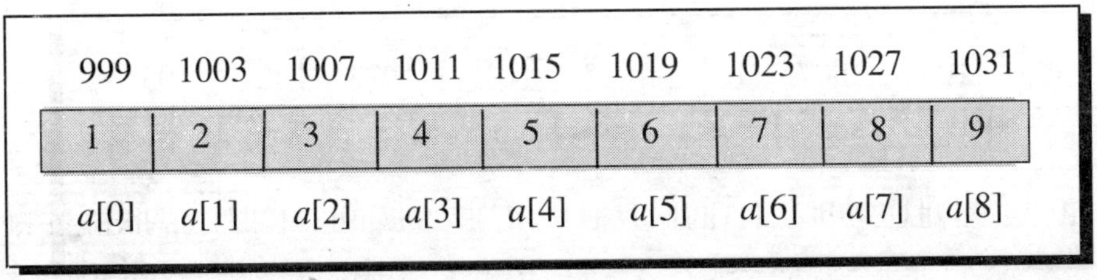

999	1003	1007	1011	1015	1019	1023	1027	1031
1	2	3	4	5	6	7	8	9
a[0]	*a*[1]	*a*[2]	*a*[3]	*a*[4]	*a*[5]	*a*[6]	*a*[7]	*a*[8]

Figure 12.4: Memory map of one-dimensional array *a* with a base address 999

Listing 12.6

```
/* Program to demonstrate the way elements of one dimensional
 *  array are accessed using pointer notation    */
#include<stdio.h>
int main()
{
   int arr[10]={12,15,20,17,25,50,11,10,8,13};
   int i, count, *ptr;
   float sum, average;
   ptr = arr;
   for ( sum = 0, i = 0; i < 10; i++ )
      sum += *(ptr+i);
   average = sum / 10;
   for ( count = 0, i = 0; i < 10; i++ ) {
      if ( *(ptr + i ) > average )
         count++;
   }
   printf( "\nNo. of elements > average = %d\n", count);
   return 0;
}
```

12.9.2 Pointers and Two-dimensional Arrays

Consider the following declaration statement

```
float a[3][3] = {{1,2,3},{4,6,9},{3,2,7}};
```

This statement set aside a contiguous block of 9 memory locations for the array named *a*. Here the array named *a* holds the address of the first element of the array, *i.e.*, base address of the array. As we know that there is no concept of rows and columns in memory. The elements of the array, irrespective of that whether it is one-dimensional, two-dimensional or three-dimensional, are stored in a continuous chain, *i.e.*, one after the other.

	Col #0	Col #1	Col #2
Row #0	1	2	3
Row #1	4	6	9
Row #2	3	2	7

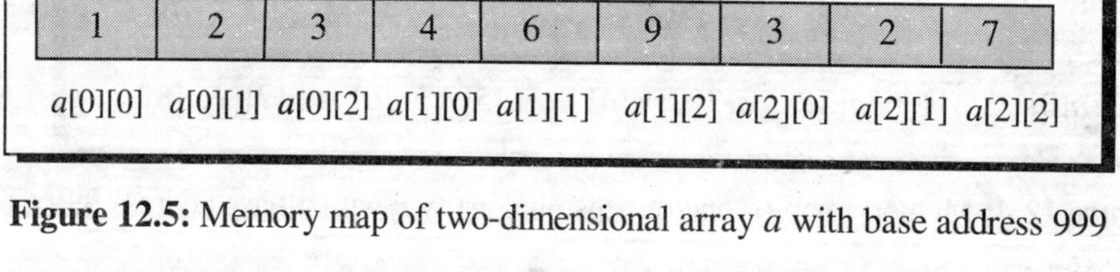

999	1003	1007	1011	1015	1019	1023	1027	1031
1	2	3	4	6	9	3	2	7
a[0][0]	*a*[0][1]	*a*[0][2]	*a*[1][0]	*a*[1][1]	*a*[1][2]	*a*[2][0]	*a*[2][1]	*a*[2][2]

Figure 12.5: Memory map of two-dimensional array *a* with base address 999

Now let us see how we can access individual elements of the above two-dimensional array using pointers. The procedure is slightly difficult. As mentioned earlier that two-dimensional array can be treated as an array of arrays, where each element of the array is one-dimensional array. For example, above array can be visualized as

a[0]	1	2	3
a[1]	4	6	9
a[2]	3	2	7

Figure 12.6: Two-dimensional array *a* as array of one-dimensional arrays

We can refer to individual elements as *a*[0], *a*[1], and *a*[2]. More specifically, *a*[0] gives the address of the first one-dimensional array, *a*[1] gives the address of the second one-dimensional array, *a*[2] gives the address of the third one-dimensional array.

Listing 12.7

```
/*  Program to illustrate 2D array as array of 1D arrays  */
#include<stdio.h>
void main()
{
    int a[3][3] = { { 1, 2, 3 }, { 4, 6, 9 }, { 3, 2, 7 } };
    int i;
    for ( i = 0; i < 3; i++ ) {
        printf("\nAddress of  1D array number %d = %u", i+1, a[i]);
    }
    printf( "\n" );
}
```

Now the question is — *how can we access the individual elements of these arrays*? Suppose we want to access the *j*th element of the *i*th row.

We know that *a*[*i*] will give address of the *i*th one-dimensional array (i.e. *i*th row). And we also know from our knowledge of one-dimensional arrays that writing *a*[*i*] in subscript notation is equivalent to *(*a*+*i*).

Further, if *a*[*i*] is the address of the *i*th row, the address of the *j*th element of *i*th row will be (*a*[*i*]+*j*), and thus the element can be accessed as *(*a*[*i*]+*j*), which on replacing *a*[*i*] by *(*a*+*i*) becomes *(*(*a*+*i*)+*j*).

Thus, all of the following expressions refer to element a_{ij}.

```
a[i][j]
*(a[i]+j)
*(*(a+i)+j)
```

Listing 12.8

```
/* Program to print elements of the 2D array using pointer notation */
#include<stdio.h>
int main()
{
   int a[3][3] = { { 1, 2, 3 }, { 4, 6, 9 }, { 3, 2, 7 } };
   int i, j;
   printf("\nElements of two-dimensional array are\n\n");
   for ( i = 0; i < 3; i++ ) {
      for ( j = 0; j < 3; j++ )
         printf ( "%4df\t", *(*(a+i)+j) );
      printf ( "\n" );
   }
   return 0;
}
```

12.9.3 Dynamic Arrays

The declaration of arrays at the compilation time (i.e. using subscripted notation) has one major limitation. The memory is allocated during the compilation time. If the number of actual data elements is very less as compared to the declared size, then a lot of memory is wasted. On the other hand, if the number of actual data elements is more than the declared size then we have to modify the source code. However, this limitation can be overcome if the memory is allocated at the execution time. The next program illustrates this concept.

Listing 12.9

```
/* Program to illustrate the allocation of memory
*  for one-dimensional array at execution time    */

#include<stdio.h>
#include<alloc.h>

int main()
{
   int *a, n, i, big;
   printf( "Enter size of one-dimensional array: " );
   scanf( "%d", &n );
   a = (int *) malloc(n*sizeof(int)); /* allocate memory for array */
   if ( a == NULL ) {
      printf ( "\nUnable to allocate memory\n" );
      return 1;
   }
   printf( "Enter %d elements of array\n", n );
   for ( i = 0 ; i < n ; i++ )
      scanf( "%d", a+i );
   for ( big = *a, i = 1 ; i < n ; i++ )
   {
      if ( *(a+i ) > big )
         big = *(a+i);
   }
```

```
    printf( "\nLargest element of the array = %d\n", big );
    free(a);    /* release the memory held by array */
    return 0;
}
```

12.10 POINTERS AND FUNCTIONS

12.10.1 Pointers as Function Arguments

If you remember, in *Chapter 9*, we had mentioned that the arguments to a function could be passed using either *call by value* or *call by address* mechanism. When we use call by value mechanism, the formal arguments of the function are declared as simple variable, and at execution time the values of the actual arguments are copied into formal arguments (formal arguments are local to the function), and the execution of the function begins.

However, when we use call by reference mechanism, the formal arguments are declared as pointers, and at execution time, the addresses of the actual arguments are copied into formal arguments and the execution of the function begins.

Listing 12.10

```
/*  Program to illustrate pointers as function arguments   */
#include<stdio.h>
int main()
{
    int a = 10, b = 20;
    void interchange( int *, int *);   /* function prototype */
    printf( "\nValue of a = %d and b = %d", a, b );
    printf( " before function call\n"  );
    interchange ( &a, &b);    /*    function call   */
    printf( "\nValue of a = %d and b = %d", a, b );
    printf( " after function call\n"  );
    return 0;
}
void interchange( int *x, int *y )   /* function definition */
{
   int temp;
   temp = *x;
   *x = *y;
   *y = temp;
}
```

Test Run

```
Value of a = 10 and b = 20 before function call
Value of a = 20 and b = 10 after function call
```

Consider the following program listing.

Listing 12.11

```
/* Program to illustrate non-pointer variables as function arguments */
#include<stdio.h>
int main()
{
    int a = 10, b = 20;
    void interchange( int, int );    /* function prototype */
    printf( "\nValue of a = %d and b = %d", a, b );
    printf( " before function call\n" );
    interchange ( &a, &b);           /* function call      */
    printf( "\nValue of a = %d and b = %d", a, b );
    printf( " after function call\n"  );
    return 0;
}
void interchange ( int x, int y )   /* function definition */
{
    int temp;
    temp = x;
    x = y;
    y = temp;
}
```

Test Run

```
Value of a = 10 and b = 20 before function call
Value of a = 10 and b = 20 after function call
```

Why this version of the program could not complete the task of interchanging the values of the arguments. In this version, values of *a* and *b* are copied into *x* and *y*, which act as local variables. Thus the values, in fact, of *x* and *y* are interchanged in the workspace of the *interchange()* function, but as soon as the function completes its execution, these variables are destroyed. Thus, the values of the variables remain un-effected. However, in the first version of the program, the addresses of variables *a* and *b* are copied into pointer variables *x* and *y*. These pointer variables, though local to the function *interchange()*, interchanges the values of *a* and *b* indirectly in the work space of the *main()* function, with the help of a local variable *temp*. When the function *interchange()* completes its execution, the variables *x*, and *y* along with temp are destroyed, but the task has been accomplished.

12.11 POINTERS AND STRINGS

As we know that a string is an array of characters, terminated with a null character '\0'. Therefore, like one-dimensional arrays, we can use individual characters in a string.

Listing 12.12

```
/* Program to illustrate access of individual characters in a string */
#include<stdio.h>
int main()
{
   char city[]= "Chandigarh is a beautiful city";
   char *ptr;
   ptr = city;
   printf( "Individual characters of the strings are" );
   printf( "\n\n" );
   while ( *ptr != '\0' )
   {
      printf ( "\n%c is stored is address %u", *ptr, ptr );
      ptr++;
   }
   printf ( "\n"  );
   return 0;
}
```

As mentioned earlier, storing a table of strings (list of strings) using a two-dimensional array is not an efficient way. We will take up that issue here again.

Consider the following declaration

```
char names[4][12] = { "Ram", "Inder Mohan", "Amanpreet", "Rajan" };
```

Where we have considered that each string can be of 20 characters, at maximum excluding the null character. This list of strings will be represented in memory as shown below:

R	a	m	'\0'								
I	n	d	e	r		M	o	h	a	n	'\0'
A	m	a	n	p	r	e	e	t	'\0'		
R	a	j	a	n	'\0'						

Total memory requirements are 48 bytes. However, if we count, first string requires 4 bytes, second string requires 12 bytes, third string requires 10 bytes, and the fourth string requires 6 bytes. Thus total memory requirements come out to be 32, wastage of 16 bytes. In practice, it has been shown that, on average, half of the allocated space is mere wastage.

However, this kind of representation of list of strings can be practical only if all strings are of equal length, but this happens rarely in practice.

Therefore, more practical and efficient way of storing a list of strings is using array of pointers to strings. The following declaration illustrates this point

```
char *names[4] = { "Ram", "Inder Mohan", "Amanpreet", "Rajan" };
```

This will be represented in memory as

R	a	m	'\0'								
I	n	d	e	r		M	o	h	a	n	'\0'
A	m	a	n	p	r	e	e	t	'\0'		
R	a	j	a	n	'\0'						

This declaration allocates only 32 bytes of memory, *i.e.*, what is actually needed.

Listing 12.13

```
/* To illustrate I/O of strings represented using array of pointers */
#include<stdio.h>
#include<string.h>
int main()
{
    char *names[30];
    int i, n;
    printf( "\nEnter number of strings : " );
    scanf( "%d", &n);
    for ( i = 0; i < n; i++ ) {
        printf ( "\nEnter string #%d : ", i+1 );
        gets(names[i]);
    }
    printf( "\nYou have entered following strings\n\n" );
    for ( i = 0; i < n; i++ )
        puts(names[i]);
    return 0;
}
```

12.12 POINTERS AND STRUCTURES

As we can use pointers with other data types, we can also use pointers for structure variables. Consider the following declarations

```
typedef struct
{
    int code;
    char name[20];
    int  dept_code;
    float  salary;
} EMPLOYEE;
EMPLOYEE emp1, *ptr;
```

These declarations create and user-defined data type and given it name *EMPLOYEE* and declares *emp1* as variable of type employee and *ptr* as pointer variable to type employee.

Now the following statement

```
ptr = &emp1;
```

can be used to assign the address of variable *emp1* to pointer variable *ptr*. Now, we can access the members of the structure variable *emp1* using pointer variable *ptr*. You may expect that the elements can be accessed as shown below:

```
*ptr.code     *ptr.name     *ptr.dept_code     *ptr.salary
```

But this approach will not work as the *dot operator* '.' has higher priority than the *indirection operator* '*'.

The correct way is to write as

```
(*ptr).code    (*ptr).name   (*ptr).dept_code   (*ptr).salary
```

There is another approach that is more elegant and simple. In this approach, the elements of structure are accessed using *arrow operator* '–>' which is a combination of two characters, *i.e.*, hyphen '–' and greater-than '>'. The syntax for accessing members of structure using arrow operator is:

```
ptr->vname
```

where *ptr is* pointer variable pointing to a structure, and *vname* is the name of element of the structure.

Using arrow operator, the elements of structure of type *employee* pointed to by pointer variable *ptr* can be accessed as

```
ptr->code     ptr->name     ptr->dept_code     ptr->salary
```

12.12.1 Self-referential Structures

The self-referential structures are structures that include an element that is a pointer to another structure of the same type. These structures find their application in building complex data structure such as *linked lists*, *trees* and *graphs*.

Following are some examples of self-referential structures.

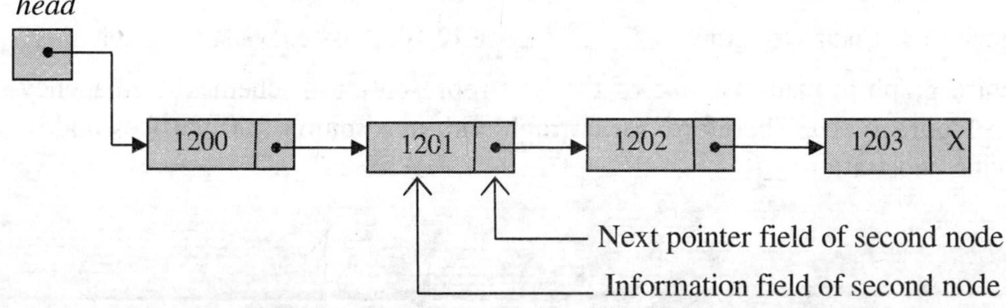

Figure 12.7: Linear linked list of integer values with nodes 4

To represent the above linear linked list in memory, we need following declarations

```
struct NODE
{
   int info;
   struct NODE *next;
};
struct NODE *start;
```

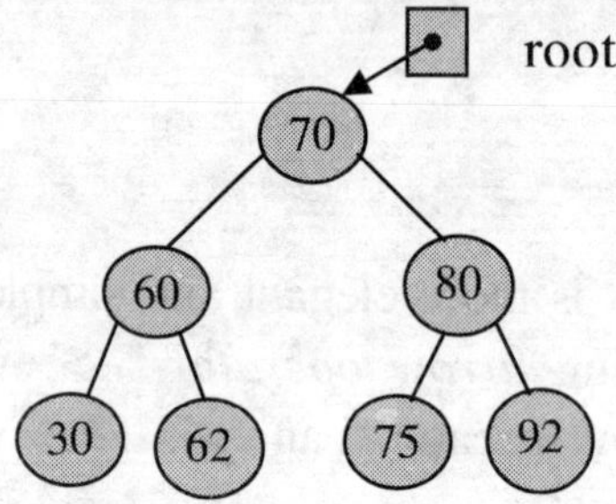

Figure 12.8: Binary tree

To represent a binary tree in memory, we need following declarations

```
struct NODE {
   struct NODE *left;
   int info;
   struct NODE *right;
};
struct NODE *root;
```

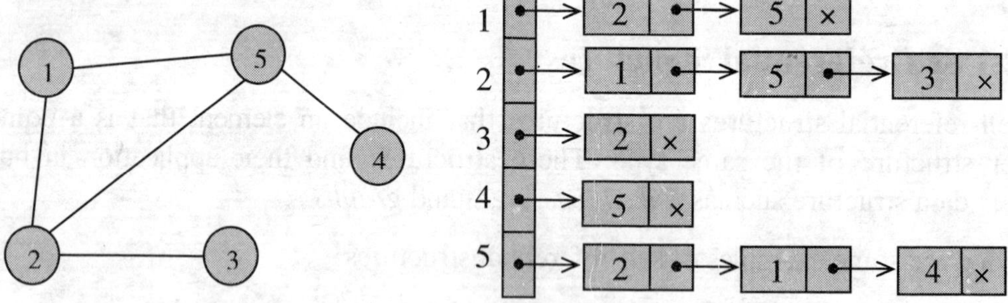

Figure 12.9: Undirected graph **Figure 12.10:** Adjacency list for graph

To represent a graph in memory, one of the best representation schemes is adjacency list as shown in Figure 12.10. Therefore, for a graph with maximum of 50 vertices/nodes, we need following declarations

```
#define MAX 50
struct NODE {
   int vertex;
   struct NODE *link;
};
struct NODE *adj[MAX];
```

12.12.2 Passing Structure by Reference to a Function

When we pass a structure by address, only address of the first byte of the structure is passed. Using this address, the called function can access all the elements of the structure, and thus the called function can process the structure in the memory space of the calling function.

```
typedef struct {
   int day, month, int year;
} DATE;
void printDate(DATE *);
void main()
{
   DATE aDate = { 10, 6, 2008 };
   int x;
   printDate(&aDate);
   /*  more statements   */
}
```

```
/*  function definition   */
void printDate(DATE *tDate )
{

   /*  local declarations */
   /*  other statements   */

}
```

Figure 12.11: Passing entire structure by address

Listing 12.14

```
/* Program to illustrate the passing of a structure by address  */
#include <stdio.h>
typedef struct {
   int day;
   int month;
   int year;
} DATE;
void printDate( DATE *aDate );        /* function declaration   */
int main()
{
   DATE tDate = { 10, 6, 2008 };   /* initialize a structure */
   printDate( &tDate );             /* function call          */
   return 0;
}
void printDate( DATE *aDate )
{
   printf( "\nDate in format dd/mm/yy: %02d/%02d/%2d\n",
           aDate->day, aDate->month, aDate.->year );
}
```

12.13 POSSIBLE PROBLEMS WITH THE USE OF POINTERS

Pointers can be the source of mysterious and catastrophic program bugs. Common bugs related to pointers and memory management is *dangling/wild pointers*, *memory leaks*, and *allocation failures*. Each of these problems is discussed next.

12.13.1 Problem of Dangling/Wild Pointers

The most common problem with pointers is that the programmer fails to initialize a pointer with a valid address. Such an initialized pointer, referred to as *dangling/wild pointer*, can end up pointing anywhere in memory that may include the program code itself or to the code of the operating system. If the programmer assign a value to such pointer, the value will overwrite the program or operating system instructions, and the computer program may show undesirable behaviour or may even crash, *i.e.*, stop responding. When the system stop responding, we say the system has hanged up and the only option left is to reboot the system. Therefore, care must be taken to initialize pointers with valid address.

12.13.2 Problem of Null Pointer Assignment

One particular situation that happens is when the pointer points to address 0, which is called NULL. For example, this may happen that if the pointer variable is declared as global since global variables are initialized to 0. Likewise, this can also happen for a local un-initialized pointer variable, particularly for local static variables, they are also initialized to 0. When it happens, and no catastrophic situations had occurred, then the system will display a message "*Null pointer assignment*" on termination of the program.

12.13.3 Problem of Memory Leak

Another very serious problem with pointers is that of *memory leak*. Memory leak is a situation where the programmer fails to release the memory allocated at run time in a module. Note that when memory is allocated, a pointer variable is used to hold the address of the allocated block, however, when the module completes its execution, the pointer variable goes out of scope and there will be no way to reach that memory block. Therefore, care must be taken to release the allocated memory block in a module where memory was allocated.

Listing 12.15

```
/*  Program to illustrate the cause of memory leak   */
#include<stdio.h>
int main()
{
   . . . . .
   fun(); //  invoke function named 'fun'
   . . . . .
}
void fun()
{
   int *intPtr;
   . . . . .
   /*  allocate memory for 'n' integer numbers
   intPtr = ( int * ) malloc(n*sizeof(int));
   /*   work on the allocated block    */
}
```

12.13.4 Problem of Allocation Failures

Allocation failure is a situation when the program through *malloc(), calloc(), realloc(),* function request for a block of memory, and the operating could not fulfill the request of the program because the sufficient memory may not be available in the free pool of memory. In that case, these functions return a NULL pointer indicating allocation failure.

Therefore, the programmer should take care of allocation failures and provide a safe exit in their programs on such failures.

Listing 12.16

```
/* Program to illustrates the allocation failure    */
#include<stdio.h>
void main()
{
    int *intPtr;
    . . . . .
    intPtr = ( int * ) malloc(n*sizeof(int));
    if ( intPtr == NULL ) {
        printf("\nUnable to allocate memory . . .\n");
        return;
    }
    /*   work on the allocated block    */
    . . . . .
    free(intPtr);    // release the memory
}
```

ILLUSTRATIVE EXAMPLES

Example 12.1: Write a program that takes a list of numbers as its command line arguments, and prints the sum of these numbers.

The procedure to be followed is as follows: The command line arguments are received by the *main()* function as string. Since the command line arguments are just numbers, therefore the string are made up of only letters 0-9 and decimal point '.'. We use here a library function *atof()* to convert a string of representing a number to numeric value. Then we add these numeric values one by one after conversion. We need to include *stdlib.h* header file in order to use *atof()* library function.

Listing 12.17

```
/* Program to add list of integer numbers given as command line arguments
 * filename : sum.c
 */
# include <stdio.h>
# include <stdlib.h>
void main( int argc, char *argv[] )
{
    int i;
```

```
    float sum = 0;
    for ( i = 1; i < argc; i++ ) {
        sum += atof( argv[i] );
    }
    printf("\nSum of numbers on command line = %.2f\n", sum);
}
```

Test Run

```
>sum 1 2 3 4 5 6 7 8 9 10
Sum of numbers on command line = 55.00
```

Example 12.2: Consider the following declaration:

```
char *names[]={"Vimal", "Amit", "Anuj", "Rohit", "Abhijit"};
```

Write a program to produce the following output:

```
tijihbA, tihoR, junA, timA, lamiV
```

Listing 12.18

```
/*  Program to reverse each string of array of strings  */
# include <stdio.h>
# include <stdlib.h>
void reverseString(char *p);
void main()
{
    int i;
    char *names[] = {"Vimal", "Amit", "Anuj", "Rohit", "Abhijit"};
    printf( "\n" );
    for ( i = 4; i >= 1; i-- ) {
        reverseString(names[i]);
        printf( "%s,", names[i]);
    }
    reverseString(names[i]);
    printf( "%s,", names[i]);
    printf( "\n" );
}
void reverseString(char *p) {
    char t;
    char *sStr, *eStr;
    sStr = p;
    eStr = p + strlen(p) - 1;
    while ( eStr > sStr ) {
        t = *sStr;
        *sStr = *eStr;
        *eStr = t;
        sStr++;
        eStr--;
    }
}
```

Test Run

```
tijihbA, tihoR, junA, timA, lamiV
```

Example 12.3: Write a function that takes an array as its argument and returns the largest, smallest and range of its elements. Using this function, write a program using pointers to find the largest, smallest and range of elements of the array.

Listing 12.19

```
/* Program to find the largest, smallest, and range of elements
*  using pointer notation only  */
#include<stdio.h>
#include<stdlib.h>
void compute(int *,int,int *,int *,int); /*function declaration*/
void main()
{
   int *a, i, n, largest, smallest, range;
   printf( "\nEnter size of array: "  );
   scanf( "%d", &n );
   a = (int *) malloc(n*sizeof(int)); /* allocate memory for array */
   if ( a == NULL )
   {
      printf ( "\nUnable to allocate memory\n" );
      exit(1);
   }
   printf( "\nEnter %d elements of array\n\n", n );
   for ( i = 0 ; i < n ; i++ )
      scanf( "%d", a+i );
   compute(a, n, &largest, &smallest, &range); /*function call*/
   printf( "\nLargest element of the array   = %d", largest );
   printf( "\nSmallest element of the array  = %d", smallest );
   printf( "\nRange of elements of the array = %d\n", range );
   free ( a );       /* release the memory held by array */
}
void compute(int *a, int n, int *big, int *small, int *range)
{
   int i;
   *big = *small = *a;
   for ( i = 1; i < n; i++ )
   {
      if ( *(a+i) > *big )
         *big = *(a+i);
      if ( *(a+i) < *small )
         *small = *(a+i);
   }
   *range = *big - *small;
}
```

Test Run

```
Enter size of array: 10
Enter 10 elements of array
```

```
12 30 21 34 9 59 70 54 44 33
Largest element of the array    = 70
Smallest element of the array   = 9
Range of elements of the array = 61
```

MULTIPLE CHOICE QUESTIONS

1. The address of a variable can be obtained using ___ operator.
   (*a*) * (*b*) &
   (*c*) ? (*d*) ->
2. Which of the following operator is known as *indirection* (*dereference*) operator?
   (*a*) & (*b*) <<
   (*c*) ^ (*d*) *
3. Which of the following operator is known as *address of* operator?
   (*a*) & (*b*) *
   (*c*) -> (*d*) **
4. Identify the correct declarations of pointer variables *ptr1*, *ptr2*.
   (*a*) `int ptr1, *ptr2;` (*b*) `int *ptr1, ptr2;`
   (*c*) `int ptr1, ptr2;` (*d*) `int *ptr1, *ptr2;`
5. The operators exclusively used with pointers are
   (*a*) * and / (*b*) & and *
   (*c*) & and ^ (*d*) * and +
6. The operand of *addressof* operator can be a/an
   (*a*) constant (*b*) array element
   (*c*) expression (*d*) None
7. Pointer variables may be assigned
   (*a*) address value represented in hexadecimal notation
   (*b*) address value represented in octal notation
   (*c*) address of another variable
   (*d*) address value represented in binary notation
8. Operand of *indirection* (*dereference*) operator is
   (*a*) pointer variable (*b*) ordinary variable
   (*c*) integer constant (*d*) none of above
9. Identify the operator that is not used with pointers
   (*a*) -> (*b*) &
   (*c*) * (*d*) >>
10. The function used for dynamic memory allocation is
    (*a*) `memalloc()` (*b*) `blockalloc()`
    (*c*) `malloc()` (*d*) `mcalloc()`

11. Pointer can be used to achieve
    (*a*) call by function (*b*) call by name
    (*c*) call by reference (*d*) call by procedure
12. Which of the following statements about functions is not correct?
    (*a*) function can return a pointer
    (*b*) function cannot return a pointer
    (*c*) function can have its formal arguments declared as pointer
    (*d*) function can be called using a pointer to a function
13. When *addressof* operator (&) is prefixed to a variable, it yields
    (*a*) variable's value (*b*) variables data type
    (*c*) variable's address (*d*) none of above
14. Which of the following operation is not permitted on pointers?
    (*a*) incrementing a pointer variable
    (*b*) adding a number to a pointer variable
    (*c*) division of a pointer variable by a number
    (*d*) difference of two pointer variables
15. Consider the following statements

```
char *ptr;
ptr = "abba";
putchar(*ptr);
```

    What will be printed?
    (*a*) first letter (*b*) entire string
    (*c*) it is a syntax error (*d*) last letter
16. What does the following declaration mean?

```
int (*ptr)[10];
```

    (*a*) *ptr* is a pointer to 10 arrays (*b*) *ptr* is a pointer to array of 10 integers
    (*c*) *ptr* is a pointer to a function (*d*) *ptr* is a generic pointer
17. What will be the output of the following program?

```
void change (int *, int);    /* function declaration  */
void main()
{
     int  i, a[5]={3,4,5,6,7};
     change(a,5);
     for(i=0 ; i<5 ; i++)
        printf("%3d",a[i]);
}
void change (int *p, int n)
{
     int i;
     for ( i = 0 ; i < n ; i++ )
        *(p+i) += i+1;
}
```

    (*a*) 3 5 7 9 11 (*b*) 4 6 8 10 12
    (*c*) 7 8 9 10 11 (*d*) 3 4 5 6 7

18. What is the difference between the following declarations?

```
const int *intptr;
int *const intptr;
```

(*a*) First statement declares pointer to integer constant while the second declares constant pointer to an integer.
(*b*) First statement declares constant pointer to an integer while the second declares pointer constant to an integer.
(*c*) Both declares pointer to an integer.
(*d*) Both declares constant pointer to an integer.

19. What is the difference between the following declarations?

```
int *ptr(int arg);
int (*ptr)(int arg);
```

(*a*) First declares a function that returns a pointer to an integer, the second one declares a pointer to a function.
(*b*) First declares a pointer to a function, the second one declares a function that returns a pointer.
(*c*) Both declares function that return a pointer to an integer.
(*d*) Both declares pointer to a function.

20. What is the difference between the following declarations?

```
int (*a)[];
int *a[];
```

(*a*) First declares a pointer an integer array, the second one declares array of integer pointers.
(*b*) First declares array of integer pointer, the second one declares a pointer an array.
(*c*) Both declares pointer to an array.
(*d*) Both declares array of pointers.

21. What is the value of *x* after executing the following segment?

```
int x = 5, *p;
p = &x;
*p = 7;
```

(*a*) 5 (*b*) 7
(*c*) Undefined (*d*) None

22. What will be the output of the following segment?

```
int a[5] = {1,2,3,4};
int *p = a;
printf("%d",*(p+2));
```

(*a*) 4 (*b*) 0
(*c*) 3 (*d*) 2

23. Consider the following statement

```
int p= 20, *q=&p, **k = &q;
```

then the variable *k* is called __________.

(*a*) pointer to an integer (*b*) pointer to pointer to an integer
(*c*) pointer to pointer to an integer pointer (*d*) just double pointer

24. Consider the following program segment:

```
int *p, a[] = { 1, 2, 3, 4, 5 };
p = &a[2];
printf("%d", p[-1]);
```

Which of the following is correct about above program segment?

(*a*) Compiler time error because *p* is not an array
(*b*) Output will be 2
(*c*) Compiler time error because subscript cannot be negative
(*d*) None of above

25. Consider the following program segment:

```
const int i = 5;
int *p;
i = 20;
p = &i;
printf("\n%d, %d\n", *p, i);
```

Which of the following is correct about above program segment?

(*a*) Output will be 20, 20
(*b*) Output will be 5, 20
(*c*) Output will be 20, 5
(*d*) Compile time error because the value of variable *i* cannot be changed.

26. Consider the following program segment:

```
char *s1 = "Mumbai";
char *s2 = "Delhi";
char *ps1 = s1, *ps2 = s2;
while ( *ps1++ = *ps2++ ) ;
puts(s1);
```

Which of the following is correct about above program segment?

(*a*) Mumbai
(*b*) Delhi
(*c*) MumbaiDelhi
(*d*) None

REVIEW EXERCISE

1. How can we get the address of a variable?
2. What is a pointer variable?
3. How is a pointer variable declared?
4. How can the values pointed to by a pointer variable be accessed?
5. What are the various types of pointers?
6. Pointers are double-edged weapons. Comment.
7. What are the operations permissible on pointers?
8. What are the operations, which are not permissible on pointers?
9. What is the difference between *static* and *dynamic* memory allocation?
10. What do you mean by a *dangling/wild pointer*? What type of problems it may cause?
11. What is a NULL pointer assignment?
12. What is a memory leak? What can be the possible reason for it? How can memory leaks be avoided?
13. What is an allocation failure?

PROGRAMMING EXERCISE

1. Write a program that dynamically allocates memory for an array of size *n*, reads elements of array and sorts them in ascending order.
2. Write a function *substr()* whose prototype is given as

```
char *substr(char *str1, char *str2);
```

that scan the first string for the occurrence of a second string, and returns a pointer to the element in the first string where the second string begins. However, if the second string does not occur in the first string, the function should return NULL. Write a program to test the above function.
3. Write a function *replace()* whose prototype is given as

```
int replace(char *str, char c1, char c2);
```

that replace every occurrence of character c1 with character c2, and returns the total replacements made. Write a program to test the above function.
4. Consider the following structure definition:

```
struct BOX
{
    char make[25];
    float length, breadth, height;
};
```

Write a function that takes address of the *BOX* structure and returns its volume.
5. Consider the following declaration:

```
struct EMPLOYEE
{
    int code;
    char name[31];
    float salary;
};
```

Given that there are *n* (≤50) employees in an organization. Write a program, using pointer notation to access the data of an employee that output the details of the employee having highest salary.
6. Write a function *substr()* whose prototype is given as

```
char *substr(char *str1, char *str2);
```

that scan the first string for the occurrence of a second string, and returns a pointer to the element in the first string where the second string begins. However, if the second string does not occur in the first string, the function should return NULL. Write a program to test the above function.
7. Write a function *replace()* whose prototype is given as

```
int replace(char *str, char c1, char c2);
```

that replace every occurrence of character c1 with character c2, and returns the total replacements made. Write a program to test the above function.

❑ ❑ ❑

Chapter 13

File Handling

Learning Outcomes

After reading this chapter, students will be able to

- comprehend the concept of data file and its usefulness
- explain the different steps in processing a file
- explain the opening, reading and writing of files
- describe the variety of library functions available for doing input/output on files
- write efficient programs for file handling

13.1 INTRODUCTION

The programs described so far, accepts the input data from the keyboard at the time of execution. The main limitation of this approach is that it is very difficult to handle large volume of data. As an example, suppose there are about 100 data items to enter. Some of the problems that may arise are enumerated below:

- It takes lot of time to enter the entire data.
- If the user makes a mistake while entering the data, he has to start from the beginning.
- If the same data is to be processed again at some later stage, again we have to enter it again. That is, for every program run, we have to enter the same data again and again.

Therefore, if the data is entered in computer's memory, checked for its correctness, and finally stored in a disk file, a program can read it directly from the disk and that too at very high speed.

In addition, the output of one program may become input for another program. In such a situation, it is more efficient to store the output of a program in a disk file, which can subsequently be read by other program(s).

A program developed in C language can read and write files in a variety of ways. Each of these ways is described in this chapter in a very simple manner and illustrated with well-designed programming examples.

13.2 TEXT MODE VERSUS BINARY MODE

Depending on the way a file is opened for processing, a file can be classified as a *text file* or a *binary file*.

The mode of opening a particular file determines how various details of the file are handled.

For example, how *newline character* ('\n') is stored and how *end-of-file* (*EOF*) is indicated.

In addition, the numbers in the text format are stored as string of characters, whereas in binary format they are stored the same way as they are stored in computer's main memory.

Newline Character **-** In C language, the end of a line is signaled by a single character; the *newline character* '\n' with ASCII value 10 decimal. Whereas for storage in the disk, the end of line is marked by two characters, the *carriage return* (CR) with ASCII value 13 decimal and the *linefeed* (LF) with ASCII value 10 decimal which is same as of C's *newline* character.

In a text mode, *newline* character is automatically translated into CR/LF combination before writing to the file. Similarly, while reading a file, the CR/LF combination on the disk is automatically translated back into a *newline* character.

However, in a binary mode, this translation will not take place. Therefore, the *newline character* '\n' will be written as only *linefeed* (LF) character. Similarly, while reading, the combination of CR/LF will be treated as two separate characters as '\r' and '\n' characters.

End-of-File **-** It is important to understand that *EOF* is not a character. It is actually an integer value sent to the program by the operating system, and is defined in a header file *stdio.h* to have a value of –1. No character with this value is stored in a file on the disk. While creating a file, when the operating system finds the last character to the file has been sent, it transmits the *EOF* signal.

Both in text as well as in binary mode, the system keeps track of the total length of the file and will signal an *EOF* when this length has been reached.

However, in a text mode, a special character, 1A hex (decimal 26) is inserted after the last character in the file. This character can also be used to indicate *EOF*. A character 1A hex can be generated by holding down *Ctrl* and *Z* keys simultaneously (often represented as ^Z) or by typing key F6. If this character is encountered at any point in the file, the read function will return the *EOF* signal to the program.

But in a binary mode, the numbers are stored in binary format. Therefore a number stored in memory may happen to be 1A hex. Thus, if a file is created in binary mode, and your program opens it in text mode for reading, the reading may terminate prematurely.

In text mode, character 1A hex indicates end-of-file, where as in binary mode, it is just as another number.

DIFFERENCES BETWEEN TEXT FILE & BINARY FILE	
1. Data is stored using human readable characters.	1. Data is stored in the same format as it is stored in memory.
2. Each line of data ends with a *newline* character.	2. There are no lines or *newline* character.
3. There is a special character called end-of-file (EOF) marker at the end of the file.	3. There is an end-of-file marker at the end of the file, which is usually implementation dependent.

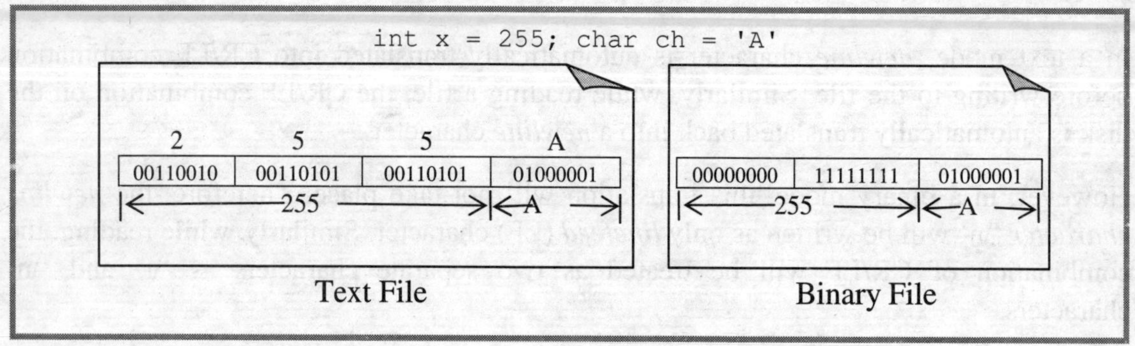

Figure 13.1: Text and Binary Files

13.3 STEPS IN PROCESSING A FILE

The processing of a file involves following steps:

1. **Opening a file** — First step in file processing is to open the file in appropriate mode. If you want to read from a file in your program, the file must be opened in the read/input mode. Similarly, if you want to write the output of your program to a file, the file must be opened in the write/output mode. However, there may be cases when you want to read from and also write onto the same file, then file must be opened in read-write mode.
2. **Reading from or writing onto a file** — Once a file is opened successfully, data can be read from a file or written to a file in a variety of ways.
3. **Closing the file** — This is the final step in file processing. Once we have completed the reading and writing of the files, each file must be closed.

13.3.1 Opening a File

Before a program can write to a file or read from a file, the program must open it. Opening a file establishes an understanding between the program and the operating system. This provides the operating system with the *name* of the file and the *mode* in which the file is to be opened, *i.e.*, whether for reading or writing and that too in text mode or binary mode. Communication areas are set up between the file and the program. One of these areas is a structure of type *FILE*, declared in header file named *stdio.h* that holds information about the file.

For example, the statement

```
fptr = fopen( "test.dat", "w" );
```

request the operating system to open a file named *test.dat* in text mode and prepare it for writing.

Table 13.1: File Opening Modes

Mode	Purpose
r	Open a file for reading. The file must already exist.
w	Open a file for writing. If the file already exists, its contents will be discarded. If it does not exist, it will be created.
a	Open a file for extending. Data will be added to the end of the existing file, if file already exists. If it does not exist, it will be created.
r	Open a file for both reading and writing. The file must already exist.
w+	Open a file for both reading and writing. If it does not exist, it will be created.
a+	Open a file for both reading and appending. If the file does not exist, it will be created.

When the program makes a request to open a file, it receives back a pointer to this particular structure, if the request is granted; otherwise it receives a sentinel value, such as a *NULL* value defined *stdio.h*. Each file opened will have its own *FILE* structure, with a pointer to it. Any program can open any number of files.

Possible Trouble with Opening a File

As mentioned earlier, the request for opening a file may be granted or rejected. In case of rejection, the program will not run and will terminate abnormally. There may be many reasons for the rejection of program request.

For example

- Trying to open a file for reading that does not exist.
- Trying to open a file for writing, but sufficient space is not available on the disk.
- Trying to open a file that has been corrupted, etc.

Thus, it is important for the program that uses a disk file to check whether a file has been opened successfully before trying to read from or write onto the file. If the file cannot be opened, *fopen()* function returns a NULL value.

Therefore, a program can check the value returned by the *fopen()* function to determine whether the file is opened successfully or not. Hence, it is always better to write program segment responsible for opening a file as in next code segment.

```
fptr = fopen ( "test.dat", "w" );
if ( fptr == NULL )
{
   printf( "\nCan't open file test.dat.\n" );
   return 1;
}
```

13.3.2 Closing a File

When a program has finished with reading/writing of a file, it must be closed. This task is carried out by using *fclose()* library function, whose syntax is

```
fclose( fptr );
```

where *fptr* is a file pointer associated with a file which is to be closed.

Closing a file explicitly has two effects. These are

- Any data remaining in the buffer is written to the file. It may be noted that the buffer used is invisible to the programmer.
- Frees the communication areas used by the particular file so that they are available for other files. These areas include the *FILE* structure and the buffer itself.

For all possible ways of reading & writing, way of opening and closing a file is same.

Why buffer is required?

A *buffer* is a temporary storage area in main memory. A buffer is necessary even if it is invisible. To visualize its importance, let us consider a particular example. Suppose our program accesses the disk to read/write one character at a time. The disk system takes some time to position the head on the correct track and then further waits for the correct sector of the track to come around.

For every read/write operation, the motor that rotates the disk actually has to start from a stand still. If you type several characters repeatedly, and each one requires a separate disk access, some of the characters would probably be lost. So this approach will be inefficient. To solve this problem, the buffer is used.

When a program writes data to a file, it is actually stored in a buffer rather than being written immediately to the disk. When the buffer is full, its contents are written to the disk all at once, and subsequent writes continue onto the buffer. Similarly, when a program reads data from a file, a block of data, equal to the size of buffer except the last block, is read from the disk file and is placed in the buffer. The input functions subsequently read data from a buffer. When a program attempts to read a data item that is not in buffer, it forces to read another block of data containing the requested data item.

13.3.3 Reading and Writing of Files

There is a variety of ways of reading data from a file and writing data to a file. These are enumerated below:

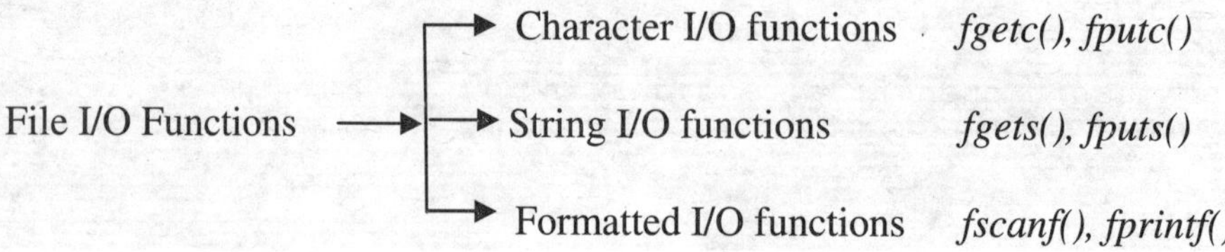

13.3.3.1 Reading and Writing using Character I/O Functions

Using character I/O functions, data can be read or written one character at a time. This is analogous to the way functions *putchar()* and *getchar()* write data to screen and read data from the keyboard.

Writing to a file

Once the program has established a line of communication with a particular file by opening it, then it can write to the file. The syntax of function that writes one character at a time is

```
fputc( ch, fptr );
```

where *ch* is a character variable or constant, and *fptr* is a file pointer.

The *fputc()* function is similar to *putchar()* function, the only difference is that a *putchar()* function always writes to screen unless redirection is employed. The following listing shows the example of a program that accepts a line of text from keyboard, one character at a time, and writes into disk file named *file1.dat*.

Listing 13.1

```
/* Program to demonstrate writing of one character at a time to file  */
#include<stdio.h>
int main()
{
   FILE *fptr;
   char ch;
   fptr = fopen( "file1.dat", "w" );
   if ( fptr == NULL )
   {
      printf( "\nUnable to open file1.dat\n" );
      return 1;
   }
   printf( "\nType a line of text, when finished" );
   printf( ", when finished hit Enter key\n" );
   while ( ( ch = getche() ) != '\r' )
      fputc ( ch, fptr );
   fclose( fptr );
   return 0;
}
```

When the above program is executed, it prompts the user to type a line of text. When you have finished, you hit the Enter (↵) key to terminate the program.

Test Run

```
C is a powerful procedural language. It provides both low-level as well
as high level features. ↵
```

Reading from a file

If the program can write to a file, it should also be able to read from a file. The syntax of the function that reads and returns one character at a time is

```
ch = fgetc( fptr );
```

where *ch* is a character variable, and *fptr* is a file pointer.

The *fgetc()* function returns the character read from the file or the end-of-file (EOF) character if it has reached the end of the file. The *fgetc()* function is similar to *getchar()* function, the only difference is that the later functions always read from the keyboard unless redirection is employed. The following listing shows the example of a program that reads one character at a time from file, and writes onto the screen.

Listing 13.2

```
/*  Program to demonstrate reading of character at a time from a file  */
#include<stdio.h>
int main()
{
   FILE *fptr;
   char ch;
   fptr = fopen( "file1.dat", "r" );
   if ( fptr == NULL ) {
      printf( "\nUnable to open file1.dat\n" );
      return 1;
   }
   printf( "\nContens of file are:\n\n" );
   while ( ( ch = fgetc (fptr) ) != EOF )
      putchar( ch );
   fclose( fptr );
   return 0;
}
```

When the above program is executed, it reads the contents of file *file1.dat*, one character at a time and writes onto screen till it encounters the end-of-file (EOF).

Test Run

```
Contents of file are:

C is a powerful procedural language. It provides both low-level as well
as high level features.
```

13.3.3.2 Reading and Writing using String I/O functions

Using string I/O functions, data can be read or written in the form of string of characters. Reading and writing strings of characters is as easy as individual characters. This is analogous to *puts()* and *gets()* functions that write data to screen and read data from the keyboard.

Writing to a file

The syntax of the function that writes a string of characters at a time is

```
fputs( str, fptr );
```

where *str* is a array of characters or a string constant, and *fptr* is a file pointer.

The *fputs()* function is similar to *puts()* function, the only difference is that the *puts()* function always writes to the screen unless redirection is employed.

The following listing shows the example of a program that accepts a series of strings from keyboard, and writes them onto a disk file.

Listing 13.3

```
/*  Program to demonstrate writing of strings to a file   */
#include<stdio.h>
#include<string.h>
int main()
{
    FILE *fptr;
    char str[81];
    fptr = fopen( "file2.dat", "w" );
    if ( fptr == NULL )
    {
        printf( "\nUnable to open file2.dat\n" );
        return 1;
    }
    printf( "\nEnter a set of strings, press just" );
    printf( " Enter key to finish\n" );
    while ( strlen( gets( str ) ) > 0 )
    {
        fputs( str, fptr );
        fputs( "\n", fptr );
    }
    fclose( fptr );
    return 0;
}
```

Test Run

```
C is a powerful procedural language.↵
It is a middle level language.↵
All Programs are tested using Turbo C Compiler.↵
↵
```

When the above program is executed, it accepts set of strings terminated by hitting *Enter* key. When you have finished, hit the *Enter* key in the beginning without entering anything, which is taken as string of length 0, *i.e.*, *null string*, to terminate the program.

In the above program, we have set up an array of characters to store the string. The *fputs()* function then writes the contents of the array to the disk. Since *fputs()* function does not add a *newline* character automatically to the end of the string, we must do this explicitly to make it easier to read the string back from the file.

Reading from a file

The syntax of the function that reads strings from a file is

```
fgets( str, n, fptr );
```

where *str* is an array of characters and specifies the address where the string is to be stored, *n* is the maximum length of the input string, and *fptr* is a file pointer.

The *fgets()* function is similar to *gets()* function, the only difference is that the *gets()* function always reads from keyboard unless redirection is employed. The *fgets()* function returns a *NULL* value when it reads end-of-file *EOF*. Following listing shows example of a program that reads one string at a time from a file, and writes onto the screen.

Listing 13.4

```
/*  Program to demonstrate reading of strings from a file   */
#include<stdio.h>
#include<string.h>
int main()
{
   FILE *fptr;
   char str[81];
   fptr = fopen( "file2.dat", "r" );
   if ( fptr == NULL ) {
      printf( "\nUnable to open file2.dat\n" );
      return 1;
   }
   printf( "\nContens of file are:\n\n" );
   while ( fgets( str, 80, fptr ) != NULL ) {
      puts( str );
   }
   fclose( fptr );
   return 0;
}
```

When the above program is executed, it reads the contents of *file3.dat* one string at a time and writes them onto the screen till it encounters EOF.

Test Run

```
Contents of file are:

C is a powerful procedural language.
It is a middle level language.
All Programs are tested using Turbo C Compiler.
```

13.3.3.3 Reading and Writing using Formatted I/O functions

So far we have considered reading and writing of characters, strings, and integer numbers. *What about the real numbers*? and *What about the mixed type of data*? For example, suppose that we want to store information about an agent comprising his name (a string), code number (an integer number), and height (a real number). We want to create a data file for a given list of agents. This can be done using formatted I/O functions.

Writing to a file

The syntax of the function that writes formatted data to a file is

```
fprintf( fptr, "format-string" , ditems );
```

where *fptr* is a file pointer, and *ditems* is a lists of variables to be written to a file. The *fprintf()* is similar to *printf()* function, the only difference is that *printf()* function writes formatted data onto the screen.

Listing 13.5

```
/*
 *  Program to demonstrate writing of formatted data to a file
 */
#include<stdio.h>
int main()
{
   FILE *fptr;
   char yes_no;
   char name[41];
   int code;
   float height;
   fptr = fopen( "file3.dat", "w" );
   if ( fptr == NULL ) {
      printf( "\nUnable to open file3.dat\n" );
      return 1;
   }
   while(1)
   {
      printf( "\nEnter name, code number, height" );
      scanf( "%s,%d,%f", name, &code, &height );
      fprintf( fptr,"%s,%d,%f", name, code, height );
      printf( "Any more input y/n?: " );
      yes_no = tolower( getche() );
      fflush ( stdin );
      if ( yes_no == 'n' )
         break;
   }
   fclose( fptr );
   return 0;
}
```

Test Run

```
Enter name, code number, height
Geetu, 10, 160.25
Any more input y/n?: y
Enter name, code number, height
Shivani, 11, 158.5
Any more input y/n?: y
Enter name, code number, height
Vijay, 12, 165.5
Any more input y/n?: y
Enter name, code number, height
Gurpreet, 13, 166.25
Any more input y/n?: n
```

The above information is now in a file named *file3.dat*. If we attempt to look at it using TYPE command, the entire output will be on the same line as there are no *newlines* in the data. To format the output more conveniently, we can write a program specifically to read the above file using formatted input.

Reading from a file

The syntax of the function that reads formatted data to a file is

```
fscanf( fptr, "format-string" , ditems );
```

where *fptr* is a file pointer, and *ditems* is a lists of address where the values read from a file are to be stored.

The *fscanf()* is similar to *scanf()* function, the only difference is that *scanf()* function reads formatted data from the keyboard.

Listing 13.6

```
/*  Program to demonstrate writing of formatted from a file   */
#include<stdio.h>
int main()
{
   FILE *fptr;
   char yes_no, char name[41];
   int code;
   float height;
   fptr = fopen( "file3.dat", "r" );
   if ( fptr == NULL )
   {
      printf( "\nUnable to open file3.dat\n" );
      return 1;
   }
   printf( "\nContens of file are:\n\n" );
   while(fscanf(fptr,"%s,%d,%f", name, &code, &height) != EOF) {
      printf( "%s %d %f\n", name, code, height );
```

```
    }
    fclose( fptr );
    return 0;
}
```

Test Run

```
Contents of file are:
Geetu 10 160.25
Shivani 11 158.5
Vijay 12 165.5
Gurpreet 13 166.25
```

13.4 FILE STATUS FUNCTIONS

The C language provides following three functions to handle file status queries:

- Function *eof()* to test end of file.
- Function *ferror()* to test error.
- Function *clearerr()* to clear error.

13.4.1 Test End of File: *feof()* Function

The *feof()* function is used to check if the end of file has been reached. If the file pointer is at the end, *i.e.*, all data have been read, the function returns 1. If the end of file is not reached, it returns value 0. The prototype for *feof()* is

```
int feof(FILE *stream);
```

where *stream* is a file pointer for the currently opened file.

13.4.2 Test Error: *ferror()* Function

The *ferror()* function is used to test the error status of file. As mentioned earlier, errors can be created for many reasons, ranging from bad media (disk, CD, etc.) to illegal operations such as trying to read a file in the write state. If the *ferror()* function returns 1 if an error has occurred after a file operation. If no error has occurred, the file ferror() returns value 0. The prototype for *ferror()* is

```
int ferror(FILE *stream);
```

where *stream* is a file pointer for the currently opened file.

It is important to note here that testing for an error does not reset the error condition. Once an error has occurred, it can only return to normal read or write state after clearing the error state using *clearer()* function, described next.

13.4.3 Clear Error: *clearerr()* Function

When an error occurs, the subsequent calls to the *ferror()* function return 1, until the error status of the file is reset. The *clearerr()* function is used for this purpose. The prototype for *clearerr()* is

```
void clearerr(FILE *stream);
```

where *stream* is a file pointer for the currently opened file.

It is important to note here that even though we have cleared the error, we have not necessarily cured the problem. We may find that subsequent read or write operation may return to error state.

13.5 FILE POSITIONING FUNCTIONS

The C language provides following three functions to handle file positioning operations:

- Function *rewind()* to set the file pointer to the beginning of the file.
- Function *ftell()* to know the current position of the file pointer in the file.
- Function *seek()* to change the position of the file pointer in the file.

13.5.1 Rewind File: *rewind()* Function

One way of positioning the file pointer to the beginning of the file is to close the file and then re-open it again. However, we can accomplish the same task without closing the file using *rewind()* function. This function positions the file pointer in the beginning of the file. This function sounds very much like rewinding the audio or video cassette in order to listen or watch the cassette from the beginning.

The prototype of *rewind()* function is

```
void rewind(FILE *stream);
```

where *stream* is a file pointer for the currently opened file.

13.5.2 Current Location: *ftell()* Function

In some situations, it may be required to find the current location of the file pointer within the file. The *ftell()* function lets us know the current position of the file pointer.

The prototype of the *ftell()* function is

```
long int ftell(FILE *stream);
```

where *stream* is a file pointer for the currently opened file. Note that the *ftell()* function returns a long integer. This is necessary because many files may have more that 32767 bytes of data.

Recall that the C I/O system considers files as streams of bytes of data. It measures the position in the file by the number of bytes, relative to zero, *i.e.*, from the beginning of the file. When file pointer is at the beginning of the file, the *ftell()* function returns 0. If the file pointer is at the second byte, the *ftell()* function returns value 1.

13.5.3 Set Position: *fseek()* Function

In order to read a data item from anywhere in a file, we have to move the file pointer to the beginning of that data item. To accomplish this task, we can use *fseek()* function.

The prototype of using *fseek()* function is

```
int fseek(file *stream, long offset, int wherefrom );
```

The *fseek()* function takes three arguments, where the first argument *stream* is a file pointer, the second argument *offset* is variable of type *long integer* that specifies the number of bytes by which file pointer is to moved, and third argument *wherefrom* specifies from which position the *offset* is measured. The various values for argument *wherefrom* are listed in Table 13.2.

Table 13.2: Various Values of *wherefrom* for *fseek()* function

Mode	Offset is measured from
0	Beginning of the file
1	Current position of the file
2	End of the file

Table 13.3: Some examples illustrating the use of *fseek()* functions

Seek Call	Action performed
`fseek(fptr, 0, 0);`	Moves the file pointer *fptr* to the beginning of the file. If file pointer is currently at beginning of file, it results in no action.
`fseek(fptr, n, 0);`	Moves the file pointer *fptr* forward by *n* bytes, *i.e.*, move the file pointer to (*n*+1)th bytes in the file.
`fseek(fptr, -n, 1);`	Moves the file pointer *fptr* backward by *n* bytes from current position.
`fseek(fptr, 0, 2);`	Moves the file pointer *fptr* to the end of the file. If file pointer is currently at end of file, it results in no action.
`fseek(fptr, 0, 2);`	Moves the file pointer *fptr* to the end of the file. If file pointer is currently at end of file, it results in no action.
`fseek(fptr, n, 1);`	Moves the file pointer *fptr* forward by *n* bytes from current position.
`fseek(fptr, 1, 1);`	Moves the file pointer *fptr* to the next byte.
`fseek(fptr, -1, 1);`	Moves the file pointer *fptr* to the previous byte.
`fseek(fptr, m, 2);`	Moves the file pointer *fptr* backward by *m* bytes from end of file.

ILLUSTRATIVE EXAMPLES

Example 13.1: Program to find the size of a given file, where the name of the file is given as the command-line argument.

To know the seize of the specified file, position the file pointer at the end of the file using *fseek()* function, and access the value of the file pointer using *ftell()* function that gives the size of the file.

Listing 13.7

```
/* Program to find the size of a given file
*  filename : fsize.c                            */
#include <stdio.h>
int main ( int argc, char *argv[] )
{
   FILE *fptr;
   fptr = fopen(argv[1], "r");
   if (!fptr) {
      printf("\nFile %s does not exist\n",   argv[1]);
      return 1;
   }
   fseek(fptr, 0L, 2);
   printf("\nSize of file = %ld bytes.\n", ftell(fptr));
   fclose(fptr);
   return 0;
}
```

Test Run

```
>fszie fsize.c↵
Size of file = 387 bytes.
```

The above test run shows that the size of the program file (source code) created in Listing 13.7 is 387 bytes.

Example 13.2: rite a program that read some text from the keyboard and write it into a file named TEXT. Program then reads this file and displays its contents on screen.

Listing 13.8

```
/*  Program that creates a file and then reads its contents
*   and display them on the computer screen         */
#include <stdio.h>
int main()
{
   char ch;
   FILE *fp1;
   fp1 = fopen("TEXT", "w");   /*  open file for writing  */
   if ( fp1 == NULL ) {
      printf("\nUnable to open file TEXT for writing\n");
      return 1;
```

```
    }
    printf("\nType some text and terminate");
    printf(" the input by Enter key\n\n");
    while ( ( ch = getche() ) != '\r' )
       fputc(ch, fp1);
    fclose(fp1);   /*  close file  */
    fp1 = fopen("TEXT", "r");    /*  open file for reading  */
    if ( fp1 == NULL ) {
       printf("\nUnable to open file TEXT for reading\n");
       return;
    }
    printf("\n\nContents of file TEXT are\n\n");
    while ( !feof(fp1) ) {
       ch = fgetc(fp1);
       putchar(ch);
    }
    fclose(fp1);
    return 0;
}
```

Test Run

```
Type some text and terminate the input by Enter key
Don't do any thing with others which you wish other should not
do with you.↵
Contents of file TEXT are
Don't do any thing with others which you wish other should not
do with you.
```

Example 13.3: Write a program to copy the contents of one file to another file byte-by-byte. The names of the files are taken as command line arguments.

Listing 13.9

```
/*  Program to copy the contents of a given file to another file.
*   filename : kopy1.c              */
#include <stdio.h>
int main(int argc, char *argv[])
{
    char ch;
    FILE *fp1, *fp2;
    fp1 = fopen(argv[1], "r");
    if ( fp1 == NULL ) {
       printf("\nUnable to open file: %s for reading\n", argv[1]);
       return 1;
    }
    fp2 = fopen(argv[2], "w");
    if ( fp2 == NULL ) {
       printf("\nUnable to open file: %s for writing\n", argv[2]);
       return 1;
    }
    while ( !feof(fp1) ) {
```

```
        ch = fgetc(fp1);
        fputc(ch, fp2);
    }
    printf("\nFile copied successfully...\n");
    fclose(fp1);
    fclose(fp2);
    return 0;
}
```

Test Run

```
>kopy1 kopy1.c kk
File copied successfully...
>
```

You can verify that the contents of file *kk* will be same as that of *kopy1.c*.

Example 13.4: Write a C program to reverse the first *n* characters in a file. The *file name* and *n* are specified on the command line.

Listing 13.10

```
/*  Program to copy contents of a given file to another file
*   filename : kopy2.c          */
#include <stdio.h>
int main(int argc, char *argv[])
{
    char *str, ch;
    int i, n;
    FILE *fp1;
    fp1 = fopen(argv[1], "r+");
    if ( fp1 == NULL ) {
      printf("\nUnable to open file: %s for reading\n", argv[1]);
       return 1;
    }
    i = 0;
    n = atoi(argv[2]);
    str = ( char * ) malloc ( n * sizeof(char) );
    while ( ( !feof(fp1) ) && ( i < n ) )
    {
       ch = fgetc(fp1);
       str[i] = ch;
       i++;
    }
    rewind(fp1); /* reposition the file pointer to the beginning */
    i = n-1;
    while ( i >= 0 )
    {
       fputc( str[i], fp1);
       i--;
    }
    fclose(fp1);
    return 0;
}
```

Test Run

```
Original Contents of file test.dat
1234567890abcdefghijklmnopqrstuvwxyz
>kopy2 test.dat 5
>
New Contents of file test.dat
5432167890abcdefghijklmnopqrstuvwxyz
```

You can verify that the contents of the file after program execution by using type command on the command prompt.

In the above program, first the given file is opened in read/write mode. Next, first *n* bytes of the file are copied in dynamically allocated character array. Next we rewind the given file, *i.e.*, reposition the file pointer at the beginning of the file. Finally, we write the contents of the character array in reverse order onto the file. Thus, we are able to accomplish the desired task.

MULTIPLE CHOICE QUESTIONS

1. Which of the following is a default file pointer?
   (*a*) stdin (*b*) stdout
   (*c*) stderr (*d*) All
2. The function *fopen ()* when fails to open file, then it returns value
   (*a*) NULL (*b*) –1
   (*c*) Null (*d*) void
3. The function *flose()* is, usually, used to
   (*a*) delete a file (*b*) disconnect file from program
   (*c*) read data from a file (*d*) write data to a file
4. Which of the following values is not a value of *from* in function `seek(fptr,offset,from)`?
   (*a*) 2 (*b*) 0
   (*c*) 1 (*d*) EOF
5. The task performed by function `fseek(fptr,0,0)` is
   (*a*) fclose (fptr) (*b*) ftell (fptr)
   (*c*) search (fptr) (*d*) rewind (fptr)
6. Which of the following is not a valid file opening mode?
   (*a*) r+ (*b*) +r
   (*c*) r (*d*) rt
7. If letter *b* is suffixed with the file opening mode, what does it convey?
   (*a*) file is to be opened for reading only
   (*b*) file is to be opened for writing only
   (*c*) file is to be opened for both read-write
   (*d*) file is a binary file

8. Consider the following code segment
```
FILE *fp;
fp =  fopen("inventory.dat", "rb");
```
What is the role of letter *b* in "*rb*"?
(*a*) Open *inventory.dat* file in binary mode for reading.
(*b*) Open *inventory.dat* file reading and writing.
(*c*) Create a new file *inventory.dat* for writing.
(*d*) Open *inventory.dat* file in binary mode for writing and reading.
9. Consider the following code segment
```
FILE *fp;
fp =  fopen("notes.txt", "r+");
```
Which of the following operations can be performed on *notes.txt* file?
(*a*) Reading (*b*) Writing (*c*) Appending (*d*) All

REVIEW EXERCISE

1. Name the various systems for performing file I/O in C.
2. Is it is advisable to close a file explicitly opened for writing?
3. Name the function to close a file.
4. What information system gets by opening a file?
5. What you mean by a *file pointer*?
6. How a file opened with *fopen()* function is referred to in a program?
7. Suppose a program wants to examine every byte of a file, which mode will be more appropriate?
8. What do you mean by term *offset* in reference to a file?
9. What is the task performed by the *fseek()* function?
10. What is the task performed by the *ftell()* function?
11. Which value is returned by the *fopen()* if some error occurs in opening a file?
12. What is the difference between *w+* and *r+* modes?
13. What a *rewind()* function does?

PROGRAMMING EXERCISE

1. Given a text file containing some text. Write a program that prompts the user to input name of a text file, reads this file and outputs the number of vowels and number of words in the text.
2. Given a text file containing some text. Write a program that prompts the user to input name of a text file, reads this file and replaces each occurrence of character 'x' with uppercase 'X'.
3. A beginner to the C language has typed the entire program in uppercase letter. As you know that the normal convention in C language is to type the source code using lowercase letters only except for macros and defined constants. Write a program that converts the source code into lowercase letter. Comments should be left unchanged.

❑ ❑ ❑

Chapter 14

Preprocessor Directives

Learning Outcomes

After reading this chapter, students will be able to

- comprehend the actual compilation process, and the preprocessing of the source code prior to compilation
- comprehend the usage of preprocessing directives
- use preprocessing directives for efficient program handling and making then generalized programs

14.1 INTRODUCTION

The C source code may include instructions for the compiler. These instructions are known as *preprocessor commands or directives*. Each directive begins with character '#' (hash character). Although, these directives are not part of the C language, they expand the scope of the C programming environment.

The C compiler is made up of two functional parts – *a preprocessor* and *a translator.* The preprocessor processes all the directives and as a result produces an intermediate code known as *translation unit*, which is then translated by the translator to machine code and places it in object file.

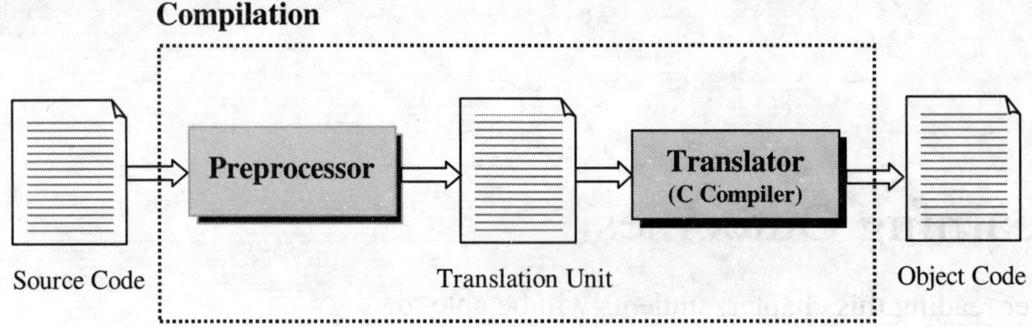

Figure 14.1: Compilation process

In this chapter, we will learn about various preprocessor directives and built-in macros.

14.2 THE #include DIRECTIVE

The *#include* directive, also known as *file inclusion,* instructs the preprocessor to copy the contents of specified file into the program. The files to be included are generally header files, though any file with valid C statements can also be included.

This directive has following forms:

```
#include <filename.h>
#include "filename.h"
```

The first form directs the preprocessor to include the header file from the standard directory while the second form directs the preprocessor to look for the specified header file in the user directory.

14.3 THE #define DIRECTIVE

The *#define* directive, is used to define an identifier and a character string that is substituted for the identifier each time it is encountered in the source file. The identifier is called a *macro name*, and the replacement process is called *macro-substitution*. The general form of the directive is

```
#define identifier string
```

Please note that the semicolon is not used after the *#define* directive. There can be any number of spaces between the identifier and the string, but once the string begins, it is terminated only by a new line.

For example, if we want to use *true* for a value 1 and *false* for a value 0, then we can do this as shown below:

```
#define true  1
#define false 0
```

Once an *identifier* has been defined, it can be used as part of the definition of other macro-names.

For example, the following segment defines ONE, TWO, and THREE to their respective values:

```
#define ONE      1
#define TWO      ONE+ONE
#define THREE    TWO+ONE
```

It is important to understand that macro-substitution simply replaces an identifier with its associated string. However, no text substitutions occur within a string.

For example

```
#define XYZ  "This is test string"
     :
printf( "XYZ" );
```

does not print *This is a test string* rather prints *XYZ.*

If a string is longer than one line, it can be continued on the next line by placing a backslash character \ at the end of the line, as shown in the following example:

```
#define LONG_STRING "This is a long string, \
      and is used to illustrate this feature"
```

It is a common practice to use capital letters for the defined identifiers. With this convention, anyone reading the program can understand at a glance that a macro-substitution will take place. It is always better to put all *#define* directives at the beginning of the program file or perhaps in a separate include file, rather than sprinkling them throughout the program.

The maximum use of macro-substitution is to define names for *magic numbers* that occur in a program. For example, you may have a program that defines an array and have several routines that access that array. Instead of *hard coding* the array's size with a constant, it is better to define a size and use that name whenever the size of the array is needed. If the size of the array changes, you only have to change it at one place in the file and recompile. For example,

```
#define MAX_SIZE 100
float a[MAX_SIZE];
```

The *#define* directive has another powerful feature — a macro-name can have arguments. Each time macro-name is encountered, the arguments associated with it are replaced by the actual arguments found in the program. This is illustrated in the following program.

```
#include<stdio.h>
#define MIN(x, y)   ( x < y ) ? x : y
void main()
{
    int a = 5, b = 20;
    printf( "\nMinimum of %d and %d is %d\n", a, b, MIN(a,b) );
}
```

When this program is compiled, the expression defined by *MIN*(*x*,*y*) is substituted except that *a* and *b* are used as the operands. That is, the *printf()* statement will look like

```
printf("\nMinimum of %d and %d is %d\n",a,b,(a<b)?a:b);
```

If you are not very careful in defining macros that take arguments, there can be some surprising results. For example

```
#include<stdio.h>
#define EVEN(x)   x % 2 ? 0 : 1
void main()
{
    int a = 5;
    if ( EVEN(a + 1) )
        printf( "\n%d is even\n", a+1 );
    else
        printf( "\n%d is odd\n", a+1 );
}
```

does not work correctly because of the way the macro-substitution is made.

When this program is compiled, the *EVEN*(*a*+1) is expanded to

```
a + 1 % 2 ? 0 : 1
```

Because the % (modulus) operator has higher precedence over the + (plus) operator, the modulus operation is performed on 1 first and then the result is added to *a*, which of course does not equal 0. To fix the trouble, there must be parentheses around *a* in the macro definition of EVEN as shown in this corrected version of the program:

```
#include<stdio.h>
#define  EVEN(x)    (x) % 2 ? 0 : 1
void main()
{
    int a = 5;
    if ( EVEN(a + 1) )
        printf( "\n%d is even\n", a+1 );
    else
        printf( "\n%d is odd\n", a+1 );
}
```

Now the expression *a*+1 is evaluated before the modulus operation. In general, it is good idea to surround macro parameters with parentheses to avoid troubles like the one just described.

The use of macro-substitution in place of real function is one major benefit: *it increases the speed of the code because no overhead for a function call is incurred.* However, this increased speed is paid for with an increase in the size of the program because of duplicated code.

14.4 THE #error DIRECTIVE

The *#error* directive forces the C compiler to stop compiling when it is encountered. It is used primarily for debugging. The general form of the directive is

```
#error error-message
```

The error-message is not between double quotes. When the compiler encounters this directive, it displays the following information and terminates compilation

```
fatal: filename linenum Error directive: error-message
```

14.5 THE #if DIRECTIVE

If the *constant expression* following *#if* is true, the code between it and *#endif* is compiled; otherwise the code is ignored by the compiler. The *#endif* is used to mark the end of a *#if* block.

The general form of *#if* directive is

```
#if constant-expression
   statement(s)
#endif
```

Consider the program given on next page.

```
#include<stdio.h>
#define MAX 100
void main()
{
   #if MAX > 99
      printf( "Compiled for small array" );
   #endif
}
```

This program displays the message on the screen because, as defined in the program, MAX is greater than 99.

This example illustrates an important point -- the expression that follows the *#if* is evaluated at compile time. Therefore, it must contain only identifiers that have been previously defined and constants; the use of variables is not permitted.

14.6 THE #else DIRECTIVE

The *#else* directive works the same way as the *else* statement that forms a part of the C language. It provides an alternative to the *#if* directive. The previous example can be expanded as shown here:

```
#include<stdio.h>
#define MAX 10
void main()
{
   #if MAX > 99
      printf( "Compiled for array greater than 99");
   #else
      printf( "Compiled for small array" );
   #endif
}
```

In this case, MAX is defined to be less than 99, so the part of the code between *#if* directive and *#else* directive is not compiled, but the part of the code between *#else* directive and *#endif* directive is compiled.

14.7 THE #elif DIRECTIVE

The *#elif* directive means *else if* and is used to establish an *if–else–if* ladder for compilation options. The *#elif* is followed by a *constant expression*. If the expression is true, that block of code is compiled and no other *#elif* expressions are tested. Otherwise, the next in the series is checked. The general form is

```
#if expression-1
   statement sequence-1
#elif expression-2
   statement sequence-2
#elif expression-3
   statement sequence-3
      :
#elif expression-N
   statement sequence-N
#endif
```

As an example, consider the following segment

```
#define US 0
#define UK 1
#define INDIA 2
#if COUNTRY == US
   char curreny[] = "Dollar";
#elif COUNTRY == UK
   char curreny[] = "Pound";
#elif COUNTRY == INDIA
   char curreny[] = "Rupee";
#endif
```

This segment uses the value of COUNTRY to define the currency sign.

14.8 THE #ifdef AND #ifndef DIRECTIVE

The directive *#ifdef* means *if defined*, and its general form is

```
#ifdef macro-name
   statement sequence
#endif
```

If the macro-name has been previously defined in a *#define* statement, the statement sequence between the *#ifdef* and *#endif* is compiled.

The directive *#ifndef* means *if not defined*, and its general form is

```
#ifndef macro-name
   statement sequence
#endif
```

If the macro-name is currently not defined by a *#define* statement, the statement sequence between the *#ifndef* and *#endif* is compiled.

Both the *#ifdef* and *#ifndef* directives can use *#else* directive but not the *#elif* directive.

14.9 THE #undef DIRECTIVE

The *#undef* directive is used to remove a previous definition of the macro-name that follows it. Its general form is

```
#undef macro-name
```

As an example, consider the following segment

```
#define  LEN 100
#define WIDTH 100
char a[LEN][WIDTH];
#undef LEN
#undef WIDTH
```

Both LEN and WIDTH are defined until the *#undef* statements are encountered. The principal use of *#undef* is to allow macro-names to be localized only to those sections of code that needed them.

14.10 THE #line DIRECTIVE

The *#line* directive is used to change the contents of __LINE__ and __FILE__, which are predefined macro-names. The __LINE__ macro contains the line number of the line currently being compiled and __FILE__ contains the name of the file being compiled. The general form of the *#line* directive is

```
#line number "filename"
```

where *number* is any positive integer, and the *filename* is any valid file identifier, which is optional. The line number is the number of the current source line and the file name is the name of the source file. The *#line* directive is used primarily for debugging purposes and special applications. As an example, consider the following program

```
#include<stdio.h>
#line 100
void main(void)
{
   printf( "\n%d\n", __LINE__);
}
```

This program specifies that line count will begin from 100. The *printf()* statement displays the number 102 because it is the third line in the program after #line 100 statement.

14.11 THE #pragma DIRECTIVE

The *#pragma* directive is defined by ANSI standard to be an implementation-defined directive that allows various instructions, defined by the compiler's creator to be given to the compiler. We will not be discussing it as it is used in advanced applications.

MULTIPLE CHOICE QUESTIONS

1. What are the types of C Preprocessor Directives.?
   (*a*) Macros (*b*) Conditional Compilation
   (*c*) File Inclusion (*d*) All the above
2. Processor Directive in C language starts with - - - - .
   (*a*) $ symbol (dollar) (*b*) @ symbol (at the rate)
   (*c*) & symbol (ampersand) (*d*) # symbol (HASH)
3. Preprocessor in C language works on?
   (*a*) DOTC file (.c) (*b*) DOTEXE file (.exe)
   (*c*) DOTH file (.h) (*d*) DOTCP file (.cp)
4. What is another name for .C file?
   (*a*) Executable code (*b*) Source Code
   (*c*) Distributable Code (*d*) Macro code
5. What is the keyword used to define a C macro?
   (*a*) def (*b*) definition
   (*c*) define (*d*) defy
6. What is the output of C program with #define.?
```
#define CVV 156
void main()
{
    int a=10;
    a = a*CVV;
    printf("CVV=%d",a);
}
```
   (*a*) 0 (*b*) 1650
   (*c*) 1560 (*d*) Compiler error
7. What is the C keyword used to create global constants?
   (*a*) constant (*b*) definition
   (*c*) def (*d*) define
8. Choose a correct C statement about #define statement.
```
#define CVV 156
```
   (*a*) CVV is called Macro Expansion. 156 is called Macro Template.
   (*b*) CVV is called Macro Expansion. 156 is also called Macro Expansion.
   (*c*) CVV is called Macro Template. 156 is called Macro Expansion.
   (*d*) None of the above

9. What is the output file generated after processing a .C file?
   (*a*) .h file
   (*b*) .exe file
   (*c*) .obj file
   (*d*) .bak file
10. What is the output of C program with macros?
```
#define ERRMSG printf("Some error.");
void main()
{
    printf("JAR.");
    ERRMSG;
}
```
   (*a*) JAR.
   (*b*) JAR.ERRMSG
   (*c*) JAR.Some error.
   (*d*) Compiler error
11. What is the output of C program with macros?
```
#define ERRMSG(a) printf("Error=%d",a);
void main()
{
    ERRMSG(10);
}
```
   (*a*) ERRMSG(10)
   (*b*) Error=
   (*c*) Error=10
   (*d*) Compiler error
12. What is the output of C program?
```
#define LOGIC(a,b) (a==b)
void main()
{
    if(LOGIC(5,5))   {
        printf("SAME ");
    }
}
```
   (*a*) SAME
   (*b*) LOGIC(5,5)
   (*c*) 5==5
   (*d*) Compiler error
13. How do you separate a multiline macro in C language?
   (*a*) Using * operator
   (*b*) Using % operator
   (*c*) Using \ operator
   (*d*) Using + operator
14. What is the output of C program.?
```
#define LOOP(a) for(int i=1;i<=a;i++) \
{printf("%d ",i);}
void main()
{
    LOOP(5);
}
```
   (*a*) 5
   (*b*) 5 5 5 5 5
   (*c*) 1 2 3 4 5
   (*d*) Compiler error

15. What is the output of C program?

```
#define TANK(a) a*10+2
void main()
{
    int a = TANK(2)*2;
    printf("%d",a);
}
```

(*a*) 44 (*b*) 22
(*c*) 24 (*d*) Compiler error

16. What is the file extension of expanded source code of .C file after preprocessing?

(*a*) .e file (*b*) .h file
(*c*) .i file (*d*) .p file

17. Choose a correct statement about C Macro?

(*a*) Macro template (e.g. PI or function) will be replaced by Macro Expansion (3.1428) as many number of times as it appears in the C program increasing Source Code size in bytes.
(*b*) Macros increase program speed when compared to functions.
(*c*) Functions use less memory as the program code is written and placed only one place in source code. Macros put function code everywhere it is called again and again.
(*d*) All the above

18. What is the C Preprocessor directive to be used to add a header file or any file to existing C program?

(*a*) #add (*b*) #present
(*c*) #include (*d*) $include

REVIEW EXERCISE

1. What is a translation unit?
2. What are the components of a C compiler? Explain the function of each component.
3. What is the difference in use when the name of the file is enclosed in angle brackets and double quotes with the file inclusion directives?
4. Explain the use of *define* directive with suitable examples.
5. What will be the output of the following program?

```
#define A 1
#define C 3
void main()
{
    #ifdef A
       printf("10");
    #endif
    #ifdef C
       printf("20")
     #endif
     #ifndef B
        printf("30");
     #endif
}
```

6. What will be the output of the following program?

```
#define  EVEN(x)    x % 2 ? 1 : 0
void main()
{
    int a = 5;
    if ( EVEN(a + 1) )
        printf( "\n%d is even\n", a + 1 );
    else
        printf( "\n%d is odd\n", a + 1 );
}
```

7. What will be the output of the following code segment?

```
#define XYZ   "This is test string"
printf( "XYZ" );
```

8. What will be the output of the following program?

```
#define MAX(a, b)    a > b ? b : a
void main()
{
    int max;
    max = MAX(2+3, 3+6);
    printf("%d", max);
}
```

9. Given a choice, would you prefer to use *const* over *#define* preprocessor in order to use a constant?

❑ ❑ ❑

Chapter 15

Introduction to Algorithms

Learning Outcomes

After reading this chapter, students will be able to

- explain the notion of algorithm complexity
- explain the time-space tradeoff among algorithms
- explain Big Oh notation
- explain various categories of algorithms
- write programs for various searching algorithms
- write programs for various sorting algorithms

15.1 INTRODUCTION

Basically, an algorithm is a description of the various steps involved in finding the solution of a particular problem. The development of algorithms is a major task in problem solving.

In general, a problem can have a *direct solution, iterative solution* or a *recursive solution.* In the direct solution, there are fixed steps, in iterative solution, there certain steps within the algorithm that may have to be repeated (iterated), and in recursive solution, the algorithm calls itself to solve a problem of reduced size.

This chapter describes the algorithms, the algorithmic complexity, Big Oh (O) notation to express algorithmic complexity, and few basic algorithms.

15.2 ALGORITHM DEFINED

An *algorithm* is a finite set of steps defining the correct solution of a particular problem. An algorithm can be expressed in English like language, called *pseudocode,* in a programming language, or in the form of a flowchart. In this book, we will describe the algorithms by numbering the steps which are executed in increasing order unless otherwise specified by the decision making for selecting particular step(s) or by repeating certain step(s).

Each algorithm must meet the following criteria:

- ***Definiteness*****:** Each step must be clear and unambiguous.
- ***Finiteness*****:** If we trace the steps of an algorithm, then for all cases, the algorithm must terminate after a finite number of steps.
- ***Effectiveness*****:** Each step must be sufficiently basic that it can in principle be carried out by a person using only paper and pencil. In addition, not only each step is definite, it must also be feasible.

15.2.1 Algorithm Complexity

There are basically two aspects of computer programming:

- First is the data organization, *i.e.*, the data structures to represent the data of the problem in hand.
- Second is choosing the appropriate algorithm to solve the problem in hand.

As an algorithm is a sequence of steps to solve a problem, there may be more than one algorithm to solve a problem.

The choice of particular algorithm depends on following considerations:

- Performance requirements, *i.e., time complexity*
- Memory requirements, *i.e., space complexity*

Performance requirements are usually more critical than memory requirements; hence, in general, it is not necessary to worry about memory unless they grow faster than performance requirements. Therefore, in general, the algorithms are analyzed only on the basis of performance requirements, *i.e.*, running-time efficiency.

15.2.1.1 Space Complexity

The space complexity of an algorithm, hence program, is the amount of memory it needs to run to completion. Some of the reasons for studying space complexity are:

- If the program is to run on multi-user system, it may be required to specify the amount of memory to be allocated to the program.
- We may be interested to know in advance that whether sufficient memory is available to run the program.
- There may be several possible solutions with different space requirements.
- Can be used to estimate the size of the largest problem that a program can solve.

The measure the space complexity, we compute the amount of memory needed for instructions, for constants, simple variables and fixed size structured variables.

15.2.1.2 Time Complexity

The time complexity of an algorithm is the amount of time it needs to run to completion. Some of the reasons for studying time complexity are:

- We may be interested to know in advance that whether the program will provide a satisfactory real-time response. For example, an interactive program, such as an editor, must provide such a response. If it takes even a few seconds to move the cursor one page up or down, it will not be acceptable to the user.
- There may be several possible solutions with different time requirements.

To measure the time complexity accurately, we can count the all sort of operations performed in an algorithm. If we know the time for each one of the primitive operations performed on a given computer, we can easily compute the time taken by an algorithm to complete its execution. This time will vary from system to system.

Our intention is to estimate the execution time of an algorithm irrespective of the computer on which it will be used. Hence, the more reasonable approach is to identify the key operation and count such operations performed till the program completes its execution. A key operation in our algorithm is an operation that takes maximum time among all possible operations in the algorithm. The time complexity can now be expressed as a function of number of key operations performed.

15.2.1.3 Time-Space Trade-off

The best algorithm, hence best program, to solve a given problem is one that requires less space in memory and takes less time to complete its execution. But in practice, it is not always possible to achieve both of these objectives. As said earlier, there may be more than one approach to solve a same problem. One such approach may require more space but takes less time to complete its execution while the other approach requires less space but takes more time to complete its execution. Thus, we may have to sacrifice one at the cost of the other. That is what we can say that there exits a *time-space* trade among algorithms.

Therefore, if space is our constraint, then we have to choose a program that requires less space at the cost of more execution time. On the other hand, if time is our constraint such as in real time systems, we have to choose a program that takes less time to complete it execution at the cost of more space.

In the analysis of algorithms, we are interested in the *average case*, the amount of time a program might be expected to take on typical input data, and in the *worst case*, the amount of time a program would take on the worst possible input configuration.

15.2.2 Expressing Space and Time Complexity

Space and/or time complexity is usually expressed in the form of a function $f(n)$, where n is the input size for a given instance of the problem being solved.

Expressing space and/or time complexity as a function $f(n)$ is important because of following reasons:

- We may be interested to predict the rate of growth of complexity as the size of the problem increases.
- To compare the complexities of two or more algorithms solving the same problem in order to find which is more efficient.

The most important notation used to express this function $f(n)$ is *Big Oh* notation, which provides the upper bound for the complexity.

Since in modern computers, the memory is not a severe constraint, therefore, our analysis of algorithms will be on the basis of time complexity.

15.2.2.1 Big Oh Notation

Big Oh is a characterization scheme that allows measuring the properties of algorithms such as performance and/or memory requirements in a general fashion.

The algorithm complexity can be determined ignoring the implementation dependent factors. Eliminating constant factors in the analysis of the algorithms does this. Basically, these are the constant factors that differ from computer to computer. Clearly, complexity function $f(n)$ of an algorithm increases as n increases. It is rate of increase of $f(n)$ that we want to examine.

Suppose, $f(n)$ and $g(n)$ are functions defined on positive integer number n, then function $f(n) = O(g(n))$, read as "f of n is big oh of g of n", or as "$f(n)$ is of the order of $g(n)$", if there exists positive constants c and n_0, such that $f(n) \le c \times g(n)$ for all values of $n \ge n_0$. That is, for all sufficiently large amount of input data ($n > n_0$), $f(n)$ will grow no more than a constant factor than $g(n)$. Thus g is an upper bound, except for a constant factor c on the value of f for all suitably large n, *i.e.*, $n \ge n_0$. While providing an upper bound function g for f, we will use only simple functional forms. These typically contain a single term in n with a multiplicative constant of one.

Table 15.1: Some examples

$g(n)$	$f(n) = O(g(n))$
20	O(1) *i.e.*, Constant
$0.5n^2 + 3$	$O(n^2)$
$500n^2 - 25$	$O(n^2)$
$n^3 + n^2 - 1$	$O(n^3)$
$5n^3 + 3n + 6$	$O(n^3)$

Note that exact constant values do not matter and that relationship $f(n) \le c \times g(n)$ may not hold for small input sizes. Since purpose of *Big Oh* notation is to compare the algorithms in a general fashion, the anomalies that appear for small input sizes are ignored.

Table 15.2: Rate of growth of some standard functions

n \ $f(n)$	$\log_2 n$	n	n^2	n^3	2^n	$n\log_2 n$
5	3	5	25	125	32	15
10	4	10	100	10^3	10^3	40
100	7	100	10^4	10^6	10^{30}	700
1000	10	10^3	10^6	10^9	10^{300}	10^4

Observe that logarithmic function $\log_2 n$ grows most slowly, whereas the exponential function 2^n grows most rapidly, and the polynomial function n^k grows according to the exponent k.

15.2.2.2 Categories of Algorithms

Based on *Big Oh* notation, the algorithms can be categorized as follows:

- Constant time (O(1)) algorithms
- Logarithmic time (O(log$n$)) algorithms
- Linear time (O(n)) algorithms
- Polynomial time (O(n^k), for $k > 1$) algorithms
- Exponential time (O(k^n), for $k > 1$) algorithms

Many algorithms are O($n$log$n$).

15.2.2.3 Limitations of Big Oh Notation

Big Oh notation has two basic limitations:

- It contains no consideration of programming effort.
- It masks (hides) potentially important constants.

As an example of later limitation, imagine two algorithms, one using $500000n^2$ time, and the other n^3 time. The first algorithm is O(n^2), which implies that it will take less time than the other algorithm which is O(n^3). However, the second algorithm will be faster for $n < 500000$, and this would be faster for many applications.

15.3 SEARCHING ALGORITHMS

Searching is the process of finding the location of given element in an array. The search is said to be successful if the given element is found, i.e., the element does exists in the array; otherwise unsuccessful.

There are two approaches to search operation:

- Linear search
- Binary search

The algorithm that one chooses generally depends on organization of the array elements. If the elements are in random order, then one have to use linear search technique, and if the array elements are sorted, then it is preferable to use binary search technique. These two search techniques are described in the next sections.

15.3.1 Linear Search

Given no information about the array *a*, the only way to search for given element *item* is to compare *item* with each element of *a* one by one. This method, which traverses *a* sequentially to locate *item* is called *linear search* or *sequential search*.

Listing 15.1

```
/*
*   Program to demonstrate the use of linear search to search
*   a given element in an un-sorted array
*/
#include <stdio.h>
int main()
{
   int a[50], i, n, item, flag;
   printf("\nEnter size of array : " );
   scanf("%d", &n );
   printf("\nEnter %d elements of array\n", n);
   for ( i = 0; i < n; i++ )
      scanf( "%d", &a[i] );
   printf( "\nEnter element to search : " );
   scanf( "%d", &item );
   flag = 0;

   for ( i = 0; i < n; i++ ) {
      if( item == a[i] )     /* match found  */
      {
         flag = 1;
         break;
      }
   }
  if ( flag == -1 )
    printf( "\nElement found at index: %d\n", i );
  else
    printf( "\nElement not found...\n" );
  return 0;
}
```

Test Run 1

```
Enter size of array : 10
Enter 10 elements of array
12
5
18
9
11
10
25
36
22
100
Enter element to search : 11
Element found at index: 4
```

Test Run 2

```
Enter size of array : 10
Enter 10 elements of array
12
5
18
9
11
10
25
36
22
100
Enter element to search : 55
Element not found...
```

Complexity Analysis

In the best possible case, the *item* may occur at first position. In this case, the search operation terminates in success with just one comparison. However, the worst case occurs when either the item is present at last position or missing from the array. In the former case, the search terminates in success with n comparisons. In the later case, the search terminates in a failure with n comparisons. Thus, we find that in worst case the linear search is $O(n)$ operations.

15.3.2 Binary Search

Suppose the elements of the array *A* are sorted in ascending order (if the elements are numbers) or dictionary order (if the elements are strings). The best searching algorithm, called *binary search,* is used to find the location of the given element. We do use this approach in our daily life. For example, suppose we want to find the meaning of the term *modem* in a computer dictionary. Obviously, we don't search page by page. We open the dictionary in the middle (roughly) to determine which half contains the term being sought. Then for the subsequent search one half is discarded, and we search in the other half. This process is continued till either we have located the required term or that term is missing from the dictionary, which will indicated by the fact that at the end we will be left with only one page.

Example 15.1: To illustrate the working of the binary search technique, consider the following sorted array *A* with 7 elements

3, 10, 15, 20, 35, 40, 60

and we want to search element 15.

Solution:

Given Array a

$a[0]$	$a[1]$	$a[2]$	$a[3]$	$a[4]$	$a[5]$	$a[6]$
3	10	15	20	35	40	60

To start with, we take $beg = 0$, $end = 6$, and compute location of middle element as

$$mid = (beg + end) / 2 = (0 + 6) / 2 = 3$$

Since $a[mid]$, *i.e.*, $a[3] \neq 15$, and $beg < end$. We start next iteration.

As $a[mid] = 20 > 15$, therefore, we take $end = mid - 1 = 3 - 1 = 2$, where as beg remains unchanged. Thus

$$mid = (beg + end) / 2 = (0 + 2) / 2 = 1$$

Since $a[mid]$, *i.e.*, $a[1] \neq 15$, and $beg < end$. We start next iteration.
As $a[mid] = 10 < 15$, therefore, we take $beg = mid + 1 = 1 + 1 = 2$, where as end remains unchanged.

Since $beg = end$, again compute location of the middle element as

$$mid = (beg + end) / 2 = (2 + 2) / 2 = 2$$

Since $a[mid]$, *i.e.*, $a[2] = 15$, the search terminates on success.

The element is found at index 2.

Listing 15.2

```
/*
 *   Program to demonstrate the use of linear search to search
 *   a given element in a sorted array
 */
#include <stdio.h>
int main()
{
    int a[50], i, n, item, beg, end, mid, flag;
    printf("\nEnter size of array : " );
    scanf("%d", &n );
    printf("\nEnter %d elements of array in ascending order\n", n);
    for ( i = 0; i < n; i++ )
       scanf( "%d", &a[i] );
    printf( "\nEnter element to search : " );
    scanf( "%d", &item );
    flag = 0;
    beg = 0;
    end = n - 1;
    while ( beg < end )
    {
       mid = ( beg + end ) / 2;
       if( item == a[mid] ) {        /* match found  */
          flag = 1;
          break;
```

```
        }
        if( item < a[mid] )
            end = mid - 1;
        else
            beg = mid + 1;
    }
    if ( flag == 1 )
        printf( "\nElement found at index: %d\n", mid );
    else
        printf( "\nElement not found...\n" );
    return 0;
}
```

Test Run 1

```
Enter size of array : 10
Enter 10 elements of array in ascending order
5
9
10
11
12
18
25
36
44
100
Enter element to search : 44
Element found at index: 8
```

Test Run 2

```
Enter size of array : 10
Enter 10 elements of array in ascending order
5
9
10
11
12
18
25
36
44
100
Enter element to search : 35
Element not found...
```

Complexity Analysis

In each iteration, the search is reduced to one half of the array. Therefore, for n elements in the array, there will be $\log_2 n$ iterations or recursive calls. Thus, the complexity of binary search is $O(\log_2 n)$. This complexity will be same irrespective of the position of the element, even if it is not present in the array.

15.4 SORTING ALGORITHMS

Sorting is the process of arranging the elements in some logical order. This logical order may be ascending or descending in case of numeric values or dictionary order in case of alphanumeric values.

We do, almost, use this process daily in our routine activities. For example, we do arrange our class notes in increasing order of date in order to refer them quickly later on. Like this there are many situations, which you can enumerate.

In this section, we sill discuss following sorting algorithms:

- Bubble sort
- Insertion sort
- Selection sort

15.4.1 Bubble Sort

The Bubble sort method requires $(n-1)$ pass to sort an array. In each pass every element $a[i]$ is compared with $a[i+1]$, for $i = 0$ to $(n-k)$, where k is the pass number and if they are out of order, *i.e.*, if $a[i] > a[i+1]$, they are swapped. This will cause the largest element move or bubble up.

Thus, after the end of the first pass the largest element in the array will be placed in nth position, and on each successive pass, the next largest element is placed at position $(n-1)$, $(n-2)$, ..., 2, respectively.

For more clarity, carefully examine the following steps.

Pass 1:

Step 1. If $a[0] > a[1]$ then swap $a[0]$ and $a[1]$.
Step 2. If $a[1] > a[2]$ then swap $a[1]$ and $a[2]$.
⋮
Step $n-1$. If $a[n-2] > a[n-1]$ then swap $a[n-2]$ and $a[n-1]$.

Pass 2:

Step 1. If $a[0] > a[1]$ then swap $a[0]$ and $a[1]$.
Step 2. If $a[1] > a[2]$ then swap $a[1]$ and $a[2]$.
⋮
Step $n-2$. If $a[n-3] > a[n-2]$ then swap $a[n-3]$ and $a[n-2]$.

⋮

Pass *k*:

Step 1. If $a[0] > a[1]$ then swap $a[0]$ and $a[1]$.
Step 2. If $a[1] > a[2]$ then swap $a[1]$ and $a[2]$.
⋮
Step $n-k$. If $a[n-k+1] > a[n-k]$ then swap $a[n-k+1]$ and $a[n-k]$.

⋮

Pass *n*–1:

Step 1. If $a[0] > a[1]$ then swap $a[0]$ and $a[1]$.

After $(n-1)$ passes, the array will be sorted in ascending order.

Example 15.2: To illustrate the working of bubble sort method, consider the sorting of the following array in ascending order.

12 40 3 2 15

Solution: Note that, the output of a given pass becomes the input for the next pass. The sorting process is self explanatory. In each pass, one element, shown as shaded, reaches to its final position. Given array can be shown as

a[0]	a[1]	a[2]	a[3]	a[4]
12	40	3	2	15

Pass-1:

a[0]	a[1]	a[2]	a[3]	a[4]
(12)	(40)	3	2	15

(*a*) Compare $a[0]$ and $a[1]$. Since $a[0] < a[1]$ $(12 < 40)$, no action is taken.

a[0]	a[1]	a[2]	a[3]	a[4]
12	(40)	(3)	2	15

(*b*) Compare $a[1]$ and $a[2]$. Since $a[1] > a[2]$ $(40 > 3)$, swap these elements.

a[0]	*a*[1]	*a*[2]	*a*[3]	*a*[4]
12	3	(40)	(2)	15

(*c*) Compare *a*[2] and *a*[3]. Since *a*[2] > *a*[3] (40 > 2), swap these elements.

a[0]	*a*[1]	*a*[2]	*a*[3]	*a*[4]
12	3	2	(40)	(15)

(*d*) Compare *a*[3] and *a*[4]. Since *a*[3] > *a*[4] (40 > 15), swap these elements.

a[0]	*a*[1]	*a*[2]	*a*[3]	*a*[4]
12	3	2	15	40

Thus, after first pass, the largest element, 40, has moved to its final position. However, the rest of the numbers, which may have changed their positions, are yet to be sorted. The shaded part shows the elements that have been sorted.

For the remaining passes, only comparisons and swaps, if required, are shown.

Pass-2:

a[0]	*a*[1]	*a*[2]	*a*[3]	*a*[4]	Action
(12)	(3)	2	15	40	Swap
3	(12)	(2)	15	40	Swap
3	2	(12)	(15)	40	No swap
3	2	12	15	40	

After second pass, the second largest element, 15, has moved to its final position.

Pass-3:

a[0]	*a*[1]	*a*[2]	*a*[3]	*a*[4]	Action
(3)	(2)	12	15	40	Swap
2	(3)	(12)	15	40	No swap
2	3	12	15	40	

After third pass, the third largest element, 12, has moved to its final position.

Pass-4:

$a[0]$	$a[1]$	$a[2]$	$a[3]$	$a[4]$	Action
2	3	12	15	40	No swap
2	3	12	15	40	

After fourth pass, the fourth largest element, 3, has moved to its final position. Thus, the array is completely sorted in 4 passes.

The above sorting process can also be visualized as shown below:

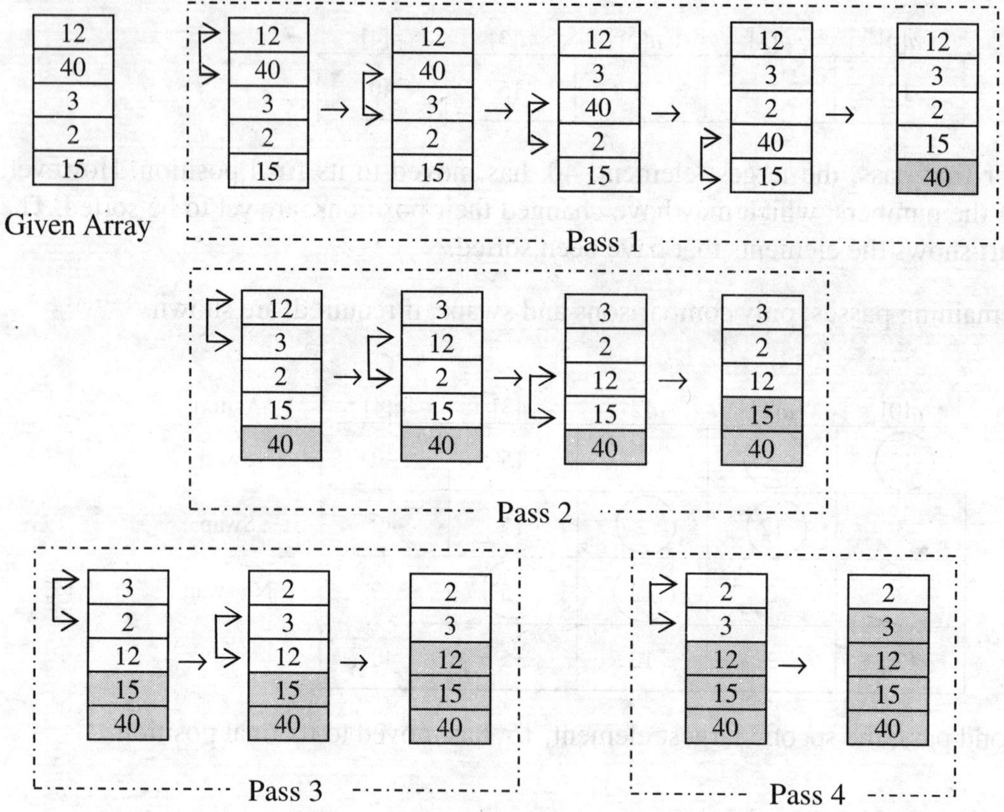

Figure 15.1: Illustration of Bubble sort method

Complexity Analysis

After $(n-1)$ passes, the array will be sorted in ascending order. As you must have noticed, the first pass requires $(n-1)$ comparisons, the second pass requires $(n-2)$,..., kth pass requires $(n-k)$, and the last pass requires only one comparison. Therefore, total comparisons are

$$f(n) = (n-1) + (n-2) + (n-3) + \ldots + (n-k) + \ldots + 3 + 2 + 1 = n(n-1)/2 = O(n^2)$$

Listing 15.3

```
/*
*   Program to demonstrate the use of Bubble sort method to sort a
*   a given array in ascending order
*/
#include <stdio.h>

int main()
{
   int a[50], i, j, n, temp;
   printf("\nEnter size of array : " );
   scanf("%d", &n );
   printf("\nEnter %d elements of array\n", n);
   for ( i = 0; i < n; i++ )
      scanf( "%d", &a[i] );
   for ( i = 0; i < n-1; i++ )
   {
      for ( j = 0; j < n-i-1; j++ )
      {
         if ( a[j] > a[j+1] ) {
             temp = a[j];
             a[j] = a[j+1];
             a[j+1] = temp;
         }
      }
   }
   printf( "\nArray after sorting...\n\n" );
   for ( i = 0; i < n; i++ )
       printf( "%d ", a[i] );
   printf( "\n" );
   return 0;
}
```

Test Run

```
Enter size of array : 10
Enter 10 elements of array
5
12
10
15
22
18
2
-10
7
16

Array after sorting...

-10 2 5 7 10 12 15 16 18 22
```

15.4.2 Selection Sort

Selection sort method also requires $(n-1)$ pass to sort an array. In the first pass, find the smallest element from elements $a[0]$, $a[1]$, $a[2]$, . . ., $a[n-1]$ and swap with the first element, *i.e*, $a[0]$. In the second pass, find the smallest element from elements $a[1]$, $a[2]$, $a[3]$, . . ., $a[n-1]$ and swap with $a[1]$. And so on.

For more clarity, carefully examine the following steps.

Pass 1:

- Find the location *loc* of the smallest element in the entire array, *i.e.*, $a[0]$, $a[1]$, $a[2]$,..., $a[n-1]$.
- Interchange $a[0]$ & $a[loc]$. Then $a[0]$ is trivially sorted.

Pass 2:

- Find the location *loc* of the smallest element in the sub array $a[1]$, $a[2]$, $a[3]$,. . ., $a[n-1]$.
- Interchange $a[1]$ & $a[loc]$. Then elements $a[0]$ and $a[1]$ are sorted, since $a[0] \le a[1]$.

⋮

Pass *k*:

- Find the location *loc* of the smallest element in the sub array $a[k]$, $a[k+1]$, $a[k+2]$,. . ., $a[n-1]$.
- Interchange $a[k]$ & $a[loc]$. Then elements $a[0]$, $a[1]$, $a[2]$, . . ., $a[k]$ are sorted.

⋮

Pass *n*–1:

- Find the location *loc* of the smaller of the elements $a[n-2]$, $a[n-1]$.
- Interchange $a[n-2]$ & $a[loc]$. Then elements $a[0]$, $a[1]$, $a[2]$, . . ., $a[n-1]$ are sorted.

After $(n-1)$ passes, the array will be sorted in ascending order.

Example 15.3: To illustrate the working of the selection sort method, consider the following array *a* with 7 elements as

20, 35, 40, 100, 3, 10, 15

Solution: In each iteration *i*, we find the location *loc* of the smallest element in the unsorted part of the array. If $loc \ne i$ then element at *i* and *loc* are swapped.

	$a[0]$	$a[1]$	$a[2]$	$a[3]$	$a[4]$	$a[5]$	$a[6]$
Given Array *a*	20	35	40	100	3	10	15

Pass 1:

a[0]	a[1]	a[2]	a[3]	a[4]	a[5]	a[6]
20	35	40	100	3	10	15

$i = 0$
$loc = 4$

Interchange elements $a[0]$ & $a[4]$, *i.e.*, 20 & 3 to obtain following array.

Pass 2:

a[0]	a[1]	a[2]	a[3]	a[4]	a[5]	a[6]
3	35	40	100	20	10	15

$i = 1$
$loc = 5$

Interchange elements $a[1]$ & $a[5]$, *i.e.*, 35 & 10 to obtain following array.

Pass 3:

a[0]	a[1]	a[2]	a[3]	a[4]	a[5]	a[6]
3	10	40	100	20	35	15

$i = 2$
$loc = 6$

Interchange elements $a[2]$ & $a[6]$, *i.e.*, 40 & 15 to obtain following array.

Pass 4:

a[0]	a[1]	a[2]	a[3]	a[4]	a[5]	a[6]
3	10	15	100	20	35	40

$i = 3$
$loc = 4$

Interchange elements $a[3]$ & $a[4]$, *i.e.*, 100 & 20 to obtain following array.

Pass 5:

a[0]	a[1]	a[2]	a[3]	a[4]	a[5]	a[6]
3	10	15	20	100	35	40

$i = 4$
$loc = 5$

Interchange elements $a[4]$ & $a[5]$, *i.e.*, 100 & 35 to obtain following array.

Pass 6:

a[0]	a[1]	a[2]	a[3]	a[4]	a[5]	a[6]
3	10	15	20	35	100	40

$i = 5$
$loc = 6$

Interchange elements $a[5]$ & $a[6]$, *i.e.*,100 & 40 to obtain following sorted array.

a[0]	a[1]	a[2]	a[3]	a[4]	a[5]	a[6]
3	10	15	20	35	40	100

Figure 15.2: Successive passes of selection sort method

The shaded part shows the part of the array sorted so for after each iteration. In each iteration, one element is added on the right of the sorted part.

To implement the selection sort algorithm, we need a routine to find the location *loc* of the smallest element among the elements $a[k-1]$, $a[k+1]$, $a[k+2]$,. . ., $a[n-1]$, during *k*th pass.

Listing 15.4

```
/*
*   Program to demonstrate the use of selection sort method to sort a
*   a given array in ascending order
*/
#include <stdio.h>
int main()
{
   int a[50], i, j, n, temp, min, loc;
   printf("\nEnter size of array : " );
   scanf("%d", &n );
   printf("\nEnter %d elements of array\n", n);
   for ( i = 0; i < n; i++ )
      scanf( "%d", &a[i] );
   for ( i = 0; i < n-1; i++ )
   {
      min = a[i];
      loc = i;
      for ( j = i+1; j < n; j++ )
      {
         if ( a[j] < min ) {
             min = a[j];
             loc = j;
         }
      }
      if ( loc != i ) {
         temp = a[i];
         a[i] = a[loc];
         a[loc] = temp;
      }
   }
   printf( "\nArray after sorting...\n\n" );
   for ( i = 0; i < n; i++ )
       printf( "%d ", a[i] );
   printf( "\n" );
   return 0;
}
```

Test Run

```
Enter size of array : 6
Enter 6 elements of array
12
5
10
15
22
18
Array after sorting...
5 10 12 15 18 22
```

Complexity Analysis

As you must have noticed, the first pass requires $(n-1)$ comparisons to find the location *loc* of smallest element, the second pass requires $(n-2)$, ..., kth pass requires $(n-k)$, and the last pass requires only one comparison. Therefore total comparisons are

$$f(n) = (n-1) + (n-2) + (n-3) + \ldots + (n-k) + \ldots + 3 + 2 + 1$$
$$= n(n-1)/2 = O(n^2)$$

15.4.4 Insertion Sort

This algorithm is very popular with bridge players when they first sort their cards.

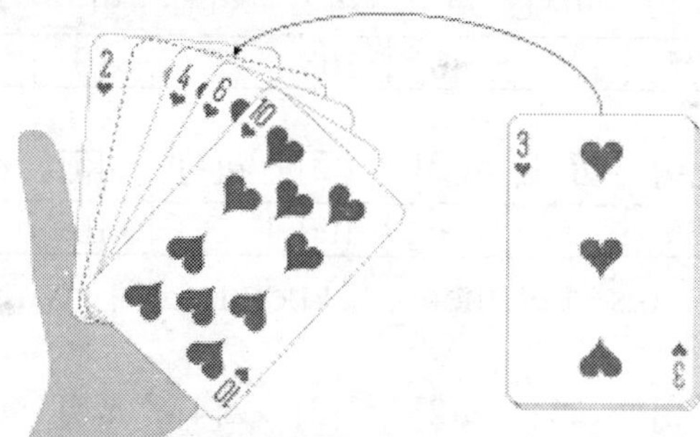

In this procedure, we pick up a particular value and then insert it at the appropriate place in the sorted sub list, *i.e.*, during kth iteration the element $a[k]$ is inserted in its proper place in the sorted sub array $a[1], a[2], a[3], \ldots, a[k-1]$.

This task is accomplished by comparing $a[k]$ with $a[k-1]$, $a[k-2]$, $a[k-3]$, and so on until the first element $a[j]$ such that $a[j] \leq a[k]$ is found. Then each of the elements $a[k-1]$, $a[k-2], \ldots, a[j+1]$ are moved one position up, and then element $a[k]$ is inserted in $(j+1)$st position in the array.

For more clarity, carefully examine the following steps.

Pass 1: $a[1]$ is inserted either before or after $a[0]$ so that $a[0]$ and $a[1]$ are sorted.

Pass 2: $a[2]$ is inserted either before $a[0]$ or between $a[0]$ and $a[1]$ or after $a[1]$ so that the elements $a[0], a[1], a[2]$ are sorted.

Pass 3: $a[3]$ is inserted either before $a[0]$ or between $a[0]$ and $a[1]$ or between $a[1]$ and $a[2]$ or after $a[2]$ so that the elements $a[0], a[1], a[2], a[3]$ are sorted.

⋮

Pass *k*: $a[k]$ is inserted in proper place in the sorted sub array $a[0], a[1], a[2], \ldots, a[k-1]$ so that after insertion, the elements $a[0], a[1], a[2], \ldots, a[k-1], a[k]$ are sorted.

⋮

Pass n–1: $a[n-1]$ is inserted in proper place in the sorted sub array $a[0]$, $a[1]$, $a[2]$, . . ., $a[n-2]$, so that after insertion, the elements $a[0]$, $a[1]$, $a[2]$, . . ., $a[n-1]$ are sorted.

Thus, after (n–1) passes, the array will be sorted in ascending order.

Example 15.4: To illustrate the working of the insertion sort method, consider the following array a with 7 elements as

35, 20, 40, 100, 3, 10, 15

Solution: Given array a

$a[0]$	$a[1]$	$a[2]$	$a[3]$	$a[4]$	$a[5]$	$a[6]$
35	20	40	100	3	10	15

Pass 1:

$a[0]$	$a[1]$	$a[2]$	$a[3]$	$a[4]$	$a[5]$	$a[6]$
35	20	40	100	3	10	15

Since $a[1] < a[0]$, insert element $a[1]$ before $a[0]$ giving the following array.

Pass 2:

$a[0]$	$a[1]$	$a[2]$	$a[3]$	$a[4]$	$a[5]$	$a[6]$
20	35	40	100	3	10	15

Since $a[2] > a[1]$, no action is performed.

Pass 3:

$a[0]$	$a[1]$	$a[2]$	$a[3]$	$a[4]$	$a[5]$	$a[6]$
20	35	40	100	3	10	15

Since $a[3] > a[2]$, again no action is performed.

Pass 4:

$a[0]$	$a[1]$	$a[2]$	$a[3]$	$a[4]$	$a[5]$	$a[6]$
20	35	40	100	3	10	15

Since $a[4]$ is less than $a[3]$, $a[2]$, $a[1]$ as well as $a[0]$, therefore insert $a[4]$ before $a[0]$, giving the following array.

Pass 5:

$a[0]$	$a[1]$	$a[2]$	$a[3]$	$a[4]$	$a[5]$	$a[6]$
3	20	35	40	100	10	15

Since $a[5]$ is less than $a[4]$, $a[3]$, $a[2]$, and $a[1]$, therefore insert $a[5]$ before $a[1]$, giving the following array.

$a[0]$	$a[1]$	$a[2]$	$a[3]$	$a[4]$	$a[5]$	$a[6]$
3	10	20	35	40	100	15

Pass 6:

Since *a*[6] is less than *a*[5], *a*[4], *a*[3], and *a*[2], therefore insert *a*[6] before *a*[2], giving the following sorted array.

a[0]	*a*[1]	*a*[2]	*a*[3]	*a*[4]	*a*[5]	*a*[6]
3	10	15	20	35	40	100

Figure 15.3: Successive passes of insertion sort method

Listing 15.5

```
/*  Program to demonstrate the use of insertion sort method to sort a
*   a given array in ascending order  */
#include <stdio.h>
void main()
{
   int a[50], i, j, k, n, temp, min, loc;
   printf("\nEnter size of array : " );
   scanf("%d", &n );
   printf("\nEnter %d elements of array\n", n);
   for ( i = 0; i < n; i++ )
      scanf( "%d", &a[i] );
   for ( k = 1; k < n; k++ ) {
      temp = a[k];
      j = k - 1;
      while ( ( temp < a[j] ) && ( j >= 0 ) ) {
          a[j+1] = a[j];
          j--;
      }
      a[j+1] = temp;
   }
   printf( "\nArray after sorting...\n\n" );
   for ( i = 0; i < n; i++ )
       printf( "%d ", a[i] );
   printf( "\n" );
}
```

Test Run

```
Enter size of array : 7
Enter 7 elements of array
5
12
10
15
22
18
16
Array after sorting...
5 10 12 15 16 18 22
```

MULTIPLE CHOICE QUESTIONS

1. What is the worst-case time for linear search finding a single item in an array?

   (*a*) Constant time (*b*) Logarithmic time
   (*c*) Linear time (*d*) Quadratic time

2. What is the worst-case time for binary search finding a single item in an array?

   (*a*) Constant time (*b*) Logarithmic time
   (*c*) Linear time (*d*) Quadratic time

3. What additional requirement is placed on an array, so that binary search may be used to locate an entry?

   (*a*) The array elements must form a heap
   (*b*) The array must have at least 2 entries
   (*c*) The array must be sorted
   (*d*) The array's size must be a power of two

4. The best way to find an item in an unsorted implemented using an array list is with ___________ .

   (*a*) Linear search (*b*) Binary search
   (*c*) Random search (*d*) Direct search

5. Using Big O notation, the number of comparisons required by a binary search is

   (*a*) $O(\log_2 n)$ (*b*) $O(n)$
   (*c*) $O(n^2)$ (*d*) $O(n\log_2 n)$

6. In a selection sort of n elements, how many times, at most, the swap function is called in the complete execution of the algorithm?

   (*a*) 1 (*b*) $n-1$
   (*c*) $n\log_2 n$ (*d*) n^2

7. Suppose that a selection sort of 100 items has completed 42 iterations of the main loop. How many elements are now guaranteed to be in their final spot?

   (*a*) 21 (*b*) 41
   (*c*) 42 (*d*) 43

8. Suppose we are sorting an array of eight integers in ascending order using some quadratic sorting algorithm. After four iterations of the algorithm's main loop, the array elements are ordered as shown here:

   52 54 55 57 58 51 53 56

Which statement is correct?

(*a*) Algorithm might be either selection sort or insertion sort
(*b*) Algorithm might be selection sort, but it is not insertion sort
(*c*) Algorithm is not selection sort, but it might be insertion sort
(*d*) Algorithm is neither selection sort nor insertion sort

9. Suppose we are sorting an array of ten integers in ascending order using some quadratic sorting algorithm. After four iterations of the algorithm's main loop, the array elements are ordered as shown here:

11 22 33 42 55 50 66 87 98 80

Which statement is correct?

(*a*) Algorithm might be either selection sort or insertion sort
(*b*) Algorithm might be selection sort, but it is not insertion sort
(*c*) Algorithm is not selection sort, but it might be insertion sort
(*d*) Algorithm is neither selection sort nor insertion sort

10. Which of the following algorithm can be modified to improve its performance?

(*a*) Selection sort (*b*) Bubble sort
(*c*) Insertion sort (*d*) All of above

REVIEW EXERCISE

1. What is an *algorithm*? What is the criterion that every algorithm must satisfy?
2. What is *algorithm complexity*? What is the need of measuring the complexity of an algorithm?
3. What do you mean by *time-space trade-off* among algorithms?
4. Which notation is popularly used to express the complexity of an algorithm?
5. Name the different categories of algorithms based on *Big Oh* notation.
6. What are the limitations of *Big Oh* notation?
7. What is sorting? What is the need of sorting?
8. If the elements of an array are given as

   12 7 13 9 10 77 2 8

   What will be the arrangement of elements after one pass of the bubble sort method?
9. If the elements of an array are given as

   11 70 13 99 5 17 21 38

   What will be the arrangement of elements after one pass of the selection sort method?
10. If the elements of an array are given as

    12 7 11 92 13 71 21 38

    What will be the arrangement of elements after one pass of the insertion sort method?

11. If the elements of an unsorted array are given as

    12 7 11 92 13 71 21 38

    Describe the steps in order to search elements 71 and 100.
12. If the elements of a sorted array are given as

    82 70 61 52 43 31 24 18

    Describe the steps in order to search elements 24 and 99.

PROGRAMMING EXERCISE

1. What a program to search an element using binary search when the given elements are sorted in descending order.
2. You are given list of names of students in your class. What a program to sort the names in dictionary order using a sorting algorithm of your choice.
3. Write a program to sort letters of your name in reverse dictionary order?
4. Write a program to sort a list of numbers in descending order using bubble sort method. Incorporate the ability in the algorithm that if the list gets sorted before exhausting all the phases, the algorithm must stop.

❑ ❑ ❑

Appendix **A**

Answers of MCQs

Chapter 3: Overview of C Language

1.	(*c*)	2.	(*b*)	3.	(*c*)	4.	(*b*)	5.	(*a*)
6.	(*b*)	7.	(*d*)	8.	(*c*)	9.	(*a*)	10.	(*c*)
11.	(*c*)	12.	(*d*)	13.	(*b*)	14.	(*d*)	15.	(*d*)
16.	(*b*)	17.	(*c*)	18.	(*d*)	19.	(*c*)	20.	(*d*)

Chapter 4: Data Types, Variables & Constants

1.	(*d*)	2.	(*a*)	3.	(*d*)	4.	(*b*)	5.	(*c*)
6.	(*c*)	7.	(*a*)	8.	(*c*)	9.	(*d*)	10.	(*a*)
11.	(*c*)	12.	(*a*)	13.	(*d*)	14.	(*d*)	15.	(*b*)
16.	(*a*)	17.	(*c*)						

Chapter 5: Operators and Expressions

1.	(*b*)	2.	(*a*)	3.	(*c*)	4.	(*b*)	5.	(*a*)
6.	(*b*)	7.	(*b*)	8.	(*a*)	9.	(*b*)	10.	(*b*)
11.	(*c*)	12.	(*b*)	13.	(*a*)	14.	(*c*)	15.	(*a*)
16.	(*d*)	17.	(*c*)	18.	(*c*)	19.	(*b*)	20.	(*c*)
21.	(*d*)	22.	(*d*)	23.	(*c*)	24.	(*d*)	25.	(*b*)

Chapter 6: Console Input/Output

1.	(*a*)	2.	(*b*)	3.	(*a*)	4.	(*b*)	5.	(*c*)
6.	(*a*)	7.	(*a*)	8.	(*b*)	9.	(*d*)	10.	(*d*)
11.	(*b*)	12.	(*a*)	13.	(*a*)	14.	(*b*)	15.	(*c*)

Chapter 7: Decision Making & Branching

1.	(*c*)	2.	(*a*)	3.	(*d*)	4.	(*a*)	5.	(*c*)
6.	(*c*)	7.	(*c*)	8.	(*b*)	9.	(*a*)	10.	(*c*)
11.	(*d*)	12.	(*c*)	13.	(*b*)	14.	(*d*)	15.	(*c*)

Chapter 8: Decision Making & Looping

1.	(*c*)	2.	(*d*)	3.	(*c*)	4.	(*d*)	5.	(*b*)
6.	(*a*)	7.	(*d*)	8.	(*c*)	9.	(*b*)	10.	(*d*)
11.	(*c*)	12.	(*c*)	13.	(*a*)	14.	(*d*)	15.	(*c*)

Chapter 9: User-Defined Functions

1.	(*d*)	2.	(*c*)	3.	(*a*)	4.	(*a*)	5.	(*c*)
6.	(*b*)	7.	(*c*)	8.	(*d*)	9.	(*c*)	10.	(*d*)
11.	(*d*)	12.	(*a*)	13.	(*b*)	14.	(*c*)	15.	(*b*)

Chapter 10: Arrays & Strings

1.	(*b*)	2.	(*a*)	3.	(*c*)	4.	(*d*)	5.	(*c*)
6.	(*b*)	7.	(*d*)	8.	(*d*)	9.	(*a*)	10.	(*c*)
11.	(*d*)	12.	(*a*)	13.	(*b*)	14.	(*d*)	15.	(*a*)
16.	(*c*)	17.	(*b*)	18.	(*c*)	19.	(*b*)	20.	(*c*)
21.	(*b*)	22.	(*a*)	23.	(*d*)	24.	(*c*)	25.	(*b*)
26.	(*a*)	27.	(*b*)	28.	(*d*)	29.	(*c*)	30.	(*a*)
31.	(*c*)	32.	(*a*)	33.	(*a*)				

Chapter 11: User-Defined Data Types: Structure, Union & Enumeration

1.	(*d*)	2.	(*c*)	3.	(*b*)	4.	(*b*)	5.	(*c*)
6.	(*b*)	7.	(*a*)	8.	(*d*)	9.	(*a*)	10.	(*c*)
11.	(*c*)	12.	(*c*)	13.	(*a*)	14.	(*c*)	15.	(*c*)
16.	(*b*)	17.	(*d*)						

Chapter 12: Pointers & Dynamic Memory Management

1.	(*b*)	2.	(*d*)	3.	(*a*)	4.	(*d*)	5.	(*b*)
6.	(*b*)	7.	(*a*)	8.	(*a*)	9.	(*d*)	10.	(*c*)
11.	(*c*)	12.	(*b*)	13.	(*c*)	14.	(*c*)	15.	(*a*)
16.	(*b*)	17.	(*b*)	18.	(*a*)	19.	(*a*)	20.	(*a*)
21.	(*b*)	22.	(*c*)	23.	(*b*)	24.	(*b*)	25.	(*d*)
26.	(*b*)								

Chapter 13: File Handling

1.	(*d*)	2.	(*a*)	3.	(*b*)	4.	(*d*)	5.	(*d*)
6.	(*b*)	7.	(*d*)	8.	(*d*)	9.	(*d*)		

Chapter 14: Pre-processor Directives

1.	(*a*)	2.	(*d*)	3.	(*a*)	4.	(*b*)	5.	(*c*)
6.	(*c*)	7.	(*d*)	8.	(*c*)	9.	(*c*)	10.	(*c*)
11.	(*c*)	12.	(*a*)	13.	(*c*)	14.	(*d*)	15.	(*c*)
16.	(*c*)	17.	(*d*)	18.	(*c*)				

Chapter 15: Introduction to Algorithms

1.	(*c*)	2.	(*b*)	3.	(*c*)	4.	(*a*)	5.	(*d*)
6.	(*b*)	7.	(*c*)	8.	(*c*)	9.	(*a*)	10.	(*b*)

❑ ❑ ❑

Appendix **B**

FAQs in Technical Interviews

Question #1

Whether C language is *low-level*, *high-level*, or *middle-level* language?

Answer

The C language is a *middle-level* language. It is called a *middle- level* language because it has features of *high-level* language as well as *low-level* language.

Question #2

List the key characteristics of C language.

Answer

The C language was and will remain as one of the popular programming languages. Some of the key characteristics that add to its popularity are:

- It is a general purpose programming language, and therefore, can be used to solve vide variety of problems.
- It is supports structuring programming, and therefore, programs developed in it can easily be understood.

- It is free form, *i.e.*, no rigid format for writing programs. Any instruction can start from anywhere and end anywhere. In addition, a single instruction can span (can be written in) many lines.
- It is case sensitive. It distinguishes between lowercase (small) and uppercase (capital) alphabets.
- It was the first language to provide rich set of operators.
- It allows you, through pointers, to access any storage location and the devices connected to your computer through your program.
- It allows you to manipulate internal processor registers, and thus it is very useful for low-level programming.
- It is portable, that means any program written in it can be used on any computer without or with little modification(s).
- It allows you to develop your own library of functions that can be linked to your any program like standard library functions.

Question #3

What are various forms that *main()* function can take?

Answer

The *main()* function is a part of every C program. The C language permits following forms of *main()* function:

- `main()`
- `main(void)`
- `void main()`
- `int main()`
- `void main(void)`
- `int main(void)`

- The empty pair of parenthesis indicates that the *main()* function takes no arguments. This may be explicitly indicated by using the keyword *void* inside the parenthesis.
- We may also specify the keyword *void* or *int* before the word main.
- When the keyword *void* is used, it indicates that the function *main()* does not return any value to the operating system.
- When the keyword *int* is used, it indicates that the function *main()* will return an integer value to the operating system. Value 0 is returned to the operating system to indicate that the program terminated on completion of its intended task whereas the value 1 is used to indicate that the program terminated prematurely (could not complete its intended task).

Question #4

The output produced by a C compiler is an object code, *i.e.*, machine code, but not in a form that can be executed. Why?

Answer

Recall that a program can contain calls to library function as well as user-defined functions that are in other disk files and are not included using *file inclusion* pre-processor directive (*#include*).

The compiler doesn't have access to the definitions of these functions, and hence can't bind the function calls to function definitions. Hence, function calls are left unresolved. In addition, some more information is required by the operating system to run the program.

Question #5

Basically, what does the header files contain?

Answer

The header files contain declarations of library functions (prototypes of library functions) and useful enumerations and macros for the convenience of the programmers.

Question #6

What is the need of having large number of header files? Why not just one?

Answer

Note that header files are read by the compiler, which helps it in the process of compilation of the program. If all header files are combined into a single header file, this file will become very large and lot of compiler's time will be wasted. Further, note that in a program, we are not going to use diverse library functions in a single program, hence relevant declarations are put in different header files and relevant and meaningful names are given to these header files that show its intent.

Question #7

What is the difference between #include <...> and # include "..."?

Answer

There are two ways to include a header file:

```
#include <stdio.h>      and      #include "stdio.h"
```

The first form directs the preprocessor to include the header file from the standard directory while the second form directs the preprocessor to look for the specified header file in the user directory.

Question #8

What happens during linking?

Answer

The next step after compilation is linking, *i.e.*, to resolve the unresolved references to functions. This task is accomplished by another program known as *linker*. The linker takes object code(s) as its inputs and searches for unresolved references. For every unresolved reference, it searches libraries for their definitions and adds them to the end of final code being generated and fills its address at point of call. If search of any function is unsuccessful, the linker will generate a message like unresolved references and also gives name of the function called. Output of a linker is a machine language program that is in executable form better known as executable code or run file.

Question #9

What is the difference between declaration and definition of a variable?

Answer

Declaration only identifies the data type of a variable whereas definition causes the space to be reserved for the variable.

Thus, declaration is a place where the nature of the variable is stated but no storage is allocated whereas definition is the place where the variable is created or assigned storage.

Question #10

What is the difference between the following statements?

```
extern int k;    and    int k;
```

Answer

The first is a declaration indicated that *k* is a variable of type *int* and is defined externally, *i.e.*, in some another source file. The second is a definition and causes 2-bytes to be reserved for it.

Question #11

What is the restriction on the declarations and/or definitions of variables?

Answer

The restriction on the declarations and/or definitions is that they must come before the executable statements in a block, and hence they appear in the beginning of a block.

Question #12

Suppose a global variable and a local variable have the same name. Is it possible to access a global variable from a block where local variable is defined?

Answer

No. It is not possible in C. It is always the most local variable that gets preference.

Question #13

Write a "hello world" program in C without using a semicolon.

Answer

```
#include<stdio.h>
void main()
{
    if ( printf("hello world\n") ) { }
}
```

Explanation: The *printf*() is a function that returns the number of characters it is able to print. This return value can be used in the *if* conditional expression. With this combination, we can print the required string without using a semicolon.

Question #14

What is the difference between *static* and *global* variables?

Answer

While the life of an object determines whether the object is still in the memory (of the process), scope of the object decides whether the variable can be accessed at that point in the program. Static variables are local in scope to the block or file in which they are defined, but their lifespan is throughout the program. For instance, a static variable inside a function cannot be called from outside the function (because it's not in the scope) but is alive and exists in the memory. It retains its old value between function calls.

Global variables persist throughout the program; scope is also throughout the program. It means that global variables can be accessed from any function, any file (unless declared static) in the program.

Question #15

What is the use of *register* keyword in the context of variables?

Answer

The register keyword specifies that the variable is to be stored in the CPU register, if possible. Variables are usually stored in stack and passed to and from the computer processor, as and when required, the speed the data is sent at is pretty but can be improved on. Almost all computer processors contain CPU registers, these are memory slots on the actual processors, and storing data there gets rid of the overhead of the retrieving data from the stack. This memory (in registers) is quite small compared to normal memory, therefore, only a few variables can be stored there. Having a register keyword for variables and parameters also reduces code size, which is important in embedded systems. The number of available CPU registers is strictly dependent on the architectural design of the CPU itself. The number of CPU registers is 32 in most cases.

Question #16

How do we make a global variable accessible across files? Explain the *extern* keyword.

Answer

If a variable is declared (with global scope) in one file but referenced in other, the extern keyword is used to inform the compiler of variable's existence. Note that the keyword extern is for declarations, not definitions. An extern declaration does not create any storage. Variable initializations should not be done with extern.

Question #17

What is a function prototype?

Answer

A function prototype is a declaration similar to variable declarations. Function declarations are generally placed in header files. Specifically, it tells the compiler

- The name of the function
- The type of the output returned by the function.
- The number of types of inputs (called arguments), if any, the function expects to receive and the order in which it expects to receive them.

Question #18

What is the difference between declaration and definition of a function?

Answer

Like variables, declaration of a function only identifies the return type of the function and the type of format arguments. The declaration of a function is usually referred as a *function prototype*. On the other hand, definition of a function contains the statements to accomplish the intended task. Like variables, functions also have an address associated with its name.

Question #19

Is the name of the formal arguments in a function prototype mandatory?

Answer

No. Only data type is mandatory. However, the name of formal arguments can be useful to know the kind of arguments that a function will be taking. This is explained with the help of the following example:

The *displayAt()*function accepts three arguments: row number, column number, and a string. It display the string starting from the given row and column.

The prototype of this function can be

```
void displayAt(int, int, char *);
```

But from this prototype, it is not clear whether row is to passed as first argument or column is to be passed as first argument. However, if its prototype is taken as

```
void displayAt(int row, int col, char *charPtr);
```

then it is clear that first argument will be row while the second will be column.

Question #20

What is the significance of storage class?

Answer

The storage class, in general, determines the scope and lifetime of a variable, which in turn helps to economize the memory usage.

Question #21

What is a dangling pointer?

Answer

It is a pointer that is not initialized with a valid address. Such a pointer can end up pointing anywhere in memory that may include the program code itself or to the code of the operating system. The use of such pointer can lead to undesirable results. Therefore, care must be taken to initialize pointers with valid address before use.

Question #22

What is a memory leak?

Answer

It is a situation where the programmer fails to release the memory allocated at run time in a module. Remember that when memory is allocated, a pointer variable is used to hold the address of the allocated block, however, when the module completes its execution, the pointer variable goes out of scope and there will be no way to reach that memory block. Therefore, care must be taken to release the allocated memory block in a module where memory was allocated.

Question #23

What do you mean by *Lvalue* and *Rvalue*?

Answer

There are two values associated with a variable:

1. Its data value that is stored at some memory address. This sometimes referred to as variable's *Rvalue* (pronounced as "are-value"). We may think *Rvalue* as meaning *read value*. Both a literal constant and a variable can serve as *rvalue*.
2. Its address value that is its address in memory within which its data value is stored. This sometimes referred to as variable's *Lvalue* (pronounced as "ell-value"). We may think *Lvalue* as meaning *location value*. Only a variable can serve as *Lvalue*.

Question #24

Cite some cases where *switch* statement cannot be used?

Answer

The *switch* statement cannot be used when the variable or expression is not of integral type (*int*, *long* or *char*). The other situation is where conditions are of type that does not require exact match, such *amount* <= 1200.

Question #25

Among *gets()* and *fgets()* functions, which function is safe to use?

Answer

fgets() is safe to use. This is illustrated below. The prototype of *gets()* functions is

```
char *gets(char *s);
```

It reads a string of characters terminated by newline characters from *stdin* and stores in *s*. As there is no way to tell the size of the string there is always a possibility of overflow.

The prototype of *fgets()* functions is

```
char *gets(char *s, int n, FILE *stream);
```

It reads a string of characters from *stream* and stores in *s*. It stops when either it reads (*n*-1) characters or a newline character, whichever comes first. So overflow condition never occurs.

Question #26

What would be the consequence if the function prototypes are local to *main()* function?

Answer

If any other function calls these functions, the compiler will flag an error message, as these prototypes are not available in that function, hence the compiler may not be able to compile the program properly.

Question #27

When should the function prototypes be made global?

Answer

When the functions can be called from more than one function.

Question #28

Is the following nested comments are acceptable?

```
/* Outer comment /* Nested comments */ allowed/not allowed */
```

Answer

No. Note that anything that lies between /* and */ is treated as comment. Therefore, in the above case, from word *Outer* to word *comments* everything is treated as comment including /*. Hence the string "allowed/not allowed */" does not lie in the comment.

Thus, it is being improper statement the compiler will flag an error.

Question #29

When should we pass arguments by reference (address)?

Answer

Argument passing by reference (address) is useful in following situations:

- When we intend to change the values of actual arguments.
- When we want to save the memory by preventing the creation of large array and/or structure variables being passed to the function.

Question #30

Given a choice, would you prefer to use *const* over *#define* in order to use a constant?

Answer

Using *const* is always a better choice because we can control its scope, which is not possible with *#define*.

Question #31

Passing arguments by reference (address), we can prevent the creation of large array and/or structure variables in the called function. Here the called function can change the values of array and/or structure variables. Is there any way to prevent the called function from doing so?

Answer

Yes, by prefixing the formal argument by keyword *const* as shown in the following example

```
void fName(const int *a)
{
   /*   some statements   */
}
```

Question #32

In how many ways can we pass arguments to a function?

Answer

There are two ways in which we can pass arguments to a function:

- By value

```
void f(int, int)   /* declaration  */

f(a, b);           /*  call   */
```

- By reference (address)

```
void f(int *, int *)  /* declaration  */

f(&a, &b);            /*  call   */
```

Question #33

Differentiate between call by value and call by reference (address).

Answer

Call by Value	Call by Reference (Address)
1. The actual arguments can be constants, variables, or expressions.	1. The actual arguments can only be variables.
2. The values of actual arguments are substituted in formal arguments which are ordinary variables.	2. The addresses of the actual arguments are substituted in formal arguments which are pointer variable.
3. Any change made to the formal arguments will have no effect on actual arguments, since the function will only be using the local copy of the arguments.	3. Any change made to the formal arguments will have immediate effect on actual arguments, since the function will be working on actual arguments through pointers.

Question #34

What is the main difference between *malloc()* and *calloc()*?

Answer

The main difference between *malloc()* and *calloc()* is that *malloc()* function just allocates the memory whereas the *malloc()* function also initializes the memory to zeros.

Question #35

What is free store/heap?

Answer

Free store/heap is area of memory from where memory is allocated at runtime using *malloc()* or *calloc()* functions.

Question #36

Recall that when we use *free()* function to deallocate the block, we only pass pointer to the block, then how *free()* function know how many bytes it should deallocate?

Answer

When operating system allocates memory, on call to *malloc()*, *calloc()* or *realloc()* function, to a program from its free pool of memory (also called free store), it keeps track of the address of the block and the size of the block. Therefore, when the memory needs to be deallocated, program need only to tell the address of block to the operating system.

Question #37

What is a void/generic pointer?

Answer

When a variable is declared as being a pointer to a variable of type void it is known as a generic pointer. Since we cannot have a variable of type void, the pointer will not point to any data and therefore cannot be de-referenced. It is still a pointer though; to use it we have to cast it to another kind of pointer first. Hence, the term generic pointer. It is very useful when you want a pointer to point to data of different types at different times.

Question #38

Why arithmetic operation can't be performed on a void/generic pointer?

Answer

The C language doesn't allow pointer arithmetic on void pointers. The argument being that we don't know the size of what's being pointed to with a void *and therefore can't know how far to seek the pointer to get to the next valid address.

Question #39

How do you access command line arguments?

Answer

A program is started by the operating system by calling a program's *main*() function. The *main*() function is the only function in C that can be defined in multiple ways. It can take no arguments, two arguments, or three arguments. The two and three argument forms allow it to receive arguments from the shell. The two arguments takes an int and an array of strings. When defining *main*() function any name can be given but it is convention to call them *argc* and *argv*[]. The first argument (*argc*) holds a count of how many elements there are in the array of strings passed as the second argument (*argv*). The array is always null terminated so *argv* [*argc*] = null.

Here's short program demonstrating the use:

```
int main (int argc, char *argv[])
{
    int i;
    for( i = 0; i < argc; i++ ) {
        printf("argv[%d] == %s\n", i, argv[i]);
    }
    return 0;
}
```

The integer *argc* is argument count (hence *argc*). It is the number of arguments passed into program from the command line, including the name of the program.

The array of character pointers is the listing of all the arguments. The *argv*[0] is the name of the program. After that every element number less than *argc* is command line arguments. You can use each *argv* element just like a string, or *argv* as a 2D array.

Question #40

What is the value of the expression 5["CONQUER"]?

Answer

```
E
```

The string literal "CONQUER" is an array, and the expression is equivalent to "CONQUER"[5]. The inside-out expression is equivalent because *a*[*b*] is equivalent to *(*a*+*b*) which is equivalent to *(*b*+*a*) which is equivalent to *b*[*a*].

Question #41

What is a self-referential structure?

Answer

If a structure contains a member that is a pointer to the same structure, then the structure is called self- referential structure.

```
struct list
{
    int nodedata;
    struct list *next;
};
```

Question #42

Write a recursive function to find sum of elements of an array named *a* with *n* elements.

Answer

```
int sum(int a[], int n)
{
    if ( n == 0 )
        return a[0];
    else
        return ( a[n-1] + sum(a,n-1) );
}
```

Question #43

What are library functions? Where are they stored?

Answer

Library functions are a collection of predefined functions. They are stored in files with extension *lib*. These functions are shipped with the compiler for the convenience of the programmers.

❑ ❑ ❑

Appendix C

Solution of Selected Programming Exercises

Chapter 7

1. Given measure of three line segments as *a*, *b*, and *c*. Write a program to check whether these line segments can be used to form a triangle or not.

Note that by the triangle inequality, in order for three line segments to make a triangle, the sum of the lengths of any two line segments must exceed the length of the third line segment.

Listing C.1

```
#include <stdio.h>
int main()
{
   int a, b, c;
   printf("\nEnter value for side AB : ");
   scanf("%d", &a);
   printf("Enter value for side BC : ");
   scanf("%d", &b);
   printf("Enter value for side CA : ");
   scanf("%d", &c);
   if ( ( (a+b) > c ) && ( (b+c) > a ) && ( (c+a) > b ) )
      printf("\nYes, triangle can be made with these sides\n");
   else
      printf("\nNo, triangle cannot be made with these sides\n");
   return 0;
}
```

Test Run 1

```
Enter value for side AB : 2
Enter value for side BC : 3
Enter value for side CA : 5

No, triangle cannot be made with these sides
```

Test Run 2

```
Enter value for side AB : 5
Enter value for side BC : 6
Enter value for side CA : 9

Yes, triangle can be made with these sides
```

6. Given three points $A(x_1, y_1)$, $B(x_2, y_2)$ and $C(x_3, y_3)$, determine whether they are collinear, *i.e.*, lie on the same line.

Note that three points are collinear if they lie on a line. This condition can be checked by following any of the following criteria:

(*a*) area of the triangle formed by these points is zero.
(*b*) slope of the line joining any two pair of points is same.

We will implement second criteria,

Listing C.2

```
#include <stdio.h>
int main()
{
   int x1, y1, x2, y2, x3, y3;
   float m1, m2;
   printf("\nEnter point A(x1,y1) : ");
   scanf("%d,%d", &x1, &y1);
   printf("\nEnter point B(x2,y2) : ");
   scanf("%d,%d", &x2, &y2);
   printf("\nEnter point C(x3,y3) : ");
   scanf("%d,%d", &x3, &y3);
   /*  slope of line segment  AB  */
   if ( (x2-x1) != 0 )
      m1 = (float) (y2-y1)/(x2-x1);
   else
   {
      printf("\nSlope of line segment AB is infinity\n");
      return 1;
   }
   /*  slope of line segment  BC  */
   if ( (x2-x1) != 0 )
      m2 = (float) (y3-y2)/(x3-x2);
   else
   {
      printf("\nSlope of line segment AB is infinity\n");
      return 1;
   }
   /*  compare slopes of line segments AB & BC  */
   if ( m1 == m2 )
    printf("\nPoints A, B, C are collinear\n");
   else
    printf("\nPoints A, B, C are not collinear\n");
   return 0;
}
```

Test Run 1

```
Enter point A(x1,y1) : 5,5
Enter point B(x2,y2) : 9,9
Enter point C(x3,y3) : 11,11
Points A, B, C are collinear
```

Test Run 2

```
Enter point A(x1,y1) : 5,5
Enter point B(x2,y2) : 6,9
Enter point C(x3,y3) : 10,7
Points A, B, C are not collinear
```

10. Write a program to check whether the given date in format *dd/mm/yyyy* is valid or not.

Listing C.3

```
#include <stdio.h>
int isLeapYear(int);
void main()
{
   int d,m,y;
   char ch;
   printf( "\n\tEnter date as dd/mm/yyyy : ");
   scanf( "%d%c%d%c%d", &d,&ch,&m,&ch,&y );
   if ( m < 0 || m > 12 ) {
       printf( "\n\n\t\tDate %d/%d/%d is invalid\n\n", d, m, y );
 } else if ( m == 2 ) {
       if ( isLeapYear(y) ) {
          if ( d >= 1 && d <= 29 )
             printf( "\n\n\t\tDate %d/%d/%d is valid\n\n", d, m, y );
          else
             printf( "\n\n\t\tDate %d/%d/%d is invalid\n\n", d, m, y );
       } else {
          if ( d >= 1 && d <= 28 )
            printf( "\n\n\t\tDate %d/%d/%d is valid\n\n", d, m, y );
          else
            printf( "\n\n\t\tDate %d/%d/%d is invalid\n\n", d, m, y );
       }
   } else if ( m==1||m==3||m==5||m==7||m==8||m==10||m==12 ) {
       if ( d >= 1 && d <= 31 )
          printf( "\n\n\t\tDate %d/%d/%d is valid\n\n", d, m, y );
       else
          printf( "\n\n\t\tDate %d/%d/%d is invalid\n\n", d, m, y );
 } else {
       if ( d >= 1 && d <= 30 )
          printf( "\n\n\t\tDate %d/%d/%d is valid\n\n", d, m, y );
       else
          printf( "\n\n\t\tDate %d/%d/%d is invalid\n\n", d, m, y );
   }
}
```

```
int isLeapYear(int year)
{
   if ((year % 4==0)||((year % 400==0) )&&(year % 100!=0))
      return 1;
   else
      return 0;
}
```

Test Run 1

```
Enter date as dd/mm/yyyy : 23/11/2020
Date 23/11/2020 is valid.
```

Test Run 2

```
Enter date as dd/mm/yyyy : 29/02/2019
Date 29/02/2019 is invalid.
```

11. Write a program to find the day of the week on a given date.

Listing C.4

```
#include <stdio.h>
/*   function prototypes   */
int isLeapYear(int);
int validDate(int,int,int);

void main()
{
  int dd, mm, yy, year, day;
  char ch;
  char *days[]={"Sunday", "Monday", "Tuesday",
               "Wednesday", "Thursday", "Friday", "Saturday" };
  int t[] = { 0, 3, 2, 5, 0, 3, 5, 1, 4, 6, 2, 4 };
  printf( "\n\tEnter date as DD/MM/YYYY : ");
  scanf( "%d%c%d%c%d", &dd,&ch,&mm,&ch,&yy );
  if ( validDate(dd,mm,yy) ) {
     year = yy;
     year -= mm <= 2;
     day = (year + year/4 - year/100 + year/400 + t[mm-1] + dd) % 7;
     printf("\nDay on %02d/%02d/%02d is %s.\n",dd,mm,yy,days[day]);
  }
  else {
     printf( "\n\n\t\tDate %02d/%02d/%02d is invalid\n\n", dd, mm, yy );
  }
}
int isLeapYear(int year)
{
   if ( ( (year % 4==0)&&(year % 100!=0) ) || (year % 400==0) )
     return 1;
   else
     return 0;
}
```

```
int validDate(int d, int m, int y)
{
   int isvalid;
   if ( m < 0 || m > 12 )
      isvalid = 0;
   else if ( m == 2 ) {
      if ( isLeapYear(y) ) {
         isvalid = ( d >= 1 && d <= 29 ) ? 1 : 0;
      } else {
         isvalid = ( d >= 1 && d <= 28 ) ? 1 : 0;
      }
   } else if ( m==1 || m==3 || m==5 || m==7 || m==8 || m==10 || m==12 ) {
      isvalid = ( d >= 1 && d <= 31 ) ? 1 : 0;
   } else {
      isvalid = ( d >= 1 && d <= 30 ) ? 1 : 0;
   }
   return isvalid;
}
```

Test Run

```
Enter date as dd/mm/yyyy : 23/11/2020
Day on 23/11/2020 is Monday.
```

16. You know that an even number is a number which is divisible by 2. However there is another approach that can be used to check whether the given number is even or not. The approach is by inspecting the unit digit - *if the unit digit is 0, 2, 4, 6, or 8, the number is even.* So you are required to write a program to check whether the given number is even or not without performing the division by 2.

Listing C.5

```
#include <stdio.h>
int main()
{
   int num, udigit;
   printf("\n\nEnter any whole  number : ");
   scanf("%d", &num);
   udigit = num % 10;
   switch (udigit)
   {
       case 0:
       case 2:
       case 4:
       case 6:
       case 8: printf("\n%d is even number\n", num);
               break;
       default; printf("\n%d is odd number\n", num);
   }
   return 0;
}
```

Test Run 1

```
Enter any whole number : 204
204 is even number
```

Test Run 2

```
Enter any whole number : 155
155 is odd number
```

Chapter 8

1. Two prime are said to be twin prime numbers if they differ by 2, such as 11 and 13, 17 and 19, etc. Write a program to print first *n* pairs of twin prime numbers.

Listing C.6

```
#include <stdio.h>
#include <math.h>
int main()
{
   int n, cprime = 2, flag, nprime, k, m, num = 3, count = 0;
   printf("\nHow many pair of twin prime numbers to print : ");
   scanf("%d", &n);
   while ( count < n )
   {
      if ( (num >> 2) && (num%2==0) )
      {
         num++;
         continue;
      } else {
         m = sqrt(num);
         flag = 1;
         for ( k = 3; k <= m; k++ )
         {
            if ( num % k == 0 )
            {
               flag = 0;
               break;
            }
         }
         if ( flag == 1 )
         {
            nprime = num;
            if ( nprime - cprime == 2 ) {
               count++;
               printf("(%d,%d)  ", cprime, nprime);
            }
            cprime = nprime;
         }
         num++;
```

```
        }
    }
    printf("\n");
    return 0;
}
```

Test Run

```
How many pair of twin prime numbers to print : 5
(3,5)  (5,7)  (11,13)  (17,19)  (29,31)
```

4. A factorial chart from 1 to 5 looks like the following:

1! = 1
2! = 1×2
3! = 1×2×3
4! = 1×2×3×4
5! = 1×2×3×4×5

Write a program that will generate a factorial chart from 1 to *n* for any ≤ 9.

Listing C.7

```
#include <stdio.h>
#include <math.h>
int main()
{
    int i, j, n;
    printf("\nEnter value for n (<=9) : ");
    scanf("%d", &n);
    for ( i = 1; i <= n; i++ )
    {
     printf("\n%d! = 1", i);
     for ( j = 2; j <= i; j++ )
            printf("x%d", j);
    }
    printf("\n");
    return 0;
}
```

Test Run

```
Enter value for n (<=9) : 5
1! = 1
2! = 1×2
3! = 1×2×3
4! = 1×2×3×4
5! = 1×2×3×4×5
```

8. A positive integer number IJK is said to be *well-ordered* if I < J < K. For example, number 138 is called *well-ordered* because the digits in the number increase from left to right, *i.e.*, 1 < 3 < 8. Number 365 is not *well-ordered* because 6 is larger than 5. Write a program that displays all possible three digit *well-ordered* numbers.

Listing C.8

```
#include <stdio.h>
int main()
{
    int i, d1, d2, d3, t;
    printf("\n\nThree digit well-ordered numbers\n\n");
    for ( i = 100; i <= 999; i++ )
    {
        t = i;
        d3 = t % 10;
        t = t / 10;
        d2 = t % 10;
        t = t / 10;
        d1 = t % 10;
        if( (d1 < d2) && (d2 < d3) )
        {
            printf("%d  ", i);
        }
    }
    printf("\n");
    return 0;
}
```

Test Run

```
Three digit well-ordered numbers
123 124 125 126 127 128 129 134 135 136 137 138 139 145 146 147
148 149 156 157 158 159 167 168 169 178 179 189 234 235 236 237
238 239 245 246 247 248 249 256 257 258 259 267 268 269 278 279
289 345 346 347 348 349 356 357 358 359 367 368 369 378 379 389
456 457 458 459 467 468 469 478 479 489 567 568 569 578 579 589
678 679 689 789
```

Chapter 10

7. Given a matrix *A* of order *m*×*n*, write a program to find the row having the maximum number of elements which are even.

Listing C.9

```
#include<stdio.h>
int main()
{
    int a[30], i, j, k, n;
    printf("\nHow many elements? : ");
    scanf("%d", &n);
```

```
    printf("\nEnter %d elements of array\n", n);
    for ( i = 0; i < n; ++i )
        scanf("%d", &a[i]);
    for( i = 0; i < n; ++i ) {
        for( j = i+1;j < n;) {
              if(a[i]==a[j])
              {
                for( k = j; k < n-1;++k ){
                     a[k]=a[k+1];
                }
                --n;
              }
              else
              {
                ++j;
              }
        }
    }
    printf("\nArray elements after removing duplicate elements\n");
    for ( i = 0; i < n; ++i )
       printf("%d  ", a[i]);
    return 0;
}
```

Test Run

```
How many elements? 10
Enter 10 elements of array
11 11 22 33 22 56 67 33 99 100
Array elements after removing duplicate elements
11 22 33 56 67 99 100
```

8. Given a matrix *A* of order *m*×*n*, write a program to find the row having the maximum number of elements which are even.

Listing C.10

```
#include <stdio.h>
int main()
{
   int a[10][10], count[10], i, m, n, j, max, loc;
   printf("\nEnter size of matrix as m,n : ");
   scanf("%d,%d", &m, &n);
   printf("\nEnter %d elements of matrix row-wise\n", m*n);
   for ( i = 0; i < m; i++ ) {
       for ( j = 0; j < n; j++ ) {
            scanf("%d", &a[i][j] )
       }
   }
   for ( i = 0; i < m; i++ ) {
      count[i] = 0;
       for ( j = 0; j < n; j++ ) {
```

```
            if ( a[i][j] % 2 == 0 )
                count[i]++;
        }
    }
    max = count[0];
    loc = 0;
    for ( i = 1; i < m; i++ ) {
        if ( count[i] > max ) {
            max = count[i];
            loc = i;
        }
    }
    printf("\nRow having maximum number of even elements = %d\n",loc+1);
    return 0;
}
```

Test Run

```
Enter size of matrix as m,n : 4,4
Enter 16 elements of matrix row-wise
5 4 3 2
6 4 5 10
5 3 1 9
1 5 6 3
Row having maximum number of even elements = 2
```

10. A square matrix *A* of size *nxn* is said to be symmetric if element $a_{ij} = a_{ji}$ for all values of *i* and *j*. Write a function, say *int isSymmetric(int a[][], int n)* that returns value 1 if the input matrix is symmetric else returns value 0.

Listing C.11

```
#include <stdio.h>
int main()
{
    int a[10][10], i, j, n;
    printf("\nEnter size of square matrix : ");
    scanf("%d", &n);
    printf("\nEnter %d elements of matrix row-wise\n", n*n);
    for ( i = 0; i < n; i++ ) {
      for ( j = 0; j < n; j++ ) {
            scanf("%d", &a[i][j] );
      }
    }
    for ( i = 0; i < n; i++ ) {
      for ( j = i+1; j < n; j++ ) {
          if ( a[i][j] != a[i][j] ) {
              printf("\nGiven matrix is not symmetric\n");
              return 1;
          }
      }
    }
```

```
    printf("\nGiven matrix is symmetric\n");
    return 0;
}
```

Test Run 1

```
Enter size of square matrix : 3
Enter 16 elements of matrix row-wise
5 4 3
6 4 5
5 3 1
Given matrix is not symmetric
```

Test Run 2

```
Enter size of square matrix : 3
Enter 16 elements of matrix row-wise
1 2 3
2 1 4
3 4 1
Given matrix is symmetric
```

11. Given a matrix *A* of order *m*×*n*, write a program to sort the even rows in ascending order and odd rows in descending order

Listing C.12

```
#include <stdio.h>
int main()
{
    int a[10][10], count[10], i, m, n, j, k, t;
    printf("\nEnter size of matrix as m,n : ");
    scanf("%d,%d", &m, &n);
    printf("\nEnter %d elements of matrix row-wise\n", m*n);
    for ( i = 0; i < m; i++ ) {
        for ( j = 0; j < n; j++ ) {
            scanf("%d", &a[i][j] );
        }
    }
    for ( i = 0; i < m; i++ ) {
        if ( i % 2 == 0 ) {        /* even row */
            for ( k = 0; k < n-1; k++ ) {
                for ( j = 0; j < n-k-1; j++ ) {
                    if ( a[i][j] > a[i][j+1] ) {
                        t = a[i][j];
                        a[i][j] = a[i][j+1];
                        a[i][j+1] = t;
                    }
                }
            }
        } else {                /* odd row */
```

```
            for ( k = 0; k < n-1; k++ ) {
                for ( j = 0; j < n-k-1; j++ ) {
                    if ( a[i][j] < a[i][j+1] ) {
                        t = a[i][j];
                        a[i][j] = a[i][j+1];
                        a[i][j+1] = t;
                    }
                }
            }
        }
    }
    printf("\nMatrix after sorting\n\n");
    for ( i = 0; i < m; i++ ) {
        for ( j = 0; j < n; j++ ) {
            printf("%3d", a[i][j]);
        }
        printf("\n");
    }
    return 0;
}
```

Test Run

```
Enter size of matrix as m,n : 4,4
Enter 16 elements of matrix row-wise
5 4 3 2
6 4 5 10
5 3 1 9
1 5 6 3
Matrix after sorting
  2  3  4  5
 10  6  5  4
  1  3  5  9
  6  5  3  1
```

❑ ❑ ❑

Index

E

F

G

H

I

J, K

L

M

N

O

P